DISCRETE MATHEMATICS
AND
ITS APPLICATIONS

Series Editor
Kenneth H. Rosen, Ph.D.

Juergen Bierbrauer, Introduction to Coding Theory

Kun-Mao Chao and Bang Ye Wu, Spanning Trees and Optimization Problems

Charalambos A. Charalambides, Enumerative Combinatorics

Henri Cohen, Gerhard Frey, et al., Handbook of Elliptic and Hyperelliptic Curve Cryptography

Charles J. Colbourn and Jeffrey H. Dinitz, The CRC Handbook of Combinatorial Designs

Steven Furino, Ying Miao, and Jianxing Yin, Frames and Resolvable Designs: Uses, Constructions, and Existence

Randy Goldberg and Lance Riek, A Practical Handbook of Speech Coders

Jacob E. Goodman and Joseph O'Rourke, Handbook of Discrete and Computational Geometry, Second Edition

Jonathan L. Gross and Jay Yellen, Graph Theory and Its Applications, Second Edition

Jonathan L. Gross and Jay Yellen, Handbook of Graph Theory

Darrel R. Hankerson, Greg A. Harris, and Peter D. Johnson, Introduction to Information Theory and Data Compression, Second Edition

Daryl D. Harms, Miroslav Kraetzl, Charles J. Colbourn, and John S. Devitt, Network Reliability: Experiments with a Symbolic Algebra Environment

Derek F. Holt with Bettina Eick and Eamonn A. O'Brien, Handbook of Computational Group Theory

David M. Jackson and Terry I. Visentin, An Atlas of Smaller Maps in Orientable and Nonorientable Surfaces

Richard E. Klima, Ernest Stitzinger, and Neil P. Sigmon, Abstract Algebra Applications with Maple

Patrick Knupp and Kambiz Salari, Verification of Computer Codes in Computational Science and Engineering

William Kocay and Donald L. Kreher, Graphs, Algorithms, and Optimization

Donald L. Kreher and Douglas R. Stinson, Combinatorial Algorithms: Generation Enumeration and Search

Charles C. Lindner and Christopher A. Rodgers, Design Theory

Alfred J. Menezes, Paul C. van Oorschot, and Scott A. Vanstone, Handbook of Applied Cryptography

Continued Titles

DISCRETE MATHEMATICS AND ITS APPLICATIONS
Series Editor KENNETH H. ROSEN

SUMS OF SQUARES
OF INTEGERS

CARLOS J. MORENO
SAMUEL S. WAGSTAFF, JR.

CRC Press
Taylor & Francis Group
Boca Raton London New York

CRC Press is an imprint of the
Taylor & Francis Group, an **informa** business
A CHAPMAN & HALL BOOK

CRC Press
Taylor & Francis Group
6000 Broken Sound Parkway NW, Suite 300
Boca Raton, FL 33487-2742

First issued in paperback 2019

© 2006 by Taylor & Francis Group, LLC
CRC Press is an imprint of Taylor & Francis Group, an Informa business

No claim to original U.S. Government works

ISBN-13: 978-1-58488-456-9 (hbk)
ISBN-13: 978-0-367-39161-4 (pbk)

Library of Congress Cataloging-in-Publication Data

Moreno, Carlos J., 1946-
    Sums of squares of integers / Carlos J. Moreno and Samuel S. Wagstaff, Jr.
        p. cm. -- (Discrete mathematics and its applications)
    Includes bibliographical references and index.
    ISBN 1-58488-456-8 (acid-free paper)
    1. Exponential sums. 2. Integrals. 3. Forms, Modular. I. Wagstaff, Samuel S. II. Title.
III. Series.

QA246.7.M67 2005
512.7'2--dc22   2005051985

Visit the Taylor & Francis Web site at
http://www.taylorandfrancis.com

and the CRC Press Web site at
http://www.crcpress.com

CRC Press
Taylor & Francis Group
6000 Broken Sound Parkway NW, Suite 300
Boca Raton, FL 33487-2742

First issued in paperback 2019

© 2006 by Taylor & Francis Group, LLC
CRC Press is an imprint of Taylor & Francis Group, an Informa business

No claim to original U.S. Government works

ISBN-13: 978-1-58488-456-9 (hbk)
ISBN-13: 978-0-367-39161-4 (pbk)

Library of Congress Cataloging-in-Publication Data

Moreno, Carlos J., 1946-
    Sums of squares of integers / Carlos J. Moreno and Samuel S. Wagstaff, Jr.
        p. cm. -- (Discrete mathematics and its applications)
    Includes bibliographical references and index.
    ISBN 1-58488-456-8 (acid-free paper)
    1. Exponential sums. 2. Integrals. 3. Forms, Modular. I. Wagstaff, Samuel S. II. Title.
III. Series.

QA246.7.M67 2005
512.7'2--dc22
                                     2005051985

Visit the Taylor & Francis Web site at
http://www.taylorandfrancis.com

and the CRC Press Web site at
http://www.crcpress.com

DISCRETE MATHEMATICS AND ITS APPLICATIONS
Series Editor KENNETH H. ROSEN

SUMS OF SQUARES
OF INTEGERS

CARLOS J. MORENO
SAMUEL S. WAGSTAFF, JR.

CRC Press
Taylor & Francis Group
Boca Raton London New York

CRC Press is an imprint of the
Taylor & Francis Group, an **informa** business
A CHAPMAN & HALL BOOK

Preface

My system of writing mathematics, whether a research paper or a book, was to write material longhand, with many erasures, with only a vague idea of what would be included. I would see where the math led me just as some novelists are said to let the characters in their books develop on their own, and I would get so tired of the subject after working on it for a while that I would start typing the material before it had assumed final form. Thus in writing even a short paper I would start typing the beginning before I knew what results I would get at the end. Originally I wrote in ink, applying ink eradicator as needed. Feller visited me once and told me he used pencil. We argued the issue, but the next time we met we found that each had convinced the other: he had switched to ink and I to pencil.   J. L. Doob. See p. 308 of [105].

This work has its origins in courses taught by the authors many years ago, when we were both in the same math department with Joe Doob, hiked with him every Saturday afternoon and learned his system of writing mathematics.

This book is one we wished we could have read when we were graduate students. It is an introduction to a vast and exciting area of number theory where research continues today.

The first chapter is a general introduction, which gives the contents of the other chapters in greater detail. It also lists the prerequisites assumed for the reader.

The second chapter treats elementary methods having their origins in a series of eighteen papers by Liouville. Formulas are proved for the number of representations of an integer as a sum of two, three or four squares.

Chapter 3 develops useful properties of the Bernoulli numbers. Many are interesting in their own right to number theorists. Some properties and values are needed for Chapter 6.

Chapter 4 discusses the central themes of the modern theory of modular forms. It does this through several large examples from the literature.

Chapter 5 introduces the basic theory of modular forms on the modular group Γ and its subgroup $\Gamma_0(N)$. It treats Hecke operators, the Petersson inner product and Poincaré series.

Chapter 6 presents a second treatment of sums of squares. This time analytic methods are used to count the representations. The number of representations of n as the sum of s squares of integers is approximated by an Euler product, and this approximation turns out to be exact when $1 \le s \le 8$. We tell how to evaluate this Euler product in terms of generalized Bernoulli numbers, which are easy to compute.

Chapter 7 deals with arithmetic progressions. In it we prove the theorems of van der Waerden, Roth and Szemerédi. We also find all arithmetic progressions of squares.

Chapter 8 gives some applications to real life of formulas proved earlier. We give algorithms for efficiently computing numbers whose existence was proved in earlier chapters. The book concludes with applications of the theory to three problems outside of number theory. These lie in the areas of microwave radiation, diamond cutting and cryptanalysis.

More than one hundred interesting exercises test the reader's understanding of the text. The exercises range in difficulty from nearly trivial to research problems.

We are grateful to Ning Shang, Zhihong Li and the students in our classes where we taught this material for providing insightful comments on earlier versions of parts of this book and other information that made the book better. We thank Abhilasha Bhargav-Spantzel and Joerg Spantzel for drawing some exquisite figures. We thank Ning Shang who read the entire manuscript and did most of the exercises to assure that they are stated correctly and doable. We are grateful to Carl Pomerance, Hugh Montgomery and Harold Diamond for valuable discussions of parts of the manuscript. We thank Professor Anant Ramdas for a delightful and educational discussion of diamond cutting.

The first author expresses thanks for financial support from the Research Foundation of CUNY. The second author thanks the Center for Education and Research in Information Assurance and Security, CERIAS, its sponsors and its director, Professor Eugene Spafford, for support while this book was being written.

Carlos J. Moreno   Sam Wagstaff
Baruch College, CUNY   Purdue University CERIAS
New York City   West Lafayette, Indiana
Carlos_Moreno@baruch.cuny.edu   ssw@cerias.purdue.edu

We dedicate this work to our wives,

Ofelia and Cheryl.

Contents

Chapter 1

Introduction

The purpose of this introductory chapter is twofold: to list the prerequisites the reader needs to understand the book and to tell what material is in the rest of the book.

1.1   Prerequisites

It is assumed that the reader is familiar with the basic properties of divisibility, congruences, prime numbers and multiplicative functions. After Chapter 2, we assume the reader is familiar with the Legendre-Jacobi symbol (a/p). Occasionally we will mention related results and give alternate proofs that require a little algebra or analysis, but these may be omitted without loss by the reader having only minimal prerequisites. The books by Hardy and Wright [52], Niven, Zuckerman and Montgomery [79] and Uspensky and Heaslet [113] are good general references for the entire book.

We use these conventions throughout the book. A "mod" in parentheses denotes the congruence relation: $a \equiv b \pmod{m}$ means that the integer m exactly divides $b - a$. A "mod" not in parentheses is the binary operation that gives the least nonnegative remainder: $a \bmod m = a - \lfloor a/m \rfloor m$ is the integer b in the interval $0 \leq b < m$ for which $a \equiv b \pmod{m}$. And $\lfloor x \rfloor$, which we just used, is the greatest integer $\leq x$. We write $\gcd(m, n)$ for the greatest common divisor of m and n. When we write "$A \overset{\text{def}}{=} B$," we are defining A to be B. This implies that the reader should already know the meaning of B. We use $\Re(z)$ and $\Im(z)$ for the real and imaginary parts, respectively, of the complex number z. We use $\mathbb{N}$, $\mathbb{Z}$, $\mathbb{Q}$, $\mathbb{R}$ and $\mathbb{C}$ to denote the natural numbers, the rational integers, the rational numbers, the real numbers and the complex numbers, respectively. We write $\mathbb{F}_q$ for the Galois (finite) field with q elements. The notation $|A|$ means the cardinality of A if A is a set, and the absolute value of A if A is a number. The meaning will always be clear from the context. In Chapter 6 we use the notation $\text{ord}_p(n)$, where p is a prime and n is a nonzero integer, to mean the largest integer e for which p^e

divides n.

Chapters 3-6 require complex analysis at the level of Ahlfors [2]. In some parts of the book, the reader is assumed to know a bit of group theory or linear algebra, mostly about the ring $SL(2,\mathbb{R})$ of 2×2 matrices.

Here and there throughout the book we mention advanced results, like the Riemann-Roch theorem and the Riemann hypothesis for function fields, which are not proved, and which are not used in our proofs. Such references indicate possible directions for further study. We give many references to the literature as suggestions to the reader about what to read after this book. But the reader can understand virtually all of this book with only the limited background just described.

For the algorithms in Chapter 8, the reader is also assumed to know about the Euclidean algorithm and simple continued fractions. For complete understanding of the applications at the end of Chapter 8, the reader needs a first course in physics and a first course in cryptography.

Each chapter concludes with a collection of exercises to test the reader's understanding of the material. A few exercises, especially in Chapter 8, require a computer for their solution.

1.2   Outline of the Rest of the Book

This section summarizes the content of the remainder of this work. In order to describe some of the difficult material we sometimes have to use language not explained until a later chapter. Those familiar with the material can quickly learn what is in this book. The novice should not despair that he cannot understand this introduction, for all the new terms will be defined later.

1.2.1   Chapter 2

Chapter 2 is the simplest chapter of the book. In a sense, it assumes nothing more than high school mathematics. If you understand even and odd numbers and proof by mathematical induction, then you can understand this chapter. However, parts of it are a very sophisticated application of even and odd numbers, and it really is too hard for high school students.

In Chapter 2 we prove formulas for the number of ways of writing a positive integer as a sum of two, of three, and of four squares of integers in Theorems 2.5, 2.7 and 2.6, respectively. The main result of the chapter is Gauss' formula for the number $r_3(n)$ of representations of an integer n as the sum of three squares of integers. From this formula it is clear that this number is positive precisely when $n \geq 0$ and n is not of the form $4^a(8b+7)$.

The triangular numbers are the numbers

$$1, \, 1+2, \, 1+2+3, \, 1+2+3+4, \, 1+2+3+4+5, \, \ldots,$$

that is, numbers of the form $k(k+1)/2$. A corollary of Gauss' theorem, which we prove, is that every positive integer is the sum of at most three triangular numbers. More generally, we will define polygonal numbers and show that every positive integer is the sum of r r-gonal numbers, for each $r \geq 3$. Our proof of the polygonal number theorem is based on Bachmann [5].

Much of this chapter is based on Chapter XIII of Uspensky and Heaslet [113], which in turn goes back to Liouville's series of eighteen papers, "Sur quelques formules générales qui peuvent être utiles dans la théorie des nombres," *Journal de Mathematiques*, series (2), 1858–1865. See Dickson [24], Volume 2, Chapter XI for a summary of these papers. Bachmann [6], Part II, Chapter 8, deals with the same material. Part of the connection between this work and elliptic functions is given in Smith [104]. See Hardy [50] and Rademacher [87] for the standard applications of elliptic functions to sums of squares. Grosswald [42] is another excellent reference for the theory of sums of squares.

Joseph Liouville derived his results on the number of representations of an integer as the sum of a certain number of squares, as well as recurrence formulas for the sum of divisors function $\sigma(n)$, from two "Fundamental Identities," which we prove as Theorems 2.1 and 2.2. He was led to these completely elementary formulas from deep investigations in the theory of elliptic modular forms. We explain their origin in Jacobi's work in Section 6.7. The exercises at the end of Chapter 2 give many more examples of consequences of Liouville's fundamental identities.

1.2.2 Chapter 3

Bernoulli numbers are certain rational numbers that arise in many places in number theory, analysis and combinatorics. In Chapter 3 we prove a selection of important theorems about them, with emphasis on number theory. For example, we prove the von Staudt-Clausen theorem that determines the denominators of the Bernoulli numbers. We use this theorem also to investigate the distribution of the fractional parts of the Bernoulli numbers, which is quite unusual. The graph of the distribution function appears on the cover of this book. Other consequences of the von Staudt-Clausen theorem include J. C. Adams' theorem, the congruences of Voronoi and Kummer, and other congruences which may be used to compute the Bernoulli numbers modulo a prime.

Fermat's Last Theorem is the statement that the equation

$$x^n + y^n = z^n$$

has no solution in positive integers x, y, z when n is an integer > 2. Before Fermat's Last Theorem was proved by Wiles, people used to prove it computationally, one exponent at a time. Whenever a prime p divided the numerator of a Bernoulli number B_{2k} with $2 \leq 2k \leq p - 1$, there was an obstacle to the computational proof of Fermat's Last Theorem for exponent p, and p is called an *irregular prime*. Now people use irregular primes to study the structure of cyclotomic fields. We prove that there are infinitely many irregular primes in certain arithmetic progressions.

Chapter 3 touches some areas of real and complex analysis where Bernoulli numbers play a significant role. We prove the Euler-MacLaurin sum formula, Wallis' product formula for π, and Stirling's formula for $n!$. Bernoulli numbers have a deep connection to the Riemann zeta function. We prove the functional equation for the Riemann zeta function $\zeta(s)$ and its generalization to Dirichlet L-functions. We define Dirichlet characters and generalized Bernoulli numbers in this part of Chapter 3. These concepts are needed for the examples in Chapter 6.

Two old but comprehensive works on Bernoulli numbers are Nielsen [78] and Saalschütz [93]. The books by Rademacher [87] and Uspensky and Heaslet [113] each contain chapters on Bernoulli numbers.

1.2.3 Chapter 4

Chapter 4 is an introduction to three of the central themes of the modern theory of modular forms:

(i) A reciprocity law is an equality between two objects: on one side there is the solution to a diophantine problem, e.g., counting the number of solutions of polynomial equations modulo primes, while on the other side there appears a formula containing spectral information about certain types of arithmetic operators related to the operation of raising a quantity to a p-th power, e.g., the so-called Hecke operators.

(ii) The multiplicative properties of the coefficients of Dirichlet series are closely related to the automorphic properties of their Mellin transforms.

(iii) The functional equations satisfied by certain Dirichlet series, when viewed from the point of view of their Mellin transforms, serve to characterize the modular behavior of the latter.

The first theme is introduced by deriving a connection between an old result of Smith that ascertains that the cardinality of the set

$$A = \{x \in \mathbb{F}_p : x^4 - 2x^2 + 2 = 0\}$$

and the coefficients in the formal q-series expansion

$$q \prod_{n=1}^{\infty} \left(1 - q^{2n}\right) \left(1 - q^{16n}\right) = 1 + \sum_{n=1}^{\infty} a(n)q^n$$

are related, that is to say,

$$|A| = \text{Card } A = 1 + (-1)^{(p-1)/2} + a(p),$$

for p an odd prime. Such a relation is tantamount to the identification of the coefficients $a(n)$ as those of a Dirichlet series associated to an Artin L-function of a two-dimensional representation of the Galois group of the splitting extension of the fourth degree equation that appears in the definition of the set A. The principal results are Theorem 4.1, an elegant deduction of Smith in the style of Jacobi, and Theorem 4.4, the "arithmetic congruence relation," a truly diophantine statement very much in the spirit of Gauss' higher reciprocity laws. Along the same lines we derive a result of Wilton concerning the congruence properties of the Ramanujan tau function $\tau(p)$ modulo the prime 23 and the Galois properties of the polynomial equation $x^3 - x - 1 = 0$ with respect to the finite field $\mathbb{F}_p$. The principal result, given in Theorem 4.6, explicates the deep diophantine significance of a congruence such as

$$q \prod_{n=1}^{\infty} (1 - q^n)^{24} \equiv q \prod_{n=1}^{\infty} (1 - q^n)(1 - q^{23n}) \quad (\text{mod } 23).$$

The second theme, that of the connection between the Fourier coefficients of modular forms and the eigenvalues of certain Hecke operators is developed along the lines pioneered by Mordell in his seminal paper on the proof of the multiplicative property of the Ramanujan tau function $\tau(n)$. The main result is Theorem 4.5, the derivation of which will take us on a tour of the basic theory of modular forms, and will touch on the elementary theory of Hecke operators, the modular group, and fundamental domains.

Our discussion of the last theme in Chapter 4 is carried out by the development of the important theorem of Hamburger, in which the uniqueness of the Riemann zeta function $\zeta(s)$ is determined, under some natural conditions, by the type of functional equation it satisfies as a function of the complex variable s. The principal result is Theorem 4.9. The method used in the proof of this theorem can be thought of as the starting point of the theory of modular forms as developed by Hecke.

1.2.4   Chapter 5

Chapter 5 is an introduction to the basic theory of modular forms on the modular group Γ and its associated subgroup $\Gamma_0(N)$. We build on the elementary methods already discussed in Chapter 4. The main themes are:

(i) The finite dimensionality of the space of holomorphic modular forms of weight k on the modular group Γ and on $\Gamma_0(N)$.

(ii) The normality of the family of Hecke operators, a property which is essential in the proof of the existence of an orthonormal basis with respect to the Petersson inner product.

(iii) The relation between the eigenfunction property of a modular form and the existence of an Euler product representation for the Dirichlet series constructed with the Fourier coefficients of a modular form.

(iv) The elementary theory of Poincaré series, a theory which provides a constructive way of building classical cuspidal modular forms.

The dimensionality results concerning (i) are obtained as a simple case of the Riemann-Roch theorem applied to the Riemann sphere

$$\mathbb{P}^1(\mathbb{C}) \stackrel{\text{def}}{=} \mathbb{C} \cup \{\infty\}.$$

The essential idea consists in deriving a diophantine equation that depends on the weight k and on the multiplicities of the zeros of a modular form within a fundamental domain, taking care to count properly the contributions arising from the elliptic points on the boundary. The resulting equation affords only a finite number of solutions, all of which can be enumerated. With the assistance of the readily available Eisenstein series, it is then possible to verify that all solutions to the diophantine equation correspond to actual modular forms. The more general case of the group $\Gamma_0(N)$ requires an algebraic analysis of the complete Riemann surface

$$X_0(N) \stackrel{\text{def}}{=} \mathcal{H}/\Gamma_0(N) \cup \mathbb{Q}/\Gamma_0(N),$$

itself a collection of orbits of points in the upper half plane under the action of the group of linear fractional transformations $\Gamma_0(N)$, as a finite topological covering of the Riemann sphere $\mathbb{P}^1(\mathbb{C})$.

In developing the topic (ii), the essential property of the Hecke operators that comes into play is their commutativity, a property that results readily from the definition of these operators as modular correspondences whose structure becomes quite explicit when expressed in terms of the Fourier expansions of modular forms. Of course, the central idea here is the existence of the Petersson inner product, which among other things establishes that the finite dimensional vector space of cusp forms is a complete Hilbert space endowed with a commuting family of arithmetically defined operators.

As already indicated in Chapter 4, using the Dirichlet series associated to the Ramanujan modular form, the Fourier coefficients of a simultaneous cuspidal eigenform of the Hecke operators lead to Dirichlet series possessing an Euler product whose local factors are quadratic polynomials in the expression $T = p^{-s}$. When these Euler products are completed with suitable gamma factors, the local factor corresponding to the ordinary absolute value on $\mathbb{Q}$, the resulting function of the complex variable s is holomorphic and satisfies a functional equation that relates its value at s with that at $k - s$.

The treatment we give of the topic in (iv) takes place in the context of the elementary theory of Poincaré series with respect to the full modular group Γ. The main and most significant result established provides the Fourier expansion of the Poincaré series of weight k, a result with many important implications for the analytic theory of numbers. At the heart of these considerations one finds the Kloosterman sum

$$S_c(m,n) \stackrel{\text{def}}{=} \sum_{\substack{a=1 \\ ab \equiv 1 \pmod{c}}}^{c} \exp\left\{\frac{2\pi i}{c}(am + bn)\right\},$$

an object with deep roots in the applications of harmonic analysis to the study of the additive as well as the multiplicative properties of the integers. Some of the important properties of these sums are developed in the exercises.

1.2.5 Chapter 6

The crowning achievement of the modern theory of modular forms as it regards the problem of representing integers as sums of squares is the clear understanding of the role played by the theta series

$$\vartheta(z) = \sum_{n \in \mathbb{Z}} q^{n^2}, \quad \text{where} \quad q = e^{2\pi i z},$$

in the representation theory of the group $SL(2)$ of 2×2 matrices. This theory reached its highest points in the analytic work of Carl L. Siegel on quadratic forms in several variables and in the work of André Weil concerning the representation theory of the metaplectic group.

The generating function for $r_s(n)$, the number of representations of the integer n as a sum of s squares, is of course, the s-power $\vartheta(z)^s$:

$$\left(1 + 2\sum_{n=1}^{\infty} q^{n^2}\right)^s = 1 + \sum_{n=1}^{\infty} r_s(n)q^n.$$

The conditions in Fermat's theorem about the representability of integers as a sum of two squares are of a purely local nature—an odd prime p is a sum of two squares if and only if it is congruent to 1 modulo 4. Similarly, Gauss' criteria for the representability of an integer as a sum of 3 squares is essentially of a local nature. The work of Smith as well as that of Minkowski concerning the number of possible representations of an integer as a sum of five squares requires the calculation of local factors attached to all the primes. It was the great insight of Hardy and Ramanujan to have discovered how to apply to modular forms, such as the theta series $\vartheta(z)^s$, the local-to-global ideas embodied in the Hardy-Littlewood method. Along these lines, the central result of the theory is the realization that for $s = 1, 2, 3, 4, 5, 6, 7$, and 8, the modular form $\vartheta(z)^s$ is of weight $s/2$ and coincides with an Eisenstein series $E_{s/2}(z)$ whose Fourier series could be calculated by a method similar to that described in Chapter 5. We prove in this chapter the beautiful formula

$$r_s(n) = \rho_s(n), \quad \text{for } 1 \leq s \leq 8,$$

where $\rho_s(n)$ is the so-called singular series, an Euler product constructed with local data, and whose value can be made explicit as a quotient of two classical Dirichlet L-functions. The latter can in turn be evaluated using generalized Bernoulli numbers.

For $s > 8$, $\vartheta(z)^s$ is no longer an Eisenstein series. In fact, the difference $\vartheta(z)^s - E_{s/2}(z)$ is a cusp form of weight $s/2$. When s is even, the classical theory provides a representation for the difference $\vartheta(z)^s - E_{s/2}(z)$ as a sum of a finite number of cusp forms, all eigenfunctions of the Hecke operators. The comparison of the Fourier coefficients leads to an explicit formula for $r_s(n) - \rho_s(n)$ as a sum of multiplicative arithmetic functions closely related to the eigenvalues of the Hecke operators. We make this connection explicit in this chapter and provide many examples.

The case when s is odd is remarkably more difficult and at the same time more interesting. In this case $\vartheta(z)^s$ is a modular form of half-integral weight, and the modular transformation law

$$\vartheta\left(\frac{az+b}{cz+d}\right)^s = J_s(\sigma,z)\vartheta(z)^s, \qquad \text{where} \qquad \sigma = \begin{pmatrix} a & b \\ c & d \end{pmatrix},$$

provides a factor of automorphy which satisfies

$$J_s(\sigma,z)^2 = \chi_s(\sigma)(cz+d)^s,$$

with $\chi_s : \Gamma \longrightarrow \mathbb{C}^\times$ a character on the modular group. Shimura's theory of modular forms of half-integral weight and their lifts to modular forms of integral weight lead, for a fixed integer t, to a representation of the difference

$$r_s(tn^2) - \rho_s(tn^2)$$

as a linear combination of multiplicative arithmetic functions of n^2 with coefficients depending on t. The elucidation of the exact value of these coefficients comes from a beautiful theorem of Waldspurger concerning the quadratic twists of Dirichlet L-functions of modular forms at the center of their critical strip. The full treatment of Waldspurger's theorem is beyond the elementary scope this book and will not be treated in this chapter. We do establish the existence of exact formulas for the representation function $r_s(n)$ for all s.

1.2.6   *Chapter 7*

Chapter 7 contains a selection of results from additive number theory, some of which are related to sums of squares, the main subject of this book. In 1973, E. Szemerédi [109] proved a long-standing conjecture which says that if the length of the arithmetic progressions in a sequence of positive integers is bounded above, then the sequence must be very thin (have zero density). The original motivation for this conjecture was to understand better the theorem of van der Waerden, which says that if the set of positive integers is partitioned into a finite number of subsets, then one subset contains arbitrarily long arithmetic progressions. The thinking was that clearly one of the subsets in the partition must be not too thin (i.e., must have positive density), and therefore that one probably would have arbitrarily long arithmetic progressions. Roth [92] made the first dent in the conjecture when he showed that every sequence with positive density must contain infinitely many arithmetic progressions of three terms. We prove Roth's theorem, Szemerédi's theorem and some other results about arithmetic progressions such as van der Waerden's theorem. The latter is Khinchin's [62] first pearl. The other theorems about arithmetic progressions have not appeared in books, but some of them are quoted in Ostmann [81]. Halberstam and Roth [45] is another good general reference for additive number theory.

In the final section of Chapter 7, we determine all arithmetic progressions of three squares. This is equivalent to finding all ways of expressing twice a square as the sum of two squares. An exercise treats the case of four squares in arithmetic progression.

1.2.7 Chapter 8

In Chapter 8 we present a more practical side to complement the theory of the earlier chapters. We tell how to find a representation of n as a sum of s squares when such a representation exists. Finding the representations of n as the sum of two squares (when there are such representations) turns out to be equivalent to factoring n, and we show that equivalence in the first two sections of this chapter. Perhaps surprisingly, it is easy to compute the representations of a large positive integer as the sum of three or more squares (when such representations exist) without knowing its prime factorization, and we tell how this may be done in the next two sections. We consider when one may represent a positive integer as the sum of a certain number of *positive* squares. (In the rest of the book, we allow 0 as a square summand.) Then we tell how to compute the number $r_s(n)$ of representations of n as a sum of s squares for any s and n for which this is possible with reasonable effort.

We end this chapter and book with three little-known applications having no apparent connection to sums of squares or even to number theory. We show that the number of ways to write a positive integer as the sum of three positive squares determines the number of eigenfrequencies for microwave radiation in a cube-shaped resonant cavity. We use the structure of crystals to explain why the number of facets on a round brilliant-cut diamond might be related to the total number of ways of writing the integers 1, 2, 3, 4 and 5 as the sum of three squares. Finally, we tell how to compromise one variation of the RSA signature scheme, which is widely used in Internet commerce, by constructing a bogus valid signature for a message not actually signed by the alleged signer. The attacker does this by writing a certain number as the sum of two squares in two different ways.

Chapter 2

Elementary Methods

In this chapter we count the number of ways an integer can be written as the sum of two, three or four squares of integers. We also show that every positive integer is the sum of three triangular numbers, of four squares, of five pentagonal numbers, etc.

Most of the results in this chapter are proved by completely elementary means developed by Liouville. Some of the results were originally obtained by far more abstruse arguments. Uspensky [112] asserts that all results previously obtained using elliptic functions may be established as well by purely arithmetical methods of extremely elementary nature. He refers to earlier papers in which he tells how to do this. Much of this chapter is based on his book [113] with Heaslet, whose last chapter illustrates some of the elementary techniques.

2.1 Introduction

For positive integers k and n, let $r_k(n)$ denote the number of ways n can be expressed as the sum of exactly k squares of integers, counting all permutations and changes of sign as different representations. Let $r_k(0) = 1$. We have $r_3(6) = 24$, for example, because

$$6 = (\pm 2)^2 + (\pm 1)^2 + (\pm 1)^2 = (\pm 1)^2 + (\pm 2)^2 + (\pm 1)^2 = (\pm 1)^2 + (\pm 1)^2 + (\pm 2)^2,$$

and the three $\pm$ signs in each representation are independent.

Clearly,

$$r_1(n) = \begin{cases} 2 & \text{if } n \text{ is the square of a positive integer} \\ 1 & \text{if } n = 0 \\ 0 & \text{otherwise.} \end{cases}$$

For $k = 2$, 3 and 4 we will find simple formulas for $r_k(n)$. For $k = 4$ we will prove Jacobi's theorem

$$r_4(n) = \begin{cases} 8\sigma(n) & \text{if } n \text{ is odd} \\ 24\sigma(m) & \text{if } n \text{ is even and } m \text{ is its largest odd divisor.} \end{cases}$$

Here $\sigma(n) = \sum_{d|n} d$ is the sum of all positive divisors of n.

Since $\sigma(n) > 0$ for every positive integer n, it follows from Jacobi's theorem that $r_4(n)$ is always positive, that is, every positive integer is the sum of four squares of integers. However, some numbers (such as 7) are not the sum of three squares. Our formulas for $r_k(n)$ will enable us to determine which n are the sum of k squares. The answer is quite simple for $k = 3$: $r_3(n) = 0$ if and only if $n = 4^a(8b + 7)$ for some nonnegative integers a and b. This statement is hard to prove.

We will also prove recursion formulas for the divisor functions $\sigma_k(n) = \sum_{d|n} d^k$, that is, the sum of the k-th powers of all positive divisors of n. We have $\sigma(n) = \sigma_1(n)$ and $\tau(n) = \sigma_0(n) =$ the number of positive divisors of n. If the prime factorization of n is $n = p_1^{\alpha_1} \cdots p_r^{\alpha_r}$, then $\tau(n) = (1+\alpha_1)\cdots(1+\alpha_r)$ and, for $k > 0$,

$$\sigma_k(n) = \prod_{i=1}^{r} \frac{p_i^{k(1+\alpha_i)} - 1}{p_i^k - 1}.$$

The σ_k are *multiplicative* functions, that is, $\sigma_k(mn) = \sigma_k(m)\sigma_k(n)$ whenever m and n are relatively prime. For $k = 2$ and 4 we will prove that the functions $f_k(n) = r_k(n)/r_k(1)$ are multiplicative. This function is multiplicative also for $k = 1$ (obvious) and $k = 8$ (not obvious) and for no other positive integer k.

Finally, we shall prove a theorem about representing a positive integer as a sum of polygonal numbers. The partial sums of the arithmetic progressions

1	2	3	4	...	n	...
1	3	5	7	...	$2n-1$	...
1	4	7	10	...	$3n-1$	...
1	5	9	13	...	$4n-1$	...

...

are called the triangular, square, pentagonal, hexagonal, etc., numbers. In general, the n-th r-gonal number $p_n^{(r)}$ is the sum of the first n terms of the arithmetic progression with first term 1 and common difference $r - 2$. Thus,

$$p_n^{(r)} = \frac{n}{2}(2 + (n-1)(r-2)) = n + (r-2)\frac{n(n-1)}{2}.$$

Since we make the convention that the empty sum is zero, it is natural to define $p_0^{(r)} = 0$ for each $r > 2$. Figure 2.1 illustrates the reason for this geometric terminology.

We will prove that, for each $r \geq 3$, every positive integer is the sum of r r-gonal numbers. In case $r \geq 5$, all but four of the r-gonal numbers can be 0

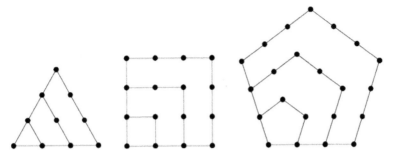

Figure 2.1 Triangular, Square and Pentagonal Numbers.

or 1. This theorem was proved by Cauchy in 1813. The special cases $r = 4$ and $r = 3$ had been proved earlier by Lagrange and Gauss, respectively.

2.2 Some Lemmas

This section contains a few lemmas we will need later.

LEMMA 2.1
If n is an odd positive integer, then n is a square if and only if $\sigma(n)$ is odd.

PROOF   Write $n = p_1^{\alpha_1} \cdots p_r^{\alpha_r}$, where the p_i are odd primes and the α_i are positive integers. We have $\sigma(p^\alpha) = 1 + p + \ldots + p^\alpha \equiv \alpha + 1 \pmod 2$. Hence, n is a square if and only if every α_i is even, that is, if and only if every $(\alpha_i + 1)$ is odd, that is, if and only if every $\sigma(p_i^{\alpha_i})$ is odd, that is, if and only if $\sigma(n)$ is odd.   ∎

LEMMA 2.2
Every square is congruent to 0 or 1 modulo 4. Every sum of two squares is congruent to 0, 1 or 2 modulo 4. Every square is congruent to 0, 1 or 4 modulo 8. Every sum of two squares is congruent to 0, 1, 2, 4 or 5 modulo 8. Every sum of three squares is congruent to 0, 1, 2, 3, 4, 5 or 6 modulo 8.

PROOF   The square of an even number $2n$ is a multiple $4n^2$ of 4. The square of an odd number $2n + 1$ is $4(n(n + 1)) + 1$, which is 1 more than a multiple of 8 because one of n, $n + 1$ must be even. Now enumerate all possible cases.   ∎

LEMMA 2.3
For every nonnegative integer n, we have $r_3(4n) = r_3(n)$ and $r_2(2n) = r_2(n)$.

PROOF If $4n = x_1^2 + x_2^2 + x_3^2$, then the integers x_i must all be even by the proof of Lemma 2.2. Thus, the equations $x_i = 2y_i$ give a one-to-one correspondence between solutions of $4n = x_1^2 + x_2^2 + x_3^2$ and solutions of $n = y_1^2 + y_2^2 + y_3^2$.

If $2n = x_1^2 + x_2^2$, then the integers x_1 and x_2 must have the same parity (both are even or both are odd). Hence, the equations

$$y_1 = \frac{x_1 + x_2}{2} \qquad y_2 = \frac{x_1 - x_2}{2}$$
$$x_1 = y_1 + y_2 \qquad x_2 = y_1 - y_2$$

give a one-to-one correspondence between solutions of $2n = x_1^2 + x_2^2$ and solutions of $n = y_1^2 + y_2^2$. ∎

LEMMA 2.4

If $n = 4^a(8b + 7)$ for some nonnegative integers a and b, then n is not the sum of three squares of integers.

PROOF The case $a = 0$ follows from Lemma 2.2. If $a > 0$, then a applications of the first part of Lemma 2.3 reduce the general case to $a = 0$. ∎

Later, as Theorem 2.7, we will prove the converse of this lemma.

2.3 Two Fundamental Identities

We will prove two identities that form the basis for Liouville's work. They involve general real-valued functions defined for integer arguments. Such a function $F(x_1, \ldots, x_n)$ is called an *even* (*odd*) function of the variables $x_{i_1}, \ldots, x_{i_\ell}$ if, for all integers $x_1, \ldots, x_n$, the value of the function is unchanged (multiplied by -1) when $x_{i_1}, \ldots, x_{i_\ell}$ are simultaneously replaced by their negatives.

For example, the function

$$F(x, y, z) = \begin{cases} 0 & \text{if } x \text{ or } z \text{ is even} \\ (-1)^{(x+z)/2+y}(2y - z) & \text{if } x \text{ and } z \text{ are both odd} \end{cases}$$

is an odd function of x and an even function of y, z.

We will use these conventions for the notation of sums. When n is a fixed positive integer, the notation $\sum_{d|n} f(d)$ means the sum of $f(d)$ where d runs over all positive divisors of n. We will often write sums like $\sum_{n=i^2+d\delta} f(d)$. This notation means the sum of $f(d)$ where i runs through all integers, positive, zero and negative, while d and δ run through all positive integers, for which $n = i^2 + d\delta$. The squared variables may be positive, zero and negative integers while the unsquared variables are positive integers. When the conditions are more complicated we will write $\sum_{(a)} f(d)$ and list the conditions as "(a) $n = i^2 + d\delta$, d and δ odd" on a separate line, for example.

The two fundamental identities are formulas relating sums of values of a function $F(x, y, z)$ over complicated sets of argument values. Let n be a positive integer and λ and μ be two real numbers, such that $\lambda > |\mu|$. Consider partitions of the following types:

$$
\begin{align}
\text{(a)} \quad & n = \lambda i^2 + \mu i + (\lambda \delta + \mu)d \\
\text{(b)} \quad & n = \lambda i^2 - \mu i + (\lambda \delta - \mu)d \\
\text{(c)} \quad & n = \lambda h^2 + \mu h + \lambda \Delta \Delta' \\
\text{(d)} \quad & n = \lambda \left(\frac{\Delta + \Delta'}{2} \right)^2 + \mu \left(\frac{\Delta - \Delta'}{2} \right) \\
\text{(e)} \quad & n = \lambda s^2 + \mu s.
\end{align}
$$

In all five partitions, n, λ and μ are fixed and the other letters are variables. In (a) and (b), i is an integer and d and δ are positive integers. In (c), h is an integer and Δ and Δ' are positive integers. In (d), Δ and Δ' are positive integers of the same parity. In (e), s is an integer. The conditions on λ and μ insure that the number of sets of parameter values is finite for each n. At most two integers satisfy (e) because it is a quadratic equation in s. Also, $s \neq 0$ because n is a positive integer.

For any function F of three variables and fixed n, λ and μ, let

$$S = \sum_{(a)} F(\delta - 2i, d + i, 2d + 2i - \delta)$$

$$S' = \sum_{(b)} F(\delta - 2i, d + i, 2d + 2i - \delta)$$

$$W = \sum_{(c)} F(\Delta + \Delta', h, \Delta - \Delta')$$

$$U = \sum_{(d)} F\left(\Delta + \Delta', \frac{\Delta - \Delta'}{2}, \Delta - \Delta' \right)$$

$$T = \sum_{(e)} \sum_{j=1}^{2|s|-1} F(2|s| - j, |s|, 2|s| - j).$$

THEOREM 2.1 First Fundamental Identity

With the notation above, if $F(x, y, z)$ is an odd function of x and an even function of y, z, then we have $S + S' = T + W - U$ for every positive integer n.

PROOF   Split S into three parts S_1, S_2, S_3 corresponding to the solutions (i, d, δ) to (a) in which the expression $2i + d - \delta$ is positive, zero and negative, respectively. Split S' into three parts S_1', S_2', S_3' in a similar fashion using (b). We shall prove the identity by showing that $S_1 = S_1' = 0$, $S_2 + S_2' = T$ and $S_3 + S_3' = W - U$.

Consider S_1 first. Given an integer i and positive integers d and δ, such that $n = \lambda i^2 + \mu(d + i) + \lambda d\delta$ and $2i + d - \delta > 0$, let $i' = \delta - i$, $d' = 2i + d - \delta$ and $\delta' = \delta$. Then i' is an integer and d' and δ' are positive integers, such that $n = \lambda i'^2 + \mu(d' + i') + \lambda d'\delta'$ and $2i' + d' - \delta' > 0$. It is easy to solve for i, d and δ in terms of the primed variables: $i = \delta' - i'$, $d = 2i' + d' - \delta'$ and $\delta = \delta'$. The six equations relating i, d, δ and i', d', δ' determine a one-to-one correspondence between solutions (i, d, δ) and (i', d', δ') of (a) with $2i + d - \delta > 0$. Hence S_1 is unchanged when (i, d, δ) is replaced by (i', d', δ'). Thus,

$$
\begin{aligned}
S_1 &= \sum F(\delta - 2i, d + i, 2d + 2i - \delta) \\
&= \sum F(\delta' - 2i', d' + i', 2d' + 2i' - \delta') \\
&= \sum F(\delta - 2(\delta - i), (2i + d - \delta) + (\delta - i), 2(2i + d - \delta) + 2(\delta - i) - \delta) \\
&= \sum F(-(\delta - 2i), d + i, 2d + 2i - \delta) \\
&= -S_1
\end{aligned}
$$

because F is an odd function of its first argument. Therefore, $S_1 = 0$.

Similarly, $S_1' = 0$. (Just replace μ by $-\mu$ in the above argument.)

In the sum S_2 we have $2i + d - \delta = 0$. Therefore, d and δ have the same parity and $i = (\delta - d)/2$. Substituting this expression for i into (a) we find

$$
n = \lambda \left(\frac{\delta - d}{2} \right)^2 + \mu \left(\frac{\delta - d}{2} \right) + (\lambda \delta + \mu)d = \lambda s^2 + \mu s,
$$

where $s = (\delta + d)/2$ is a positive integer because d and δ are positive integers with the same parity. We have $d + \delta = 2s$, $\delta - 2i = 2s - \delta$, $d + i = s$ and $2d + 2i - \delta = 2s - \delta$. Thus $S_2 = 0$ unless $n = \lambda s^2 + \mu s$, in which case (writing j for δ) we have

$$
\begin{aligned}
S_2 &= \sum_{\substack{(a) \\ 2i+d-\delta=0}} F(\delta - 2i, d + i, 2d + 2i - \delta) \\
&= \sum_{\substack{(e) \\ s>0}} \sum_{j=1}^{2s-1} F(2s - j, s, 2s - j).
\end{aligned}
$$

Similarly, $S_2' = 0$ unless $n = \lambda s^2 - \mu s$, in which case (replacing s by $-s$) we

have

$$S_2' = \sum_{\substack{n=\lambda s^2 - \mu s \\ s>0}} \sum_{j=1}^{2s-1} F(2s-j, s, 2s-j)$$

$$= \sum_{\substack{(e) \\ s<0}} \sum_{j=1}^{2|s|-1} F(2|s|-j, |s|, 2|s|-j).$$

We cannot have $s = 0$ because $n > 0$. Hence, $S_2 + S_2' = T$.

Now consider S_3. Given an integer i and positive integers d and δ, such that $n = \lambda i^2 + \mu(d+i) + \lambda d\delta$ and $2i + d - \delta < 0$, let $h = d+i$, $\Delta = d$ and $\Delta' = \delta - d - 2i$. It is easy to check that h, Δ and Δ' satisfy (c) and $\Delta' - \Delta + 2h = \delta > 0$. Conversely, given integers h, Δ and Δ' satisfying (c) and $\Delta' - \Delta + 2h = \delta > 0$, if we let $i = h - \Delta$, $d = \Delta$ and $\delta = \Delta' - \Delta + 2h$, then i, d, δ is a solution to (a) satisfying $2i + d - \delta < 0$. Thus,

$$S_3 = \sum_{\substack{(a) \\ 2i+d-\delta<0}} F(\delta - 2i, d+i, 2d + 2i - \delta)$$

$$= \sum_{\substack{(c) \\ \Delta'-\Delta+2h>0}} F(\Delta + \Delta', h, \Delta - \Delta').$$

Similarly,

$$S_3' = \sum_{\substack{n=\lambda h^2 - \mu h + \lambda \Delta \Delta' \\ \Delta'-\Delta+2h>0}} F(\Delta + \Delta', h, \Delta - \Delta').$$

Replacing h by $-h$ and interchanging Δ and Δ', we get

$$S_3' = \sum_{\substack{(c) \\ \Delta'-\Delta+2h<0}} F(\Delta + \Delta', -h, \Delta' - \Delta)$$

$$= \sum_{\substack{(c) \\ \Delta'-\Delta+2h<0}} F(\Delta + \Delta', h, \Delta - \Delta')$$

because $F(x, y, z)$ is an even function of y, z. Hence $S_3 + S_3'$ differs from W only by the terms in which $\Delta' - \Delta + 2h = 0$. In these terms, Δ and Δ' have the same parity. These terms combine to give U. Hence, $S_3 + S_3' = W - U$.

∎

The difference between the hypotheses of the two fundamental identities is that F is an *even* function of y, z in the first but an *odd* function of y, z in the second.

THEOREM 2.2 Second Fundamental Identity
Let n, λ, μ, S, S', W and U be as in the First Fundamental Identity. If $F(x,y,z)$ is an odd function of x and an odd function of y, z, then we have $S - S' = T_1 - T_2 + W - U$ for every positive integer n, where

$$T_1 = \sum_{\substack{n=\lambda s^2 + \mu s \\ s \text{ is a positive integer}}} \sum_{j=1}^{2s-1} F(2s-j, s, 2s-j)$$

and

$$T_2 = \sum_{\substack{n=\lambda s^2 - \mu s \\ s \text{ is a positive integer}}} \sum_{j=1}^{2s-1} F(2s-j, s, 2s-j).$$

PROOF   Split S and S' into three parts as in the proof of Theorem 2.1. Since $F(x,y,z)$ is an odd function of x we have $S_1 = S_1' = 0$. Also, $S_2 = T_1$ and $S_2' = T_2$. Since $F(x,y,z)$ is an odd function of y, z, it follows that $S_3 - S_3'$ differs from W only by terms in which $\Delta' - \Delta + 2h = 0$, that is, by U.   ∎

2.4   Euler's Recurrence for $\sigma(n)$

According to Section 2.1, the pentagonal numbers are the integers of the form

$$s + (5-2)\frac{s(s-1)}{2} = \frac{3s^2 - s}{2}$$

for nonnegative integers s. In the present section only, it will be convenient to allow s to be negative and say that the pentagonal numbers are all the numbers $(3s^2 + s)/2$ for any integer s. Note that distinct integers s produce distinct pentagonal numbers.

Most of the identities in this chapter follow from the First Fundamental Identity. We will make exactly one application of the Second Fundamental Identity. The function

$$F(x,y,z) = \begin{cases} 0 & \text{if } x \text{ or } z \text{ is even} \\ (-1)^{(x+z)/2+y} & \text{if } x \text{ and } z \text{ are both odd} \end{cases}$$

is an odd function of x and an odd function of y, z. We apply the Second Fundamental Identity with $\lambda = 3/2$, $\mu = 1/2$ and n a positive integer. Note that

$$U = \sum_{(d)} F\left(\Delta + \Delta', \frac{\Delta - \Delta'}{2}, \Delta - \Delta'\right) = 0$$

because $\Delta + \Delta'$ is even.

In T_1, we have $n = (3s^2 + s)/2$, and in T_2, we have $n = (3s^2 - s)/2$, where s is a positive integer in each case. Suppose $n = (3s^2 + s)/2$ for some integer s. If $s > 0$, then $T_2 = 0$ and

$$T_1 = \sum_{j=1}^{2s-1} F(2s - j, s, 2s - j)$$

$$= \sum_{\substack{j=1 \\ j \text{ odd}}}^{2s-1} (-1)^{s+j} = (-1)^{s-1}s.$$

If $s < 0$, then $T_1 = 0$ and

$$-T_2 = -\sum_{j=1}^{2(-s)-1} F(2(-s) - j, (-s), 2(-s) - j)$$

$$= -\sum_{\substack{j=1 \\ j \text{ odd}}}^{-2s-1} (-1)^{s+j} = (-1)^{s-1}s.$$

Thus,

$$T_1 - T_2 = \begin{cases} 0 & \text{if } n \text{ is not pentagonal} \\ (-1)^{s-1}s & \text{if } n = (3s^2 + s)/2 \text{ for some integer } s. \end{cases}$$

The Second Fundamental Identity becomes

$$S - S' = \underset{(a)}{\sum(-1)^i} - \underset{(b)}{\sum(-1)^i} = W + T_1 - T_2 = \underset{(c)}{\sum(-1)^{h+\Delta}} + T_1 - T_2,$$

where the partitions are

(a) $\quad n = \dfrac{3i^2 + i}{2} + \dfrac{3\delta + 1}{2}d \qquad \delta \text{ odd}$

(b) $\quad n = \dfrac{3i^2 - i}{2} + \dfrac{3\delta - 1}{2}d \qquad \delta \text{ odd}$

(c) $\quad n = \dfrac{3h^2 + h}{2} + \dfrac{3}{2}\Delta\Delta' \qquad \Delta \not\equiv \Delta' \pmod 2.$

The sum on (c) vanishes because the terms $(-1)^{h+\Delta}$ and $(-1)^{h+\Delta'}$ cancel, since Δ and Δ' have opposite parity. They must have opposite parity since $F(x, y, z) = 0$ when x or z is even, and here $x = \Delta + \Delta'$ and $z = \Delta - \Delta'$. We have proved

$$\underset{(a)}{\sum(-1)^i} - \underset{(b)}{\sum(-1)^i} = \begin{cases} 0 & \text{if } n \text{ is not pentagonal} \\ (-1)^{s-1}s & \text{if } n = (3s^2 + s)/2. \end{cases} \qquad (2.1)$$

DEFINITION 2.1   *If m is a positive integer, define $w(m)$ to be the number of divisors of m with the form $3k+1$ diminished by the number of divisors of m with the form $3k-1$. In symbols,*

$$w(m) = \sum_{\substack{d|m \\ d \equiv 1 \ (\mathrm{mod}\ 3)}} 1 - \sum_{\substack{d|m \\ d \equiv 2 \ (\mathrm{mod}\ 3)}} 1.$$

Write $\delta = 2k-1$ in (a) and $\delta = 2k+1$ in (b). Then replace i by $-i$ in (b). These partitions become

$$\text{(a}_1\text{)} \qquad n = \frac{3i^2 + i}{2} + (3k-1)d \qquad k > 0,\, d > 0$$

$$\text{(b}_1\text{)} \qquad n = \frac{3i^2 + i}{2} + (3k+1)d \qquad k \geq 0,\, d > 0$$

Recall that n is a fixed positive integer. For each fixed i, the sum $\sum_{(a_1)} 1$ counts the number of divisors of $n - (3i^2 + i)/2$ with the form $3k-1$ and the sum $\sum_{(b_1)} 1$ counts the number of divisors of $n - (3i^2 + i)/2$ with the form $3k+1$. Hence, for fixed n and i,

$$\sum_{(b_1)} 1 - \sum_{(a_1)} 1 = w(n - (3i^2 + i)/2).$$

Thus, allowing i again to vary,

$$\sum_{(a_1)} (-1)^i \; \sum_{(b_1)} (-1)^i = \sum_i (-1)^i w\left(n - \frac{3i^2 + i}{2}\right).$$

Using Equation (2.1), we have proved

THEOREM 2.3 Recursion Formula for $w(m)$
Let n be a positive integer. Then

$$\sum_{\substack{\text{integer } i \\ (3i^2+i)/2 < n}} (-1)^i w\left(n - \frac{3i^2 + i}{2}\right) = \begin{cases} 0 & \text{if } n \text{ is not pentagonal} \\ (-1)^s s & \text{if } n = (3s^2 + s)/2 \\ & \text{for some integer } s. \end{cases}$$

We now head towards a recursion formula for $\sigma(n)$. In the First Fundamental Identity, let n be a positive integer, $\lambda = 3/2$, $\mu = 1/2$ and

$$F(x, y, z) = \begin{cases} 0 & \text{if } x \text{ or } z \text{ is even} \\ (-1)^{(x+z)/2 + y}(2y - z) & \text{if } x \text{ and } z \text{ are odd} \end{cases}$$

(which is an odd function of x and an even function of y, z, as required). The identity becomes

$$\sum_{(a)}(-1)^i\delta + \sum_{(b)}(-1)^i\delta = \sum_{(c)}(-1)^{h+\Delta}(2h + \Delta' - \Delta) + T. \qquad (2.2)$$

The sum $\sum_{(c)}(-1)^{h+\Delta}2h$ vanishes as in the previous application. The partition (e) is $n = (3s^2+s)/2$, so $T = 0$ unless n is pentagonal, say $n = (3s^2+s)/2$, in which case

$$T = \sum_{j=1}^{2|s|-1} F(2|s| - j, |s|, 2|s| - j)$$

$$= \sum_{\substack{j=1 \\ j \text{ odd}}}^{2|s|-1} (-1)^{-j+|s|}j = (-1)^{s-1}s^2.$$

Adding $3/2$ times Equation (2.2) to $1/2$ times Equation (2.1) gives

$$\sum_{(a)}(-1)^i\frac{3\delta + 1}{2} + \sum_{(b)}(-1)^i\frac{3\delta - 1}{2} = \frac{3}{2}\sum_{(c)}(-1)^{h+\Delta}(\Delta' - \Delta) + V, \quad (2.3)$$

where

$$V = \begin{cases} 0 & \text{if } n \text{ is not pentagonal} \\ (-1)^{s-1}n & \text{if } n = (3s^2 + s)/2 \text{ for some integer } s. \end{cases}$$

In (a) and (b), δ is odd, so as before we let $\delta = 2k - 1$ in (a) and $\delta = 2k + 1$ in (b). Replacing i by $-i$ in (b), we find that the new partitions are

$$\text{(a)} \qquad n = \frac{3i^2 + i}{2} + (3k - 1)d \qquad k > 0, d > 0$$

$$\text{(b)} \qquad n = \frac{3i^2 + i}{2} + (3k + 1)d \qquad k \geq 0, d > 0$$

For $j = 0, 1, 2$, let $\sigma_{(j)}(m) = \sum d$, where the sum is taken over all positive integer divisors d of m with $d \equiv j \pmod 3$.

Thus, $\sigma_{(0)}(m) + \sigma_{(1)}(m) + \sigma_{(2)}(m) = \sigma(m)$,

$$\sum_{(a)}(-1)^i\frac{3\delta + 1}{2} = \sum_i(-1)^i\sigma_{(2)}\left(n - \frac{3i^2 + i}{2}\right)$$

and

$$\sum_{(a)}(-1)^i\frac{3\delta - 1}{2} = \sum_i(-1)^i\sigma_{(1)}\left(n - \frac{3i^2 + i}{2}\right),$$

where the sums on the right sides run over all integers i for which $(3i^2+i)/2 <$ n. We will prove that

$$W \stackrel{\text{def}}{=} \frac{3}{2} \sum_{(c)} (-1)^{h+\Delta} (\Delta' - \Delta) = -\sum_h (-1)^h \sigma_{(0)} \left(n - \frac{3h^2 + h}{2} \right),$$

where the sum on the right side runs over all integers h for which $(3h^2+h)/2 <$ n. Substituting the last three equations into Equation (2.3) gives Euler's recurrence formula for $\sigma(m)$:

THEOREM 2.4 Euler's Recurrence Formula for $\sigma(m)$

Let n be a positive integer. Then

$$\sum_{(3i^2+i)/2 < n} (-1)^i \sigma \left(n - \frac{3i^2 + i}{2} \right) = \begin{cases} 0 & \text{if } n \text{ is not pentagonal} \\ (-1)^{s-1} n & \text{if } n = (3s^2 + s)/2 \\ & \text{for some integer } s. \end{cases}$$

PROOF   We have

$$W = \frac{3}{2} \sum_{(3h^2+h)/2 < n} (-1)^h \sum_{\substack{\Delta, \Delta' > 0, \Delta \not\equiv \Delta' \pmod 2 \\ n - (3h^2+h)/2 = (3/2)\Delta\Delta'}} (-1)^\Delta (\Delta' - \Delta).$$

For fixed h, if $n - (3h^2+h)/2$ is not divisible by 3, then the inner sum is zero because there are no appropriate Δ, Δ'. Otherwise, write $(2/3)\left(n - (3h^2 + h)/2 \right) =$ $\Delta\Delta' = 2^{\alpha+1} M$, where M is odd and $\alpha \geq 0$. (At least one 2 must divide $\Delta\Delta'$ because Δ and Δ' have opposite parity.) The inner sum becomes

$$\sum_{\substack{\Delta\Delta'=2^{\alpha+1}M, \, \Delta, \Delta'>0 \\ \Delta \not\equiv \Delta' \pmod 2}} (-1)^\Delta (\Delta' - \Delta) = 2 \sum_{\substack{\Delta\Delta'=2^{\alpha+1}M, \, \Delta, \Delta'>0 \\ \Delta \text{ even}, \, \Delta' \text{ odd}}} (\Delta' - \Delta)$$

$$= 2 \sum_{\Delta'|M} \Delta' - 2 \sum_{\beta|M} 2^{\alpha+1}\beta \qquad (\Delta = 2^{\alpha+1}\beta)$$

$$= -2 \left(2^{\alpha+1} - 1 \right) \sigma(M)$$

$$= -2\sigma(2^\alpha M)$$

$$= -2\sigma \left(\left(n - \left(\frac{3h^2 + h}{2} \right) \right) / 3 \right)$$

$$= -\frac{2}{3}\sigma_{(0)} \left(n - \frac{3h^2 + h}{2} \right).$$

Therefore,

$$W = - \sum_{(3h^2+h)/2 < n} (-1)^h \sigma_{(0)} \left(n - \frac{3h^2 + h}{2} \right).$$

∎

Euler's recurrence formula may be stated in the alternative form

$$\sum_{(3i^2+i)/2 \le n} (-1)^i \sigma\left(n - \frac{3i^2+i}{2}\right) = 0.$$

or

$$\sigma(n) = \sigma(n-1) + \sigma(n-2) - \sigma(n-5) - \sigma(n-7) + \sigma(n-12) + \sigma(n-15) - \cdots,$$

where the term $\sigma(n-n)$, if it appears, has the value n. There are approximately $2\sqrt{\frac{2n}{3}}$ terms in this formula, which makes it useful in preparing tables of $\sigma(n)$ for n up to a few thousand.

2.5   More Identities

Let $\lambda = 1$ and $\mu = 0$. The partitions of n in the fundamental identities become

$$\begin{array}{cc} \text{(a)} & n = i^2 + \delta d \\ \text{(b)} & n = i^2 + \delta d \\ \text{(c)} & n = h^2 + \Delta\Delta', \end{array}$$

where i and h are integers and d, δ, Δ and Δ' are positive integers.
   The second fundamental identity is not interesting here since $S = S'$, $T_1 = T_2$ and $W = U = 0$, but the first fundamental identity becomes

$$2 \sum_{\text{(a)}} F(\delta - 2i, d + i, 2d + 2i - \delta) = \tag{2.4}$$

$$\sum_{\text{(c)}} F(\Delta + \Delta', h, \Delta - \Delta') + T - U,$$

where $T = U = 0$ unless $n = s^2$ for some positive integer s, in which case

$$T = 2 \sum_{j=1}^{2s-1} F(2s - j, s, 2s - j)$$

and

$$U = \sum_{j=1}^{2s-1} F(2s, j - s, 2j - 2s).$$

(We have written $2s$ for $\Delta + \Delta'$ and j for Δ in U.)
   In this section we give four special cases of (2.4). The reader should verify each time that $F(x, y, z)$ is an odd function of x and an even function of y, z, for this is assumed in (2.4).

Let $f(y)$ be an odd function of y and let

$$F(x, y, z) = \begin{cases} 0 & \text{if } x \text{ or } z \text{ is even} \\ (-1)^{(x+z)/2+y} f(y) & \text{if } x \text{ and } z \text{ are both odd.} \end{cases}$$

Since F is an odd function of y, the terms for (h, Δ, Δ') and $(-h, \Delta, \Delta')$ cancel in the sum on (c); hence that sum vanishes. Also, $U = 0$ because $x = 2s$ is even, and $T = 0$ unless $n = s^2$ for some positive integer s, in which case

$$T = 2 \sum_{\substack{j=1 \\ j \text{ odd}}}^{2s-1} (-1)^{j+s} f(s) = 2(-1)^{s-1} s f(s).$$

It is convenient to introduce the following notation. If E is an expression involving n and/or s, let $\{E\} = E$ if $n = s^2$ for some positive integer s and $\{E\} = 0$ if n is not the square of a positive integer. Using this notation the T of the preceding paragraph may be written $T = \{2(-1)^{s-1} s f(s)\}$. Then Equation (2.4), with the F just defined, becomes

$$\sum_{\text{(a), } \delta \text{ odd}} (-1)^i f(d+i) = \{(-1)^{s-1} s f(s)\}. \tag{2.5}$$

We must have δ odd because $F(x, y, z) = 0$ if $x = \delta - 2i$ is even.

Now suppose $F(x, y, z) = 0$ when x or y is odd, and let n be an odd positive integer. Then we may assume that δ and h are even. Replacing them by 2δ and $2h$, respectively, (2.4) becomes

$$2 \sum_{\text{(e)}} F(2\delta - 2i, d+i, 2d+2i-2\delta) = \sum_{\text{(f)}}{}' F(\Delta + \Delta', 2h, \Delta - \Delta') + T - U,$$

where the partitions are

$$\begin{array}{ll} \text{(e)} & n = i^2 + 2d\delta, \ d \text{ odd} \\ \text{(f)} & n = 4h^2 + \Delta\Delta'. \end{array}$$

(Since n is odd, i must be odd because $n = i^2 + 2d\delta$, and so d has to be odd because $F(x, y, z) = 0$ when $y = d+i$ is odd. Also, Δ and Δ' must both be odd because of (f).) Let $f(t)$ be an odd function of t. For

$$F(x, y, z) = \begin{cases} 0 & \text{if } x \text{ or } y \text{ is odd} \\ f(x/2) & \text{if } x \text{ and } y \text{ are even,} \end{cases}$$

we get

$$2 \sum_{\text{(e)}} f(\delta - i) = \sum_{\text{(f)}} f\left(\frac{\Delta + \Delta'}{2}\right) + T - U.$$

Since n is odd, s must be odd, too, if $n = s^2$. Therefore, $T = 0$. Also, $U = \{sf(s)\}$. Replacing i by $-i$ we have

$$\sum_{(f)} f\left(\frac{\Delta + \Delta'}{2}\right) = 2\sum_{(e)} f(\delta + i) + \{sf(s)\}. \qquad (2.6)$$

Again letting $f(t)$ be an odd function of t, define

$$F(x, y, z) = \begin{cases} 0 & \text{if } x \text{ or } y \text{ is odd} \\ (-1)^{y/2} f(x/2) & \text{if } x \text{ and } y \text{ are both even.} \end{cases}$$

Arguing as above we find

$$\sum_{(f)} (-1)^h f\left(\frac{\Delta + \Delta'}{2}\right) = \qquad (2.7)$$

$$2\sum_{(e)} (-1)^{(d-1)/2 + (i-1)/2} f(\delta + i) + \{(-1)^{(s-1)/2} sf(s)\}.$$

We make one more application of (2.4). Suppose $F(x, y, z) = 0$ when x is odd or y is even, and let $n \equiv 1 \pmod 4$. Then we may assume in (a) and (c) that $\delta - 2i$ and $\Delta + \Delta'$ are even, and $d + i$ and h are odd. This implies that i is odd and d, δ, Δ and Δ' are all even. If we replace d, δ, Δ and Δ' by $2d$, 2δ, 2Δ and $2\Delta'$, respectively, the partitions (a) and (c) become

$$(a_1) \qquad n = i^2 + 4d\delta$$
$$(c_1) \qquad n = h^2 + 4\Delta\Delta'.$$

With f an odd function, let

$$F(x, y, z) = \begin{cases} 0 & \text{if } x \text{ is odd or } y \text{ is even} \\ f(x/2) & \text{if } x \text{ is even and } y \text{ is odd.} \end{cases}$$

In T and U, s must be odd, so j must be even. Hence,

$$T = \left\{ 2\sum_{k=1}^{s-1} f(k) \right\} \text{ and } U = \{(s-1)f(s)\},$$

and Equation (2.4) becomes

$$2\sum_{(a_1)} f(\delta + i) = \qquad (2.8)$$

$$\sum_{(c_1)} f(\Delta + \Delta') + \left\{ 2\sum_{k=1}^{s-1} f(k) - (s-1)f(s) \right\}.$$

The five identities (2.4), (2.5), (2.6), (2.7) and (2.8) will be used to prove formulas for the number of ways of representing n as a sum of two, three and four squares.

2.6   Sums of Two Squares

We apply Equation (2.4) with $n \equiv 1 \pmod 4$ and

$$F(x,y,z) = \begin{cases} (-1)^{(x+z-2)/4} & \text{if } y \text{ is even, } x \equiv 2 \pmod 4 \\ & \qquad \text{and } z \equiv 0 \pmod 4 \\ 0 & \text{otherwise.} \end{cases}$$

Then F is odd in x and even in y, z, as Equation (2.4) requires. We may assume that $d+i$ and h are even, $\delta - 2i \equiv \Delta + \Delta' \equiv 2 \pmod 4$ and $2d + 2i - \delta \equiv \Delta - \Delta' \equiv 0 \pmod 4$. This implies that d, i, Δ and Δ' are all odd. Replacing h by $2h$, (2.4) becomes

$$2 \sum_{\substack{n=i^2+d\delta \\ d,i \text{ odd}}} (-1)^{(d-1)/2} = \sum_{\substack{n=4h^2+\Delta\Delta' \\ \Delta,\Delta' \text{ odd}}} (-1)^{(\Delta-1)/2} + T - U. \qquad (2.9)$$

If $n = s^2$, then s is odd, so $T = 0$ and $U = \{1\}$.

Let $\chi(d) = (-1/d)$, the Jacobi symbol. That is,

$$\chi(d) = \begin{cases} 0 & \text{if } d \text{ is even} \\ 1 & \text{if } d \equiv 1 \pmod 4 \\ -1 & \text{if } d \equiv 3 \pmod 4 \end{cases} = \begin{cases} 0 & \text{if } d \text{ is even} \\ (-1)^{(d-1)/2} & \text{if } d \text{ is odd.} \end{cases}$$

The function $\chi(d)$ is totally multiplicative, that is, $\chi(de) = \chi(d)\chi(e)$ for all integers d and e.

Define $T(0) = 1/4$ and $T(n) = \sum_{d|n} \chi(d)$, where the sum is over the positive divisors d of n. Then T is multiplicative, that is, $T(mn) = T(m)T(n)$ for all relatively prime positive integers m and n. We have

$$T(n) = \sum_{\text{odd } d|n} (-1)^{(d-1)/2} = \sum_{\substack{d|n \\ d\equiv 1 \pmod 4}} 1 - \sum_{\substack{d|n \\ d\equiv 3 \pmod 4}} 1.$$

Since T is multiplicative, it is determined by its values on prime powers. They are

$$T(p^e) = \chi(1) + \chi(p) + \cdots + \chi(p^e) = \begin{cases} 1 & \text{if } p = 2 \\ e+1 & \text{if } p \equiv 1 \pmod 4 \\ 1 & \text{if } p \equiv 3 \pmod 4 \text{ and } e \text{ is even} \\ 0 & \text{if } p \equiv 3 \pmod 4 \text{ and } e \text{ is odd.} \end{cases}$$

By the prime number theorem, any positive integer n may be written in the form

$$n = 2^\gamma p_1^{\alpha_1} \cdots p_k^{\alpha_k} q_1^{\beta_1} \cdots q_\ell^{\beta_\ell},$$

where the $p_i \equiv 1 \pmod 4$, the $q_j \equiv 3 \pmod 4$ and the exponents are nonnegative integers.

If any β_j is odd, then $T(n) = 0$. If all β_j are even, then

$$T(n) = (\alpha_1 + 1) \cdots (\alpha_k + 1) = \tau\left(p_1^{\alpha_1} \cdots p_k^{\alpha_k}\right),$$

where $\tau(m)$ is the number of positive integer divisors of m.

If $n \equiv 3 \pmod 4$, then some β_j must be odd and so $T(n) = 0$. Note that $T(2n) = T(n)$ for every positive integer n.

When $n \equiv 1 \pmod 4$, Equation (2.9) can be rewritten as (see below for the $T - U$)

$$2 \sum_{\substack{i^2 \le n \\ i \text{ odd}}} T(n - i^2) = \sum_{4h^2 < n} T(n - 4h^2).$$

If $n = i^2$, where i is odd, then also $n = (-i)^2$, and we get $2\left(\frac{1}{4} + \frac{1}{4}\right) = 1$ on the left side, which accounts for the $T - U = \{-1\}$ in Equation (2.9).

If $n \equiv 3 \pmod 4$, then $r_2(n) = 0$ by Lemma 2.2. Also, $r_2(2n) = r_2(n)$ for all n by Lemma 2.3. Now let $n \equiv 1 \pmod 4$ and let

$$R = \sum_{4h^2 < n} r_2(n - 4h^2) \quad \text{and} \quad S = \sum_{i^2 \le n,\, i \text{ odd}} r_2(n - i^2).$$

Then R is the number of solutions of $n = x^2 + y^2 + z^2$ with x even $(x = 2h)$ and S is the number of solutions of the same equation with x odd $(x = i)$. But two of x, y, z are even and one is odd since $n \equiv 1 \pmod 4$. Hence, $R = 2S$, that is

$$2 \sum_{i^2 \le n,\, i \text{ odd}} r_2(n - i^2) = \sum_{4h^2 < n} r_2(n - 4h^2).$$

A simple induction on n shows that $r_2(n) = 4T(n)$ for all $n \ge 0$, and we have proved the following theorem, due to Gauss. Much earlier, Fermat had proved Corollary 2.1 and the fact that n is not the sum of two squares if any prime $q \equiv 3 \pmod 4$ divides n to an odd power.

THEOREM 2.5 Sums of Two Squares
Let $n = 2^\gamma p_1^{\alpha_1} \cdots p_k^{\alpha_k} q_1^{\beta_1} \cdots q_\ell^{\beta_\ell}$, where the p_i and the q_j are primes $\equiv 1$ and $3 \pmod 4$, respectively. Then

$$r_2(n) = 4 \sum_{\substack{d|n, d>0 \\ d \text{ odd}}} (-1)^{(d-1)/2} = \begin{cases} 4\tau\left(p_1^{\alpha_1} \cdots p_k^{\alpha_k}\right) & \text{if all } \beta_j \text{ are even} \\ 0 & \text{if any } \beta_j \text{ is odd} \end{cases}$$

and

$$r_2(n) = 4 \left(\sum_{\substack{d|n, d>0 \\ d \equiv 1 \pmod 4}} 1 - \sum_{\substack{d|n, d>0 \\ d \equiv 3 \pmod 4}} 1 \right).$$

If p is prime and e is a nonnegative integer, then

$$r_2(p^e) = \begin{cases} 4 & \text{if } p = 2, \text{ or } p \equiv 3 \pmod 4 \text{ and } e \text{ is even} \\ 4(e+1) & \text{if } p \equiv 1 \pmod 4 \\ 0 & \text{if } p \equiv 3 \pmod 4 \text{ and } e \text{ is odd.} \end{cases}$$

COROLLARY 2.1
If $p \equiv 1 \pmod 4$ is prime, then there are unique positive integers x and y, such that x is odd, y is even and $p = x^2 + y^2$.

PROOF   When $p \equiv 1 \pmod 4$, we have $r_2(p) = 8$, so there are integers a, b with $p = a^2 + b^2$. Let $x = |a|$ and $y = |b|$. Then $p = x^2 + y^2$. The integers x and y must be positive because p is not a square. One must be even and the other odd because p is odd. Hence, the representations

$$p = (\pm x)^2 + (\pm y)^2 = (\pm y)^2 + (\pm x)^2,$$

where the $\pm$ signs are independent, give all eight ways to write p as the sum of two squares, so x and y are unique.   ∎

COROLLARY 2.2
The number of divisors d congruent to 1 modulo 4 of a positive integer n is at least as great as the number of divisors d congruent to 3 modulo 4.

PROOF   This is clear from the second displayed formula in the theorem and the fact $r_2(n) \geq 0$.   ∎

There is a simpler argument that proves Corollary 2.1 without obtaining a formula for $r_2(n)$. Let $f(x) = x$ in Equation (2.5). This leads easily to

$$R(n) - 2R(n - 1^2) + 2R(n - 2^2) - + \cdots = \{(-1)^{n-1}n\},$$

where $R(m) = \sum_{\substack{d|m \\ m/d \text{ odd}}} d$. If n is prime and $n \equiv 1 \pmod 4$, then $R(n) = n + 1$ and $\{(-1)^{n-1}n\} = 0$. We obtain

$$R(n - 1^2) - R(n - 2^2) + - \cdots = \frac{n+1}{2} \equiv 1 \pmod 2.$$

Hence, $R(n - i^2)$ is odd for some i. But $R(2^e m) = 2^e \sigma(m)$ for all odd m. Therefore, $n - i^2$ is odd and $\sigma(n - i^2)$ is odd. By Lemma 2.1, $n - i^2$ is a square, say j^2, and $n = i^2 + j^2$. The uniqueness is easy to prove directly.

The proof of Theorem 2.5 can be modified to give formulas for the number of representations of n in the forms $x^2 + 2y^2$ and $x^2 + 3y^2$. Just replace $\chi(d)$ by the Jacobi symbols $(-2/d)$ and $(-3/d)$, respectively.

2.7 Sums of Four Squares

The goal of this section is to prove Jacobi's theorem, which expresses the number $r_4(n)$ of ways of writing a nonnegative integer n as a sum of four squares of integers in terms of the sum-of-divisors function $\sigma(n)$. Many years before Jacobi's work, Lagrange, building on ideas of Euler, had proved that every positive integer is the sum of four squares of nonnegative integers, that is, that $r_4(n) > 0$ for every positive integer n.

THEOREM 2.6 (Jacobi)   Sums of Four Squares
If n is a positive integer, then

$$r_4(n) = \begin{cases} 8\sigma(n) & \text{if } n \text{ is odd} \\ 24\sigma(m) & \text{if } n \text{ is even and } m \text{ is its largest odd divisor.} \end{cases}$$

The proof will require one lemma.

LEMMA 2.5
If n is a positive integer, then

$$r_4(2n) = \begin{cases} 3r_4(n) & \text{if } n \text{ is odd} \\ r_4(n) & \text{if } n \text{ is even.} \end{cases}$$

PROOF   First suppose that n is even. Then 4 divides $2n$ and in

$$2n = x^2 + y^2 + z^2 + w^2 \tag{2.10}$$

the numbers x, y, z, w are either all even or all odd. Therefore, the integers

$$a = \frac{x+y}{2}, \quad b = \frac{x-y}{2}, \quad c = \frac{z+w}{2}, \quad d = \frac{z-w}{2}$$

satisfy

$$n = a^2 + b^2 + c^2 + d^2. \tag{2.11}$$

Conversely, given a solution to Equation (2.11), the numbers

$$x = a+b, \quad y = a-b, \quad z = c+d, \quad w = c-d, \tag{2.12}$$

satisfy Equation (2.10). We have a one-to-one correspondence between solutions of Equations (2.10) and (2.11), and this shows $r_4(2n) = r_4(n)$ when n is even.

Now suppose n is odd. In Equation (2.11) either one or three of a, b, c, d are odd, according as $n \equiv 1$ or 3 (mod 4). In Equation (2.10), two of x, y, z, w are even and two are odd. Let P be the number of solutions of (2.10), in which x and y are even and z and w are odd. Since there are six ways to choose the two even numbers in (2.10), we have $r_4(2n) = 6P$. Let Q be the number of solutions of (2.11), in which a and b have the same parity and c and d have opposite parity. In every solution of Equation (2.11), either a and b have the same parity or c and d have the same parity (and not both). Therefore, $r_4(n) = 2Q$. But Equation (2.12) gives a one-to-one correspondence between solutions of Equation (2.10) counted in P and solutions of Equation (2.11) counted in Q. Hence, $P = Q$ and $r_4(2n) = 3r_4(n)$ when n is odd. ∎

We can now prove Theorem 2.6.

PROOF   If $n \equiv 3 \pmod 4$ and the equation

$$n = x^2 + y^2 + z^2 + v^2 + w^2 \tag{2.13}$$

has solutions, then in any solution three of x, y, z, v, w are odd and two are even. Let

$$R = \sum_{4h^2 \leq n} r_4(n - 4h^2)$$

be the number of solutions in which $x = 2h$ is even, and

$$S = \sum_{i^2 < n,\ i \text{ odd}} r_4(n - i^2)$$

be the number of solutions in which $x = i$ is odd. Then $R = (2/3)S$, and this holds even if (hypothetically) Equation (2.13) has no solution. We have

$$\sum_{4h^2 \leq n} r_4(n - 4h^2) = \frac{2}{3} \sum_{i^2 < n,\ i \text{ odd}} r_4(n - i^2) \qquad (n \equiv 3 \pmod 4).) \tag{2.14}$$

It follows from Lemma 2.5 that $r_4(4m) = 3r_4(m)$ when m is odd and $r_4(4m) = r_4(m)$ when m is even. Hence,

$$r_4(j) = \begin{cases} 3r_4(j/4) & \text{if } j \equiv 4 \pmod 8 \\ r_4(j/4) & \text{if } j \equiv 0 \pmod 8. \end{cases}$$

When $4|j$, $r_4(j/4)$ tells the number of representations of j as a sum of four even squares.

Let $n \equiv 1 \pmod 4$. We claim that

$$\sum_{4h^2 \leq n} r_4(n - 4h^2) = 4 \sum_{i^2 \leq n,\ i \text{ odd}} r_4 \left(\frac{n - i^2}{4} \right) \qquad (n \equiv 1 \pmod 4).) \tag{2.15}$$

When $n \equiv 1 \pmod 8$ in Equation (2.13), one of x, y, z, v, w is odd and four of them are even. The $x = 2h$ on the left side of (2.15) is even and the $x = i$ on the right side is odd. Equation (2.15) holds because x is even four times as often as it is odd. When $n \equiv 5 \pmod 8$ in Equation (2.13) it could happen that all five variables are odd. These solutions are not counted on the left side of Equation (2.15) because $x = 2h$ is forced to be even. They are not counted on the right side either, because $r_4((n-i^2)/4)$ counts only the representations of $n - i^2$ as a sum of four even squares. In this case both sides of Equation (2.15) count the representations (2.13) with one of x, y, z, v, w odd and four of them even, and Equation (2.15) holds as for the case $n \equiv 1 \pmod 8$.

We now use Equations (2.6) and (2.8) to construct a simple function that satisfies the same identities as $r_4(n)$. In (2.6), n must be an odd positive integer and $f(x)$ must be an odd function. With $f(x) = x$ we have

$$\underset{\text{(a)}}{\sum} \frac{\Delta + \Delta'}{2} = 2 \underset{\text{(b)}}{\sum} (\delta + i) + \{ss\},$$

where the partitions are

$$\begin{array}{ll} \text{(a)} & n = 4h^2 + \Delta\Delta' \\ \text{(b)} & n = i^2 + 2d\delta, \ d \text{ odd.} \end{array}$$

Clearly, $\sum_{(b)} i = 0$ and the left side is $\sum_{4h^2 < n} \sigma(n - 4h^2)$.

Let $n \equiv 3 \pmod 4$. Then $\{s^2\} = 0$ because n is not a square. Also, $\sum_{(b)} \delta = (1/3) \sum_{i^2 < n} \sigma(n - i^2)$ because $n - i^2 \equiv 2 \pmod 4$, d is odd and $\sigma(2) = 3$. Hence, we have

$$\sum_{4h^2 \leq n} \sigma(n - 4h^2) = \frac{2}{3} \sum_{i^2 < n, \ i \text{ odd}} \sigma(n - i^2) \qquad (n \equiv 3 \pmod 4.)$$

For $n \equiv 5 \pmod 8$ we have $\delta \equiv 2 \pmod 4$ in (b) and

$$\underset{\text{(b)}}{\sum} \delta = 2 \sum_i \left(\sum_{(n-i^2)/4 = d(\delta/2)} \frac{\delta}{2} \right) = 2 \sum_i \sigma \left(\frac{n - i^2}{4} \right).$$

Therefore, we have

$$\sum_{4h^2 \leq n} \sigma(n - 4h^2) = 4 \sum_{i^2 < n} \sigma \left(\frac{n - i^2}{4} \right) \qquad (n \equiv 5 \pmod 8.)$$

If $n \equiv 1 \pmod 8$, then δ is divisible by 4. Replacing δ by 4δ, we have

$$\sum_{4h^2 \leq n} \sigma(n - 4h^2) = 4 \underset{\text{(c)}}{\sum} \left(1 - (-1)^d\right) \delta + \{n\}, \qquad (2.16)$$

with partition (c) $n = i^2 + 8d\delta$.

To simplify the right side of (2.16), we use (2.8), in which $n \equiv 1 \pmod 4$ and $f(x)$ is an odd function. With $f(x) = (-1)^x x$, Equation (2.8) becomes

$$\sum_{(d)} (-1)^{d+\delta}(d+\delta) = 2 \sum_{(d)} \left((-1)^{\delta+i}\delta + (-1)^{\delta+i}i \right)$$

$$+ \left\{ (s-1)(-1)^s s - 2 \sum_{k=1}^{s-1} (-1)^k k \right\},$$

with partition (d) $n = i^2 + 4d\delta$. Since s must be odd, the expression in braces becomes

$$(s-1)(-s) - 2(-1+2-3+4-+\cdots+(s-1)) = s - s^2 - 2(s-1)/2 = -n+1.$$

Also, since i is odd,

$$\sum_{(d)} (-1)^{\delta+i}i = -\sum_{(d)} (-1)^{\delta}i = 0$$

because terms with $+i$ and $-i$ cancel. Hence we get

$$\sum_{(d)} (-1)^{\delta} \left((-1)^d d + (-1)^d \delta + 2\delta \right) = \{-n+1\}.$$

Also,

$$\sum_{(d)} (-1)^{d+\delta}d = \sum_{(d)} (-1)^{d+\delta}\delta,$$

so

$$2\sum_{(d)} (-1)^{\delta}\delta \left((-1)^d + 1 \right) = \{-n+1\}.$$

The terms in which d is odd vanish. Change d to $2d$ and we have

$$4\sum_{(c)} (-1)^{\delta}\delta = \{-n+1\}. \tag{2.17}$$

Adding (2.16) and (2.17) gives

$$\sum_{4h^2 \le n} \sigma(n - 4h^2) = 4\sum_{(c)} \left(1 - (-1)^d - (-1)^{\delta} \right) \delta + \{1\}.$$

It is easy to see that

$$\sum_{d\delta=m} \left(1 - (-1)^d - (-1)^{\delta} \right) \delta = 3\bar{\sigma}(m),$$

where $\bar{\sigma}(m) = \sum_{d|m,\, d \text{ odd}} d$ is the sum of all odd positive divisors of m, and we have

$$\sum_{4h^2 \le n} \sigma(n - 4h^2) = 12 \sum_{i^2 < n,\, i \text{ odd}} \bar{\sigma}\left(\frac{n - i^2}{4}\right) + \{1\} \qquad (n \equiv 1 \ (\text{mod } 8).)$$

A little more notation will enable us to combine the formulas for $n \equiv 1 \ (\text{mod } 8)$ and $n \equiv 5 \ (\text{mod } 8)$. Define $\chi(0) = 1/8$ and

$$\chi(n) = (2 + (-1)^n)\bar{\sigma}(n)$$

for positive integers n. Notice that $\chi(n) = \sigma(n)$ when n is odd and $\chi(n) = 3\bar{\sigma}(n)$ when n is even. In case $n \equiv 2 \ (\text{mod } 4)$, $\chi(n) = 3\bar{\sigma}(n) = 3\sigma(n/2) = \sigma(n)$. We have

$$\sum_{4h^2 \le n} \chi(n - 4h^2) = 4 \sum_{i^2 \le n,\, i \text{ odd}} \chi\left(\frac{n - i^2}{4}\right) \qquad (n \equiv 1 \ (\text{mod } 4).)$$

(Note that if $n = i^2$, then $n = (-i)^2$, too, and $4(\chi(0) + \chi(0)) = 1 = \{1\}$.) Also, since $\chi(n) = \sigma(n)$ when $n \not\equiv 0 \ (\text{mod } 4)$, we have

$$\sum_{4h^2 \le n} \chi(n - 4h^2) = \frac{2}{3} \sum_{i^2 \le n\, i \text{ odd}} \chi\left(n - i^2\right) \qquad (n \equiv 3 \ (\text{mod } 4).)$$

These are the analogues of Equations (2.15) and (2.14). The formula

$$\chi(2n) = \begin{cases} 3\chi(n) & \text{if } n \text{ is odd} \\ \chi(n) & \text{if } n \text{ is even.} \end{cases}$$

corresponds to Lemma 2.5, and an easy induction on n shows that $r_4(n) = 8\chi(n) = 8\left(2 + (-1)^n\right)\bar{\sigma}(n)$ for all nonnegative integers n, which is Jacobi's theorem. ∎

Note that χ is multiplicative and always positive. Therefore, $r_4(n)/r_4(1) = r_4(n)/8 = \chi(n)$ is multiplicative, and every positive integer is the sum of four squares.

REMARK 2.1   Let $n = 4m$ with m odd. By Jacobi's theorem, $r_4(n) = 24\sigma(m)$. In the equation

$$n = 4m = x^2 + y^2 + z^2 + w^2,$$

either all of x, y, z, w are even or else all are odd. The number of solutions having all four of these numbers even is $r_4(m) = 8\sigma(m)$. Hence, the number of representations of $4m$ as the sum of four odd squares is $16\sigma(m)$, and the number of ways of expressing $4m$ as the sum of the squares of four positive odd numbers is $\sigma(m)$. ∎

2.8   Still More Identities

In this section we use the identities of Section 2.5 to prove some others that we will need in the proof of the three squares theorem in the next section.

Let us begin with (2.5):

$$\sum_{(c)}(-1)^i\psi(d'+i) = \{(-1)^{s-1}s\psi(s)\}, \qquad (2.5)$$

where (c) is $n = i^2 + d'\delta'$, δ' odd, d' and δ' are positive integers, and $\psi(x)$ is an odd function of x. Let d'' and δ'' be positive integers, such that $d''\delta'' < n$ and δ'' is odd. Let $f(x)$ be an even function of x, so that $\psi(x) = f(x - d'') - f(x + d'')$ is an odd function of x. Replace n by $n - d''\delta''$ and (2.5) becomes

$$\sum_{(c_1)}(-1)^i(f(d'+i-d'') - f(d'+i+d'')) = \{(-1)^{s-1}s(f(s-d'') - f(s+d''))\},$$

where the partition (c_1) is $n - d''\delta'' = i^2 + d'\delta'$, δ' odd, and the right side vanishes unless $n - d''\delta'' = s^2$ for some positive integer s. Summing this equation over all possible d'', δ'', we find (see below)

$$\sum_{(d)}(-1)^i(f(d'-d''+i) - f(d'+d''+i)) = \sum_{(e)}(-1)^i i\, f(i+d), \quad (2.18)$$

with partitions

$$\begin{array}{lll} \text{(d)} & n = i^2 + d'\delta' + d''\delta'', & \delta',\delta'' \text{ odd} \\ \text{(e)} & n = i^2 + d\delta, & \delta \text{ odd} \end{array}$$

(On the right side of Equation (2.18), $i = +s$ gives

$$(-1)^i i f(i+d) = (-1)^{s-1}s(-f(s+d''))$$

while $i = -s$ gives

$$(-1)^i i f(i+d) = (-1)^{-s}(-s)f(-s+d'') = (-1)^{s-1}sf(s-d'')$$

because $f(x)$ is an even function.)

With the same even function $f(x)$ as above, define the two functions

$$F(x,y,z) = \begin{cases} (-1)^{(x-1)/2-(z+1)/2+y}(2y-z)f(y) & \text{if both } x \text{ and } z \text{ are odd} \\ 0 & \text{if either } x \text{ or } z \text{ is even,} \end{cases}$$

and

$$F_1(x,y,z) = \begin{cases} (-1)^{(x-1)/2-(z+1)/2+y}yf(y) & \text{if both } x \text{ and } z \text{ are odd} \\ 0 & \text{if either } x \text{ or } z \text{ is even.} \end{cases}$$

The reader should check that each of F and F_1 is an odd function of x and an even function of y, z. Applying Equation (2.4) to F and then to F_1 yields the two identities

$$2\sum_{(e)}(-1)^i\delta f(d+i) = \sum_{(f)}(-1)^{\Delta'-1+h}(2h+\Delta'-\Delta)f(h)+2\{(-1)^{s-1}s^2f(s)\}$$

and

$$2\sum_{(e)}(-1)^i(d+i)f(d+i) = \sum_{(f)}(-1)^{\Delta'-1+h}hf(h)+2\{(-1)^{s-1}s^2f(s)\},$$

where the partition (f) is $n = h^2 + \Delta\Delta'$ with $\Delta + \Delta'$ odd. Now

$$\sum_{(f)}(-1)^{\Delta'-1+h}hf(h) = 0$$

and

$$\sum_{(f)}(-1)^{\Delta'-1+h}(\Delta'-\Delta)f(h) = -2\sum_{(e)}(-1)^h(d-\delta)f(i).$$

If we subtract the two identities we got from Equation (2.4), we have

$$\sum_{(e)}(-1)^iif(d+i) = \sum_{(e)}(-1)^i(\delta-d)f(d+i)+\sum_{(e)}(-1)^i(d-\delta)f(i)$$

and (2.18) becomes

$$\sum_{(d)}(-1)^i(f(d'-d''+i)-f(d'+d''+i)) = \tag{2.19}$$

$$\sum_{(e)}(-1)^i((\delta-d)f(d+i)+(d-\delta)f(i)),$$

valid for even functions $f(x)$. If $f(x)$ is an even function and i is a fixed integer, then

$$\psi_i(x) = f(x+i)+f(x-i)$$

is an even function of x. Let

$$w_i(n) = \sum_{(g)}(\psi_i(d'-d'')-\psi_i(d'+d''))-\sum_{(h)}(\delta-d)(\psi_i(d)-\psi_i(0)),$$

where the partitions are

(g)	$n = d'\delta' + d''\delta''$,	δ', δ'' odd
(h)	$n = d\delta,$	δ odd.

With this notation we can write (2.19) in the compact form

$$\omega_0(n) + 2 \sum_{i^2 < n,\ i>0} (-1)^i \omega_i(n - i^2) = 0.$$

One shows easily by induction on n that $\omega_0(n) = 0$ for every n, that is,

$$\sum_{(g)} (f(d' - d'') - f(d' + d'')) = \sum_{(h)} (\delta - d)(f(d) - f(0)) \qquad (2.20)$$

holds for every even function $f(x)$.

Now let $n = 2m$ be even. Then in (g) the numbers d' and d'' must have the same parity. Likewise, d must be even in (h). Replace d by $2d$ and change (h) to

$$(k) \qquad m = d\delta, \qquad \delta \text{ odd}.$$

When d' and d'' are even in (g) replace them by $2d'$ and $2d''$, respectively. Equation (2.20) becomes

$$\sum_{\substack{m=d'\delta'+d''\delta'' \\ \delta',\delta'' \text{ odd}}} (f(2d' - 2d'') - f(2d' + 2d'')) +$$

$$\sum_{(\ell)} (f(d' - d'') - f(d' + d'')) = \sum_{(k)} (\delta - 2d)(f(2d) - f(0)),$$

with partition

$$(\ell) \qquad 2m = d'\delta' + d''\delta'', \qquad d', d'', \delta', \delta'' \text{ odd}.$$

Replace $f(x)$ by $f(2x)$ and n by m in (2.20):

$$\sum_{\substack{m=d'\delta'+d''\delta'' \\ \delta',\delta'' \text{ odd}}} (f(2d' - 2d'') - f(2d' + 2d'')) = \sum_{(k)} (\delta - d)(f(2d) - f(0)).$$

Subtracting the last two equations yields the marvelous identity

$$\sum_{(\ell)} (f(d' - d'') - f(d' + d'')) = \sum_{(k)} d(f(0) - f(2d)). \qquad (2.21)$$

We will use (2.21) in the next section to find a formula for $r_3(n)$. We give a different application of it here. Let m be an odd positive integer and let $f(x) = x^2$, a well-known even function. Then $f(d' - d'') - f(d' + d'') = -4d'd''$ and $f(0) - f(2d) = -4d^2$. Canceling the -4's, Equation (2.21) becomes

$$\sum_{\substack{k=1 \\ k \text{ odd}}}^{2m-1} \sigma(k)\sigma(2m - k) = \sigma_3(m).$$

(Here, $d'|k$, $d''|(2m-k)$ and $d|m$.) But, for k odd, $\sigma(k)$ is the number of representations of $4k$ as a sum of the squares of four positive odd integers. The equation shows that $\sigma_3(m)$ is the number of representations of $8m$ as a sum of the squares of eight positive odd integers.

As another easy application of (2.21), one can show that every prime $p \equiv 1 \pmod 4$ can be expressed as the sum of two squares of integers by letting $f(x) = \cos(\pi x/2)$ and $m = p$ in (2.21). Note that

$$f(x) = \cos(\pi x/2) = \begin{cases} (-1)^{x/2} & \text{if } x \text{ is even} \\ 0 & \text{if } x \text{ is odd.} \end{cases}$$

Equation (2.21) becomes

$$S \stackrel{\text{def}}{=} \sum_{(\ell)} \left((-1)^{(d'-d'')/2} - (-1)^{(d'+d'')/2} \right) = \sum_{(k)} d \left(1 - (-1)^d \right).$$

As (k) in this case is, $p = d\delta$ with δ odd, the right side is just $1(1-(-1)^1) + p(1-(-1)^p) = 2(p+1)$ whenever p is an odd prime. If also $p \equiv 1 \pmod 4$, then the right side must be $\equiv 4 \pmod 8$.

On the left side (ℓ) is $2p = d'\delta' + d''\delta''$, where d', δ', d'' and δ'' are odd positive integers. It is easy to see (by trying $\pm 1 \pmod 4$ for for d' and d'') that $(d'-d'')/2$ and $(d'+d'')/2$ always have different parity so that

$$T \stackrel{\text{def}}{=} (-1)^{(d'-d'')/2} - (-1)^{(d'+d'')/2} = \pm 2.$$

If $d' \neq \delta'$ in (ℓ), then there are four solutions with the same numbers:

$$2p = d'\delta' + d''\delta'' = \delta'd' + d''\delta'' = d''\delta'' + d'\delta' = d''\delta'' + \delta'd'.$$

The first and third solutions give the same $T = \pm 2$ and so do the second and fourth ones. Hence, the contribution to S from these four solutions is either 0 or $4(\pm 2)$, a multiple of 8 in any case. The same result holds when $d'' \neq \delta''$.

The sum of multiples of 8 must be a multiple of 8. But we have seen that when p is prime and $p \equiv 1 \pmod 4$, then the right side must be $\equiv 4 \pmod 8$, not a multiple of 8. This can happen only if there is a solution to (ℓ) with $d' = \delta'$ and $d'' = \delta''$, that is, $2p = (d')^2 + (d'')^2$, and so p is the sum of two squares by Lemma 2.3.

Now we repeat the steps in the above derivation of (2.21) from (2.5) with different functions. Let $F(x)$ be an odd function. Let n be a positive integer and let d'' and δ'' be odd positive integers with $d''\delta'' < n$. In (2.5) replace n by $n - d''\delta''$ and use the odd function $\psi(x) = F(x-d'') + F(x+d'')$. Multiply by $(-1)^{(\delta''-1)/2}$ and sum over all possible d'' and δ''. We get

$$\sum_{(d)} (-1)^{i+(\delta''-1)/2} (F(d'-d''+i) - F(d'+d''+i)) = \quad (2.22)$$

$$- \sum_{(e)} (-1)^{i+(\delta-1)/2} i \, F(i+d),$$

analogous to (2.18). We use Equation (2.4) to transform the right side of (2.22). Let

$$F(x, y, z) = \begin{cases} 0 & \text{if } x \text{ or } z \text{ is even} \\ (-1)^{(x-1)/2} z F(y) & \text{if } x \text{ and } z \text{ are odd} \end{cases}$$

Equation (2.4) yields

$$\sum_{(e)} (-1)^{i+(\delta''-1)/2} (2d - \delta + 2i) F(i + d) = \{(-1)^s s F(s)\}.$$

Using (2.5) and (2.22) we have

$$\sum_{(d)} (-1)^i \ (F(d' - d'' + i) + F(d' + d'' + i))(-1)^{(\delta''-1)/2} =$$

$$\sum_{(e)} (-1)^i \left((-1)^{(\delta-1)/2} d - \frac{1 + (-1)^{(\delta-1)/2}\delta}{2} \right) F(i + d),$$

which is similar to (2.19). Repeat the same steps from (2.19) to (2.20) and find

$$\sum_{(g)} (-1)^{(\delta''-1)/2} \ (F(d' - d'') + F(d' + d'')) =$$

$$\sum_{(h)} (-1)^{(\delta-1)/2} d F(d) - \sum_{(h)} \frac{1 + (-1)^{(\delta-1)/2}\delta}{2} F(d).$$

Just as we went from (2.20) to (2.21) we go from the last equation to

$$\sum_{(\ell)} (-1)^{(\delta''-1)/2}(F(d' - d'') + F(d' + d'')) = \sum_{(k)} (-1)^{(\delta-1)/2} d \, F(2d). \quad (2.23)$$

2.9   Sums of Three Squares

We are now ready to prove the three squares theorem. Begin with (2.6) and (2.7) of Section 2.5. If $N \equiv 3 \pmod 4$, then N is not a square and these identities are

$$\sum_{(a)} F\left(\frac{d+\delta}{2}\right) = 2 \sum_{(b)} F(d' + i) \qquad (2.6)$$

and

$$\sum_{(a)} (-1)^h F\left(\frac{d+\delta}{2}\right) = 2 \sum_{(b)} (-1)^{(i-1)/2+(\delta'-1)/2} F(d' + i), \qquad (2.7)$$

where $F(x)$ is an arbitrary odd function and the partitions are

$$
\begin{array}{lll}
\text{(a)} & N = 4h^2 + d\delta & d, \delta \text{ odd} \\
\text{(b)} & N = i^2 + 2d'\delta' & d', \delta' \text{ odd.}
\end{array}
$$

Replace N by $4n+1-2d''\delta''$, where n is a positive integer and d'' and δ'' are odd positive integers with $2d''\delta'' < 4n + 1$. Note that these conditions imply $N \equiv 3 \pmod 4$. Let $f(x)$ be an even function and $F(x) = f(x-d'')-f(x+d'')$, which is an odd function. Equations (2.6) and (2.7) become

$$
\sum_{\text{(a)}} \left(f\left(\frac{d+\delta}{2} - d''\right) - f\left(\frac{d+\delta}{2} + d''\right) \right) =
$$

$$
2 \sum_{\text{(b)}} (f(d' + i - d'') - f(d' + i + d''))
$$

and

$$
\sum_{\text{(a)}} (-1)^h \left(f\left(\frac{d+\delta}{2} - d''\right) - f\left(\frac{d+\delta}{2} + d''\right) \right) =
$$

$$
2 \sum_{\text{(b)}} (-1)^{(i-1)/2+(\delta'-1)/2} (f(d' + i - d'') - f(d' + i + d'')),
$$

where the partitions are

$$
\begin{array}{lll}
\text{(a)} & 4n + 1 - 2d''\delta'' = 4h^2 + d\delta & d, \delta \text{ odd} \\
\text{(b)} & 4n + 1 - 2d''\delta'' = i^2 + 2d'\delta' & d', \delta' \text{ odd.}
\end{array}
$$

Sum each of these two equations over all possible d'' and δ''. We get the same equations as above, but with (a) and (b) replaced by

$$
\begin{array}{lll}
\text{(c)} & 4n + 1 = 4h^2 + d\delta + 2d''\delta'' & d, \delta, d'', \delta'' \text{ odd} \\
\text{(d)} & 4n + 1 = i^2 + 2d'\delta' + 2d''\delta'' & d', \delta', d'', \delta'' \text{ odd,}
\end{array}
$$

respectively.

Since i and $-i$ run through the same set of integers, we have

$$
2 \sum_{\text{(d)}} (f(d' - d'' + i) - f(d' + d'' + i)) = \sum_{\text{(d)}} (\psi_i(d' - d'') - \psi_i(d' + d'')),
$$

where $\psi_i(x) = f(x+i) + f(x-i)$. Since $\psi_i(x)$ is an even function of x we can apply (2.21) with f replaced by ψ_i, and $m = (4n + 1 - i^2)/4$ in (k) and (ℓ). Adding the equations we get for each i we find

$$
\sum_{\text{(d)}} (\psi_i(d' - d'') - \psi_i(d' + d'')) = 2 \sum_{\text{(e)}} d(f(i) - f(i - 2d)),
$$

where the partition (e) is $4n + 1 = i^2 + 4d\delta$, δ odd.

Combining the last few equations, we have

$$\sum_{(c)} \left(f\left(\frac{d+\delta}{2} - d''\right) - f\left(\frac{d+\delta}{2} + d''\right) \right) = 2 \sum_{(e)} d(f(i) - f(i - 2d)). \quad (2.24)$$

Now apply (2.23) in the same manner to the identity we obtain from (2.7). This yields

$$\sum_{(c)} (-1)^h \left(f\left(\frac{d+\delta}{2} - d''\right) - f\left(\frac{d+\delta}{2} + d''\right) \right) = \quad\quad (2.25)$$

$$2 \sum_{(e)} (-1)^{(i-1)/2 + (\delta-1)/2} \quad d\, f(i - 2d).$$

We will use these two equations with $f(x) = 1$ if $|x| = 1$ and $f(x) = 0$ for other x. On the left side of (2.24) and (2.25) the terms $f\left(\frac{d+\delta}{2} + d''\right)$ always vanish because $\frac{d+\delta}{2} + d'' \geq 2$, and $f\left(\frac{d+\delta}{2} - d''\right) = 0$ unless $d'' = \frac{d+\delta}{2} \pm 1$. Since d'' is odd this implies that $d + \delta \equiv 0 \pmod 4$. Hence nonzero terms arise in $\sum_{(c)}$ only for solutions of one of the equations

$$4n + 1 = 4h^2 + d\delta + (d + \delta - 2)\delta''$$
$$4n + 1 = 4h^2 + d\delta + (d + \delta + 2)\delta'',$$

where d, δ and δ'' are positive odd integers, such that $d + \delta \equiv 0 \pmod 4$.

For positive integers n, let $T(n)$ denote the total number of solutions to the two equations

$$4n + 1 = d\delta + (d + \delta - 2)\delta''$$
$$4n + 1 = d\delta + (d + \delta + 2)\delta'',$$

with the same restrictions on d, δ and δ''. With this notation the left sides of (2.24) and (2.25) are

$$\sum_{h^2 < n} T(n - h^2) \quad\quad \text{and} \quad\quad \sum_{h^2 < n} (-1)^h T(n - h^2),$$

respectively.

We will need to evaluate the right sides of (2.24) and (2.25) only when $n = 2m$ and $n = 4m$, where m is an odd positive integer. In (2.24), $f(i) = 0$ unless $i^2 = 1$, and we have

$$2 \sum_{(e)} d\, f(i) = 4 \sum_{n = d\delta,\ \delta \text{ odd}} d = \begin{cases} 8\sigma(m) & \text{if } n = 2m \\ 16\sigma(m) & \text{if } n = 4m. \end{cases}$$

The only terms in $-2\sum_{(e)} d\, f(i-2d)$ that do not vanish are those with $i = 2d\pm 1$. For this i, (e) is $n = d(d+\delta\pm 1)$. But δ is odd, so the two factors d, $(d+\delta\pm 1)$ have the same parity. This is impossible when $n = 2m$ with m odd, so the sum vanishes in this case. Similarly,

$$2\sum_{(e)}(-1)^{(i-1)/2+(\delta-1)/2}d\, f(i-2d) = 0$$

when $n = 2m$. Thus for $n = 2m$, m odd, we have

$$\sum_{h^2<2m} T(2m - h^2) = 8\sigma(m)$$

and

$$\sum_{h^2<2m} (-1)^h T(2m - h^2) = 0.$$

Adding and subtracting these equations we obtain

$$T(2m) + 2T(2m - 2^2) + 2T(2m - 4^2) + \cdots = 4\sigma(m) \qquad (2.26)$$
$$T(2m - 1^2) + T(2m - 3^2) + T(2m - 5^2) + \cdots = 2\sigma(m). \qquad (2.27)$$

In case $n = 4m$, the equation $4m = d(d+\delta\pm 1)$ with m and δ odd implies that d is even, say $d = 2\Delta$, and

$$-2\sum_{(e)} d\, f(i-2d) = -4 \sum_{m=\Delta(\Delta+(\delta\pm 1)/2)} \Delta.$$

Similar reasoning shows that

$$2\sum_{(e)}(-1)^{(i-1)/2+(\delta-1)/2}d\, f(i-2d)$$

has the same value. These sums cancel when we subtract (2.25) from (2.24). The result is

$$T(4m - 1^2) + T(4m - 3^2) + T(4m - 5^2) + \cdots = 4\sigma(m). \qquad (2.28)$$

Next we show that $r_3(n)$ satisfies identities like those just proved for $T(n)$, and thereby establish relations between the two functions. Let m be an odd positive integer. By Remark 2.1 at the end of Section 2.7, the number of solutions of

$$4m = x^2 + y^2 + z^2 + w^2$$

in odd numbers is $16\sigma(m)$. Therefore,

$$16\sigma(m) = 2r_3(4m - 1^2) + 2r_3(4m - 3^2) + \cdots .$$

Comparing this equation with (2.28) gives $r_3(4m-1) = 2T(4m-1)$ for all odd m, that is,

$$r_3(8n+3) = 2T(8n+3)$$

for every nonnegative integer n.

For odd m there are exactly $6\sigma(m)$ solutions to

$$2m = x^2 + y^2 + z^2 + w^2 \qquad (2.29)$$

with positive odd x, by Jacobi's theorem (Theorem 2.6). (That theorem says that $r_4(2m) = 24\sigma(m)$ when m is odd. Since two of x, y, z, w are even and two are odd, x is odd in half of the solutions, and x is positive in half of those.) Hence,

$$6\sigma(m) = r_3(2m - 1^2) + r_3(2m - 3^2) + \cdots .$$

Comparison with (2.27) yields $r_3(2m-1) = 3T(2m-1)$ for odd m, or

$$r_3(4n+1) = 3T(4n+1)$$

for every nonnegative integer n.

By Jacobi's theorem there are exactly $12\sigma(m)$ solutions to (2.29) in which x is even. Thus,

$$12\sigma(m) = r_3(2m) + 2r_3(2m - 2^2) + 2r_3(2m - 4^2) + \cdots .$$

From Equation (2.26) it follows that $r_3(2m) = 3T(2m)$ for odd m, that is,

$$r_3(4n+2) = 3T(4n+2)$$

for every nonnegative integer n.

Combining these formulas and recalling Lemmas 2.3 and 2.4, we find

$$r_3(n) = \begin{cases} 3T(n) & \text{if } n \equiv 1, 2, 5 \text{ or } 6 \pmod 8 \\ 2T(n) & \text{if } n \equiv 3 \pmod 8 \\ 0 & \text{if } n \equiv 7 \pmod 8 \\ r_3(n/4) & \text{if } n \equiv 0 \text{ or } 4 \pmod 8, \end{cases}$$

and the proof of the three squares theorem is nearly complete.

THEOREM 2.7 (Gauss [35] Art. 291) Sums of Three Squares
A positive integer n is the sum of three squares of integers if and only if it is not of the form $4^a(8b + 7)$, where a and b are nonnegative integers.

PROOF   Following the above formula for $r_3(n)$ in terms of $T(n)$ it suffices to show that $T(n)$ is always positive. Now $T(n)$ is the total number of solutions of the two equations

$$4n + 1 = d\delta + (d + \delta - 2)\delta''$$
$$4n + 1 = d\delta + (d + \delta + 2)\delta''$$

in positive odd integers, such that $d + \delta \equiv 0 \pmod 4$. If n is odd, then $d = 2n + 1$, $\delta = 1$, $\delta'' = 1$ satisfies the first equation. If n is even, then $d = 2n - 1$, $\delta = 1$, $\delta'' = 1$ is a solution to the second equation. Therefore, there is always a solution to at least one of the two equations. ∎

COROLLARY 2.3
Every positive integer is the sum of three triangular numbers.

PROOF   Let n be a positive integer. Then, by the theorem, $8n + 3 = a^2 + b^2 + c^2$ for some integers a, b and c. All three of a, b, c must be odd, and we may assume they are positive. Let $x = (a - 1)/2$, $y = (b - 1)/2$ and $z = (c - 1)/2$. Then x, y and z are nonnegative integers and

$$n = \frac{x^2 + x}{2} + \frac{y^2 + y}{2} + \frac{z^2 + z}{2}$$

is a sum of three triangular numbers. ∎

REMARK 2.2   It may be shown (Smith [14], p. 323) that $T(n)$ is four times the number of odd classes of binary quadratic forms $ax^2 + 2bxy + cy^2$ of discriminant $b^2 - ac = -n$, where classes of $ax^2 + ay^2$ are counted as $1/2$. ("Odd" means that at least one of a, c is odd; if one form in a class is odd, then so is every form in that class.) A related result is shown in the next section. ∎

Here is a short table of $T(n)$ and $r_3(n)$.

n	1	2	3	4	5	6	7	8	9	10	11	12	13	14	15
$T(n)$	2	4	4	4	8	8	4	8	10	8	12	8	8	16	8
$r_3(n)$	6	12	8	6	24	24	0	12	30	24	24	8	24	48	0

2.10   An Alternate Method

Let $\rho_k(m)$ denote the number of ways the positive integer m can be written as the sum of the squares of k *odd positive* integers. We showed in Remark 2.1 that $\rho_4(4m) = \sigma(m)$ when m is odd and positive. In this section we will prove that fact by a different elementary argument due to Jacobi and Dirichlet. A proof is also given of the theorem of Gauss (part of Theorem 2.7) that $r_3(m) > 0$ when $m \equiv 3 \pmod 8$. In this case we express $r_3(m)$ as a class number of binary quadratic forms. We shall follow in essence Weil's presentation [119] of Kronecker's elementary proof.

The congruence $\frac{m \cdot n - 1}{2} \equiv \frac{m-1}{2} + \frac{n-1}{2}$ (mod 2) for odd m and n shows that the function

$$\chi(n) = \begin{cases} (-1)^{(n-1)/2} & \text{if } n \text{ is odd} \\ 0 & \text{if } n \text{ is even} \end{cases}$$

defines a Dirichlet character modulo 4, that is to say, $\chi(m \cdot n) = \chi(m)\chi(n)$, and $\chi(m) = \chi(n)$ if $m \equiv n$ (mod 4), for all integers m and n.

In Theorem 2.5 we have shown that

$$r_2(m) = 4 \sum_{\substack{d|m,\ d>0 \\ d \text{ odd}}} \chi(d).$$

When $m \equiv 2$ (mod 4) and $m > 0$, the only way to express m as the sum of two squares is with odd squares. If we require also that the squares be squares of positive integers, the factor of 4 is not needed, and we can write this formula in the form

$$\rho_2(m) = \sum_{m=2ab} \chi(a), \qquad m \equiv 2 \text{ (mod 4)}, \qquad m > 0 \qquad (2.30)$$

because $\chi(a) = 0$ when a is even.

Note that for a and b odd we have

$$\chi(a)\chi(b) = (-1)^{\frac{a-b}{2}}. \qquad (2.31)$$

If m is positive and $m \equiv 4$ (mod 8), then by considering the possible decompositions

$$m = x_1^2 + x_2^2 + x_3^2 + x_4^2,$$

of m as the sum of the squares of four odd positive integers, we obtain when writing $r = x_1^2 + x_2^2$ and $s = x_3^2 + x_4^2$,

$$\rho_4(m) = \sum_{m=r+s} \rho_2(r)\rho_2(s), \qquad r \equiv s \equiv 2 \text{ (mod 4)}, \quad r > 0 \quad s > 0.$$

From Equations (2.30) and (2.31) we then get

$$\rho_4(m) = \sum_{m=r+s} \sum_{r=2ab} \chi(a) \cdot \sum_{r=2cd} \chi(c)$$

$$= \sum_{m=2ab+2cd} \chi(a)\chi(c)$$

$$= \sum_{m=2ab+2cd} (-1)^{\frac{a-c}{2}},$$

where a, b, c, d are odd positive integers.

Performing the change of variables

$$a = x + y, \qquad c = x - y, \qquad b = z - t, \qquad d = z + t,$$

we obtain

$$m = 2ab + 2cd = 2(x+y)(z-t) + 2(x-y)(z+t) = 4(xz - yt). \quad (2.32)$$

Since $c = x - y > 0$ and $a = x + y > 0$ we have $-x < y < x$, that is, $|y| < x$. Similarly, $b = z - t > 0$ and $d = z + t > 0$ implies $-z < t < z$, and so $|t| < z$. Also note that x and y have opposite parity because $c = x - y$ is odd. Likewise, t and z have opposite parity because $b = z - t$ is odd. Finally, it follows from Equation (2.32) that $xz - ty$ is odd because $m \equiv 4 \pmod 8$. Therefore, t and y have the same parity, and x and z have the same parity.

Let N_0 be the sum $N_0 = \sum (-1)^y$, where the sum runs over all quadruples (a, b, c, d) for which Equation (2.32) holds and also $y = (a - c)/2 = 0$. Similarly, we let N_+ (respectively, N_-) denote the same sum running over all quadruples (a, b, c, d) for which Equation (2.32) holds and also $y = (a-c)/2 > 0$ (respectively, $y = (a - c)/2 < 0$).

For N_0 we have $y = 0$, and Equation (2.32) implies

$$xz = \frac{m}{4}, \quad x \equiv z \equiv 1 \pmod 2, \quad |t| < z, \quad t \equiv 0 \pmod 2.$$

If e is an odd positive divisor of m, then there exist e solutions of Equation (2.32) for which $y = 0$, $z = e$ and $x = m/(4e)$. Therefore we have

$$N_0 = \sum_{\substack{e|m,\ e>0 \\ e \text{ odd}}} e. \quad (2.33)$$

Here we have used the fact that $d = t + z$ has solutions $t = -d + 2i$, $z = 2(d-i)$, for $i = 0, \ldots, d - 1$.

If x, y, z, t satisfy $m = 4(xz - yt)$, and we replace y by $-y$ and t by $-t$, but leave x and z unchanged, then the new values still satisfy the equation. This change of variables gives a one-to-one correspondence between terms $(-1)^y$ summed in N_+ and those summed in N_-, and shows that $N_+ = N_-$.

Consider now a solution (x, y, z, t) of Equation (2.32) with $y > 0$. We then have

$$\frac{x}{y} > 1 \quad \text{and} \quad x \not\equiv y \pmod 2,$$

which implies that there is an integer u, such that

$$2u - 1 < \frac{x}{y} < 2u + 1.$$

We put

$$x' = 2uz - t$$
$$y' = z$$
$$z' = y$$
$$t' = 2uy - x.$$

Then clearly (x', y', z', t') is a solution of Equation (2.32) with $y' > 0$ and $y' \not\equiv y \pmod 2$, which implies that $(-1)^{y'} = -(-1)^y$. This then gives $N_+ = -N_+$ and therefore

$$N_+ = N_- = 0.$$

This in turn yields

$$\rho_4(m) = N_0 = \sum_{\substack{d|m,\, d>0 \\ d \text{ odd}}} d$$

when $m \equiv 4 \pmod 8$, the result of Remark 2.1.

The proof above is essentially Dirichlet's simplified version [25] of Jacobi's elementary argument, which avoids the use of elliptic functions.

In the following we present Kronecker's solution to the problem of counting the number of representations of an integer $\equiv 3 \pmod 8$ as a sum of three squares, a solution which unlike that of Gauss, is in the same spirit as the elementary proof of Dirichlet and Jacobi just given.

We begin by proving an elementary lemma concerning rational solutions of certain linear binary equations. We place its proof here so that the flow of Kronecker's argument given later is more transparent.

Let us first note the geometric significance of the lemma by establishing a simple claim which exhibits clearly the symmetry at play.

CLAIM 2.1

Let a, b and n be positive integers. Let $f(a, b, n)$ denote the number of integer solutions of

$$ax + by = n, \quad 0 < x < b, \quad a < y, \quad y \not\equiv 0 \pmod a \tag{2.34}$$

Then we have

$$f(a, b, n) = f(b, a, n).$$

REMARK 2.3   The geometric interpretation of the claim is that the symmetric transformation which takes the segment $\left(0, \frac{n}{a}\right) - \left(\frac{n}{b}, 0\right)$ to the segment $\left(\frac{n}{a}, 0\right) - \left(0, \frac{n}{b}\right)$ maps the lattice points in the appropriate ranges into lattice points in a one-to-one fashion. See Figure 2.2 below.   ▊

PROOF   We now prove the claim. Note that $f(b, a, n)$ is the number of integer solutions of the equation

$$bx + ay = n, \quad 0 < x < a, \quad b < y, \quad y \not\equiv 0 \pmod b. \tag{2.35}$$

We want to show that there is a bijection between the sets of solutions of Systems (2.34) and (2.35).

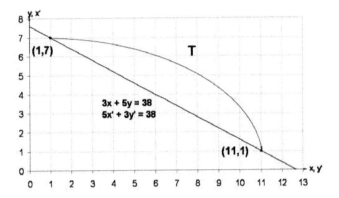

Figure 2.2 T acts on the line $3x + 5y = 38$.

$$T: \begin{pmatrix} x \\ y \end{pmatrix} \mapsto \begin{pmatrix} x' \\ y' \end{pmatrix} = \begin{pmatrix} 0 & 1 \\ 1 & 0 \end{pmatrix} \begin{pmatrix} x \\ y \end{pmatrix} + \left\lfloor \frac{y}{3} \right\rfloor \begin{pmatrix} -3 \\ 5 \end{pmatrix}.$$

First notice that $f(a, b, n) = 0$ unless

$$n = ax + by \geq a \cdot 1 + b(a + 1) = a + b + ab,$$

and n is a multiple of $\gcd(a, b)$. If n were not a multiple of $\gcd(a, b)$, then there would be no integer solutions of System (2.34) because one side would be divisible by $\gcd(a, b)$ while the other side would not be. Also note that these conditions are symmetric in a and b.

Now let (x, y) be an integer solution of System (2.34). Since $y \not\equiv 0 \pmod{a}$, there is a unique positive integer, namely, $u = \lfloor y/a \rfloor$, such that

$$u < \frac{y}{a} < u + 1,$$

or equivalently,

$$au < y < au + a,$$

that is,

$$0 < y - au < a.$$

Put $x' = y - ua$.

Since $0 < x < b$, we have $0 < \frac{x}{b} < 1$ and hence

$$u < \frac{x}{b} + u < 1 + u,$$

which implies

$$u < \frac{x + ub}{b} < u + 1.$$

Therefore $x + ub \not\equiv 0 \pmod b$. Put $y' = x + ub$.

  We now note that

$$bx' + ay' = b(y - ua) + a(x + ub) = ax + by = n.$$

Also, since $x > 0$, we have $y' = x + ub > ub \geq b$, and hence

$$y' > b.$$

Therefore, if (x, y) is a solution of System (2.34), then there exists a unique positive integer u, such that $(x', y') = (y - ua, x + ub)$ is a solution of System (2.35). This proves the claim. ∎

  We are now ready to state the main lemma. Its statement and proof are similar to those of Claim 2.1, but we restrict the parity of x and y.

LEMMA 2.6

  Let a and b be positive integers. Let m be an integer. Let α, $\beta \in \{0, 1\}$. Let $\phi(a, b, \alpha, \beta, m)$ denote the number of integer solutions of

$$ax + by = m, \ |x| < b, \ y > a, \tag{2.36}$$
$$x \equiv \alpha \text{ and } y \equiv \beta \pmod 2, \ y \not\equiv a \pmod{2a}.$$

Then $\phi(a, b, \alpha, \beta, m) = \phi(b, a, \beta, \alpha, m)$.

PROOF   Note that $\phi(b, a, \beta, \alpha, m)$ is the number of solutions of

$$bx + ay = m, \ |x| < a, \ y > b, \tag{2.37}$$
$$x \equiv \beta \text{ and } y \equiv \alpha \pmod 2, \ y \not\equiv b \pmod{2b}.$$

Let (x, y) be an integer solution of System (2.36). We then have

$$y \not\equiv a \pmod{2a}$$

and hence there exists a unique integer $u \geq 1$, such that

$$u - 1 < \frac{y - a}{2a} < u,$$

which implies

$$2au - 2a < y - a < 2au,$$

or equivalently,

$$-a < y - 2au < a,$$

that is,

$$|y - 2au| < a.$$

Denote $x' = y - 2au$.

From the inequality $|x| < b$, that is, $-b < x < b$, with the same integer u as above, we get

$$-b + 2bu < x + 2bu < b + 2bu.$$

This yields

$$-2b + 2bu < x + 2bu - b < 2bu,$$

which is the same as

$$u - 1 < \frac{x + 2bu - b}{2b} < u.$$

Therefore $x + 2bu \not\equiv b \pmod{2b}$. Denote $y' = x + 2bu$.

Observe that

$$bx' + ay' = b(y - 2au) + a(x + 2bu) = ax + by = m.$$

Also note that since $-b < x < b$, we have $-b + 2bu < x + 2bu$, and this implies $b \leq b(2u - 1) < y'$. Therefore $y' > b$.

Similarly $x \equiv \alpha \pmod{2}$ implies $x + 2bu \equiv \alpha \pmod{2}$, that is, $y' \equiv \alpha \pmod{2}$. Using $y \equiv \beta \pmod{2}$, we get $y - 2au \equiv \beta \pmod{2}$ and hence $x' \equiv \beta \pmod{2}$.

We have thus shown that there is a bijection between the solutions of Systems (2.36) and (2.37) given by $(x, y) \mapsto (x', y')$. We should note that $\phi(a, b, \alpha, \beta, m) = 0$ except if m is a multiple of $\gcd(a, b)$, $m \geq a + b$ and $m \equiv a\alpha + b\beta \pmod{2}$. These conditions for $\phi(a, b, \alpha, \beta, m) = 0$ are symmetric for a, b and α, β. This completes the proof of the lemma. ∎

We now come to the derivation of the formula for $r_3(m)$. The elementary method of Kronecker which we present requires that we know *a priori* the shape of the formula to be proved, a fact that can be taken as a weak point in the derivation, as compared to Gauss' original proof.

DEFINITION 2.2 When m is a positive integer, let $H(m)$ denote the number of solutions (a, b, c) of

$$m = 4ac - b^2, \quad b > 0, \quad b < 2a, \quad b < 2c, \quad b \equiv 1 \pmod{2}, \qquad (2.38)$$

in integers a, b, c.

Note that $H(m) = 0$ unless $-m < 0$ and $-m \equiv 1 \pmod{4}$. This follows from the fact that $b \equiv 1 \pmod{2}$ implies $b^2 \equiv 1 \pmod{4}$, and this in turn yields $m = 4ac - b^2 \equiv -1 \pmod{4}$.

If $a \leq c$, then (2.38) gives $0 < b < 2a$ and $0 < b \leq 2a - 1$, which implies $b^2 \leq 4a^2 - 4a + 1$, or $-b^2 \geq 4a - 1 - 4a^2$. Therefore,

$$m + 1 = 4ac - b^2 + 1 \geq 4ac - 4a^2 + 4a - 1 + 1 = 4a(c - a + 1).$$

This implies in particular that $H(m)$ is finite.

Example 2.1

Let $m = 3$. Then $-m < 0$ and $-m = -3 \equiv 1 \pmod 4$. Let us compute $H(3)$. We will count the reduced binary quadratic forms $ax^2 + bxy + cy^2$ with discriminant $b^2 - 4ac = -3$. The formula displayed just before this example gives

$$4 = m + 1 = 4a(c - a + 1)$$

or $a(c - a + 1) = 1$. From (2.38), a and c must be positive integers. The only possibility is $a = c = 1$. From $-3 = b^2 - 4ac = b^2 - 4$ we find that $b = 1$. Hence $x^2 + xy + y^2$ is the only reduced binary quadratic form with discriminant -3, and so $H(3) = 1$.

Note that when $k \equiv 3 \pmod 8$, the only way to write k as a sum of three squares uses odd squares, and so $\rho_3(k) = r_3(k)/8$. We can now state the main result.

THEOREM 2.8 (Gauss [35] Art. 291)

We have $\rho_3(k) = H(k)$ whenever the positive integer $k \equiv 3 \pmod 8$. Therefore, $r_3(k) = 8H(k)$ for such k.

PROOF　　Let $m = k + 1$. Then $m \equiv 4 \pmod 8$ and we have

$$\rho_4(m) = \sum_{0 < x \text{ odd}} r_3(m - x^2),$$

since we can write $m - x^2 = y^2 + z^2 + w^2$ for odd x. We have already given a formula for $\rho_4(m)$ above. If we can show that for all $m \equiv 4 \pmod 8$, we have

$$\rho_4(m) = \sum_{0 < x \text{ odd}} H(m - x^2), \tag{2.39}$$

the theorem will follow by a simple mathematical induction on m. (The base step of the induction uses $\rho_4(4) = 1$ and, for $m = 4$, the only nonzero term in the sum is $H(4 - 1) = H(3) = 1$.) Thus it suffices to prove Equation (2.39) for all $m \equiv 4 \pmod 8$.

To simplify the notation we write $m = 4n$ with n odd. (Here we use $m \equiv 4 \pmod 8$.) Let the positive integer n be fixed for the rest of this proof. In the following we will use $|A|$ to denote the cardinality of the finite set A. Let X_n denote the right side of Equation (2.39) with $m = 4n$. Then

$$X_n = \frac{1}{2} \sum_{x \text{ odd}} H(4n - x^2),$$

where the summation is now taken over all odd integers x, positive and negative.

Note that if in (2.38) we write

$$m - x^2 = 4n - x^2 = 4ac - b^2,$$

then

$$n = a \cdot c + \frac{x^2 - b^2}{4}.$$

Therefore, we can express X_n in the equivalent form $X_n =$

$$\frac{1}{2}\left|\left\{(a, b, c, x) : n = ac + \frac{x^2 - b^2}{4}, b > 0, b < 2a, b < 2c, b \equiv x \equiv 1 \ (\mathrm{mod}\ 2)\right\}\right|.$$

Since $b \equiv x \ (\mathrm{mod}\ 2)$, we can put $b + x = 2y$, $b - x = 2z$, and hence $X_n =$

$$\frac{1}{2}|\{(a, c, y, z) : n = ac - yz, y + z > 0, y + z < 2a, y + z < 2c, y \not\equiv z \ (\mathrm{mod}\ 2)\}|.$$

Note also that the conditions imposed imply that both a and c are odd, and that y and z have opposite parity.

Observe now that these conditions are symmetric in a and c and also in y and z. Therefore we can halve the size of the set on the right side of the last formula for X_n above by imposing the additional condition $c - z > a - y$. Since we already have $y + z < 2a$ and $y + z < 2c$ we find by adding these two inequalities and dividing by 2, $y + z < a + c$. Thus we have $-(c - z) < a - y < c - z$, which we write in the form

$$|a - y| < c - z.$$

We then obtain the new formula $X_n = |A|$, where $A =$

$$\{(a, c, y, z) : n = ac - yz, 0 < y + z < 2a, y + z < 2c, |a - y| < c - z, y \not\equiv z \ (\mathrm{mod}\ 2)\}.$$

Then $A \subseteq B$, where $B =$

$$\{(a, c, y, z) : n = ac - yz, 0 < y + z < 2a, |a - y| < c - z, y \not\equiv z \ (\mathrm{mod}\ 2)\},$$

and the set difference $C = B - A =$

$$\{(a, c, y, z) : n = ac - yz, 0 < y + z < 2a, y + z > 2c, |a - y| < c - z, y \not\equiv z \ (\mathrm{mod}\ 2)\}.$$

We claim that in the definition of C, the inequality $y + z < 2a$ is a consequence of the other inequalities. The other inequalities may be written

$$0 < y + z, \quad y + z > 2c, \quad z - c < y - a < c - z,$$

and these imply $2c < y + z < c + a$, from which we obtain $2c < c + a$ and $0 < y + z < c + a$. Therefore, $-a < c < a$, and hence

$$y + z < c + a < 2a,$$

as claimed.

Note that the conditions that define B also imply that both a and c are odd, and that y and z have opposite parity.

On the set B we perform the following change of variables:

$$y = a - u, \quad z = u + w, \quad c = u + v + w.$$

Thus we have

$$n = ac - yz = a(u + v + w) - (a - u)(u + w) = u^2 + au + uw.$$

The equality $y + z = a + w$ implies $0 < a + w < 2a$. Hence $-a < w < a$, or equivalently,

$$|w| < a.$$

From the inequality $|a - y| < c - z$ we obtain

$$v = u + v + w - u - w = c - z > |a - y| = |a - a + u| = |u|.$$

Finally, the incongruence $y \not\equiv z \pmod 2$ gives $a - u \not\equiv u + w \pmod 2$, which implies $a \not\equiv 2u + w \pmod 2$, or $a \not\equiv w \pmod 2$.

The change of variables $(a, c, y, z) \mapsto (a, u, v, w)$ gives a one-to-one mapping of the set B onto the set

$$D = \{(a, u, v, w) : n = u^2 + av + uw, |w| < a, v > |u|, w \not\equiv a \pmod 2)\}.$$

Note that $|B| = |D|$. Since the conditions defining D are equivalent to those defining B, we have that a is odd and therefore w is even.

Let us now consider the quadruples in D for which $u = 0$. In this case we have

$$n = av, \quad |w| < a, \quad v > 0.$$

Thus a can be any positive divisor of n. For each such a, w can be any even number with $|w| < a$, and there are a possible values for w. Therefore the total number of quadruples with $u = 0$ is

$$\sum_{\substack{d|n,\, d>0 \\ d \text{ odd}}} d = \rho_4(4n).$$

Note also that the description of D does not change if we simultaneously replace u and w by their negatives. Therefore,

$$|B| = \rho_4(4n) + 2|E|,$$

where E is the set

$$E = \{(a, u, v, w) : n - u^2 = av + uw, |w| < a, v > u > 0, a \text{ odd}, w \text{ even}\}.$$

The conditions defining E imply that v and u have opposite parity. To see this, observe that the equality

$$n - u^2 = av + uw,$$

where n and a are odd, while w is even, implies $1 + u \equiv v \pmod 2$.

With ϕ as in Lemma 2.6 we get, for fixed a and u, the number of pairs (v, w) for which $(a, u, v, w) \in E$ is $\phi(u, a, 0, u + 1, n - u^2)$. Hence the size of the set E is

$$|E| = \sum_{u,a} \phi(u, a, 0, u + 1, n - u^2),$$

where the sum runs over all pairs (u, a) for which $u > 0$, $a > 0$ and a is odd. The same argument we used in the two paragraphs after (2.33) to show that $N_+ = N_- = 0$ also shows that all the terms in the sum $\sum_{u,a}$ vanish except those for which

$$n - u^2 \geq u + a.$$

(There is no $(-1)^y$ here because we are just trying to show that there are no solutions.) This implies in particular that the set B is finite.

As we have already shown that in the set C the inequality $y + z < 2a$ is implied by the other inequalities, we will omit it from here on.

We make a further change of variables in C,

$$a = u + v + w, \quad y = u + w, \quad z = c - u,$$

and observe that

$$n = ac - yz = (u + v + w)c - (u + w)(c - u) = u^2 + vc + uw,$$

or equivalently,

$$n - u^2 = vc + uw.$$

The inequality $y + z > 2c$ implies $u + w + c - u > 2c$, and hence

$$w > |c|.$$

Similarly, the inequality $|a - y| < c - z$ can be restated as

$$u > |v|.$$

The relation $y \not\equiv z \pmod 2$, which can be written as $u + w \not\equiv c - u \pmod 2$, also yields

$$w \not\equiv c \pmod 2.$$

As before, these conditions imply that c is odd w is even. Thus we have $c \neq 0$ and the change of variables $(c, v) \mapsto (-c, -v)$ gives that $|C| = 2|F|$, where

$$F = \{(c, u, v, w) : n - u^2 = cv + uw, w > c > 0, u > |w|, c \text{ odd}, w \text{ even}\}.$$

Again with ϕ as in Lemma 2.6, for a given pair (c, u), the number of pairs (v, w) for which $(c, u, v, w) \in F$ is $\phi(c, u, u + 1, 0, n - u^2)$ and hence we see that

$$|F| = \sum_{c,u} \phi(c, u, u + 1, 0, n - u^2),$$

where the sum runs over all pairs (c, u) for which $c > 0$, $u > 0$ and c is odd. We now apply Lemma 2.6 to obtain

$$|E| = |F|.$$

This proves that

$$|C| = |B| - |A|,$$

or equivalently,

$$2|F| = \rho_4(4n) + 2|E| - |A|,$$

which shows that $X_n = |A| = \rho_4(4n)$. This establishes Equation (2.39) and completes the proof of Theorem 2.8. ∎

2.11   *Sums of Polygonal Numbers*

We have seen that every positive integer is the sum of three triangular numbers and also of four squares. With just a little more work we will show now that for every positive integer m, every positive integer is the sum of $m + 2$ polygonal numbers of order $m + 2$. Fermat claimed this was so in a note he wrote in his copy of Diophantus' *Arithmetica*. Cauchy proved this result in 1813 and in fact he proved that at most four of the polygonal numbers have to be greater than 1. We may assume that $m > 3$ because we have already dealt with the triangular numbers and squares.

THEOREM 2.9 Cauchy's Polygonal Number Theorem
Let $m \geq 3$ and n be positive integers. Then there exist nonnegative integers t, u, v and w, such that

$$n = \left(t + m\frac{t^2 - t}{2}\right) + \left(u + m\frac{u^2 - u}{2}\right) + \left(v + m\frac{v^2 - v}{2}\right) + \left(w + m\frac{w^2 - w}{2}\right) + r,$$

where $0 \leq r \leq m - 2$, that is,

$$n = \frac{m}{2}(k - s) + s + r, \tag{2.40}$$

where

$$k = t^2 + u^2 + v^2 + w^2 \quad \text{and} \quad s = t + u + v + w. \tag{2.41}$$

The proof requires one lemma, called Cauchy's lemma:

LEMMA 2.7

Let k and s be odd positive integers, such that $s^2 < 4k$ and $3k < s^2 + 2s + 4$. Then there exist nonnegative integers t, u, v and w, such that (2.41) holds.

PROOF   Since k and s are odd, we have $4k - s^2 \equiv 3 \pmod 8$. Also, $4k - s^2 > 0$ by hypothesis. By Theorem 2.7, there exist integers x, y and z so that

$$4k - s^2 = x^2 + y^2 + z^2. \tag{2.42}$$

These three integers must be odd. By reordering them and changing their signs, we may assume that $x \geq y \geq z > 0$. Choose the sign of $\pm z$ to make $s + x + y \pm z \equiv 0 \pmod 4$. Define integers $t = (s + x + y \pm z)/4$, $u = (s + x - y \mp z)/4$, $v = (s - x + y \mp z)/4$, $w = (s - x - y \pm z)/4$. They satisfy (2.41) and $t \geq u \geq v \geq w$. Using the Cauchy-Schwartz inequality, (2.42) and $3k < s^2 + 2s + 4$ we find

$$x + y + z \leq \sqrt{3(x^2 + y^2 + z^2)} = \sqrt{12k - 3s^2}$$
$$\leq \sqrt{4(s^2 + 2s + 3) - 3s^2} = \sqrt{s^2 + 8s + 12}$$
$$< \sqrt{s^2 + 8s + 16} = s + 4.$$

Hence, $s - x - y - z > -4$ and $w \geq (s - x - y - z)/4 > -1$, so $w \geq 0$. Therefore, t, u, v and w are all nonnegative integers, as required.   ∎

Now we can prove Cauchy's Theorem (Theorem 2.9).

PROOF   One verifies easily that for each odd positive integer $k \leq 9$ there is an odd positive integer s, such that $s^2 < 4k$ and $3k < s^2 + 2s + 4$. The same statement holds for $k > 9$ because then $\sqrt{4k} - (\sqrt{3k - 2} - 1) > 2$ (and $3k < s^2 + 2s + 4$ is equivalent to $s \geq \sqrt{3k - 2} - 1$).

Now let k be a fixed odd positive integer. Let s_1 be the smallest and s_2 be the largest odd integers between $\sqrt{3k - 2} - 1$ and $\sqrt{4k}$. By the lemma just proved, the theorem holds for all numbers of the form

$$\frac{m}{2}(k - s) + s + r, \tag{2.43}$$

where s is odd, $s_1 \leq s \leq s_2$ and $0 \leq r \leq m - 2$. Since $m > 2$, the smallest of these numbers is

$$C_k = \frac{m}{2}(k - s_2) + s_2$$

and the largest is

$$D_k = \frac{m}{2}(k - s_1) + s_1 + m - 2.$$

Now

$$\frac{m}{2}(k - s) + s + m - 2 = \frac{m}{2}(k - (s - 2)) + (s - 2),$$

so every number between C_k and D_k (inclusive) is of the form (2.43). Since k is an arbitrary positive integer, the theorem holds also for all numbers between

$$C_{k+2} = \frac{m}{2}((k+2) - s_2') + s_2'$$

and

$$D_{k+2} = \frac{m}{2}((k+2) - s_1') + s_1' + m - 2,$$

where s_1' and s_2' are the smallest and largest odd numbers, respectively, between $\sqrt{3(k+2) - 2} - 1$ and $\sqrt{4(k+2)}$. We have

$$\sqrt{3(k+2) - 2} - 1 - (\sqrt{3k - 2} - 1) < 2$$

and

$$\sqrt{4(k+2)} - \sqrt{4k} < 2$$

for all positive integers k. Therefore, $s_1' = s_1$ or $s_1 + 2$, and $s_2' = s_2$ or $s_2 + 2$. Hence $D_{k+2} \geq D_k + 2$ and $\lim_{k \to \infty,\ k \text{ odd}} D_k = \infty$. Note also $C_1 = \frac{m}{2}(1 - 1) + 1 = 1$. We will show that the closed intervals $[C_k, D_k]$ overlap except for a few small gaps when k is small.

We always have $s_2' \geq s_1' \geq s_1$. Suppose that $s_2' - s_1 \geq 2$. By this inequality and $m \geq 3$ we have

$$\begin{aligned}
C_{k+2} &= \frac{m}{2}k + m + s_2'\left(1 - \frac{m}{2}\right) \\
&\leq \frac{m}{2}k + m + (2 + s_1)\left(1 - \frac{m}{2}\right) \\
&= D_k + (4 - m) \leq D_k + 1.
\end{aligned}$$

This inequality shows that there is no integer between the interval $[C_k, D_k]$ and $[C_{k+2}, D_{k+2}]$ and the theorem holds for all n in $C_k \leq n \leq D_{k+2}$.

Otherwise we have $s_2' - s_1 = 0$, which implies $C_{k+2} = D_k + 2$, and also that there is at most one odd number (namely $s_1 = s_2'$) between $\sqrt{3k - 2} - 1$ and $\sqrt{4(k+2)}$. Hence $\sqrt{4(k+2)} - (\sqrt{3k - 2} - 1) < 4$, which holds only when $k \leq 105$. We have represented every positive integer in the form (2.43) except those numbers $D_k + 1$ with $k = 1, 3, 5, \ldots, 105$, for which there is only one odd s between $\sqrt{3k - 2} - 1$ and $\sqrt{4(k+2)}$. But $\sqrt{4(k+2)} - (\sqrt{3k - 2} - 1) > 2$ for every positive integer k, so there is at least one odd s between these limits. One verifies, perhaps by computer calculation, that for $1 \leq k \leq 105$, k odd, if there is only one such odd s, then there exist nonnegative integers t, u, v and w, such that either

$$k + 1 = t^2 + u^2 + v^2 + w^2 \qquad \text{and} \qquad s - 1 = t + u + v + w$$

or

$$k + 1 = t^2 + u^2 + v^2 + w^2 \qquad \text{and} \qquad s + 1 = t + u + v + w.$$

Now

$$C_k + 1 = m \left(\frac{(k+1) - (s-1)}{2} \right) + (s - 1)$$

$$= m \left(\frac{(k+1) - (s+1)}{2} \right) + (s + 1) + m - 2.$$

Hence these numbers are representable in the form (2.40), too, and we are done.   ∎

Legendre [65] proved that four or five polygonal numbers are enough for all sufficiently large n. See Nathanson [75] for a modern proof of the following version of Legendre's theorem:

THEOREM 2.10 Four or Five Polygonal Numbers Suffice
Let $m \geq 3$. If m is even, then every integer $\geq 28m^3$ is the sum of five polygonal numbers of order $m + 2$, at least one of which is either 0 or 1. If m is odd, then every integer $\geq 28m^3$ is the sum of four polygonal numbers of order $m + 2$.

2.12   *Exercises*

1. Show that distinct integers s give distinct pentagonal numbers $(3s^2 + s)/2$ in the sense of Section 2.4.

2. Make a table of the "pentagonal" numbers $(3s^2 + s)/2 < 100$. Beginning with $w(1) = \sigma(1) = 1$, use the recursion formulas in Theorems 2.3 and 2.4 to make tables of $w(n)$ and $\sigma(n)$ for all $n \leq 100$.

3. Let $\chi(d) = 0$ if d is even, and $\chi(d) = (-1)^{(d^2-1)/8+(d-1)/2}$ if d is odd. Prove that χ is totally multiplicative, that is, $\chi(de) = \chi(d)\chi(e)$ for all integers d, e. Prove that for odd positive integers n we have

$$\sum_{\substack{n=i^2+de \\ i,e \text{ odd}}} \chi(e) = \sum_{\substack{n=i^2+de \\ i \text{ even, } e \text{ odd}}} \chi(e) - \{1\}.$$

4. For positive integers n, let $r(n)$ denote the number of representations of n as the sum of a square plus twice a square, that is, $r(n)$ is the number of pairs (x, y) of integers with $n = x^2 + 2y^2$. Prove that $r(n) = 2 \sum_{d|n} \chi(d)$ with χ as in the preceding exercise.

5. Let $\bar{\sigma}(n)$ denote the sum of the odd positive divisors of n. Prove that $\bar{\sigma}$ is multiplicative and that $\bar{\sigma}(n) = \sigma(m)$, where m is the greatest odd

divisor of n. Show that

$$\sum_{\substack{de=n \\ d,e \text{ positive}}} \left(1 - (-1)^d - (-1)^e\right) e = 3\bar{\sigma}(n)$$

6. Prove that the number of representations of an integer $n \geq 0$ as the sum of four triangular numbers is $\sigma(2n+1)$.

7. In the notation of Section 2.8, prove easily by induction that $w_0(n) = 0$ for all positive integers n. This shows that Equation (2.20) follows from Equation (2.19).

8. Let $R(n) = \sum_{d|n, \, n/d \text{ odd}} d$ for positive integers n. Let m be an odd positive integer. Prove that

$$\sigma_3(m) = m\sigma(m) + 4 \sum_{i=1}^{m-1} R(i)R(m-i).$$

9. Prove that for odd positive integers m, $\sigma_2(m) = \sum \min(d', d'')$, where the sum extends over all odd positive integers $d', d'', \delta', \delta''$, such that $2m = d'\delta' + d''\delta''$.

10. Prove that for odd positive integers m,

$$\sigma_5(m) = \sum_{i=1}^{m} \sigma(2i+1)\sigma_3(2m - 2k + 1).$$

11. Prove that a positive integer n is representable in the form $n = x^2 + y^2 + 2z^2$ if and only if n is not of the form $4^a(16b + 14)$.

12. Prove that $r_3(n)/r_3(1)$ is not a multiplicative function of n.

13. Let $r(n)$ denote the number of triples (x, y, z) of integers with $n = x^2 + y^2 + 2z^2$. Prove or disprove the following statements:

   (a) $r(n)/r(1)$ is a multiplicative function of n.

   (b) $r(4n) = r(n)$ for every positive integer n.

14. Which integers are representable in the form $x^2 + 2y^2 + 2z^2$? Which are representable in the form $x^2 + 2y^2 + 4z^2$? Prove your answers.

15. Find all positive integers d, such that every positive integer is representable in the form $x^2 + y^2 + z^2 + dw^2$. Repeat for $x^2 + y^2 + 2z^2 + dw^2$ with $d \geq 2$, for $x^2 + 2y^2 + 2z^2 + dw^2$ with $d \geq 2$ and for $x^2 + y^2 + 4z^2 + dw^2$ with $d \geq 4$.

16. Find the smallest positive integer k, such that every positive integer is the sum of k or fewer odd squares.

17. Find the smallest positive integer k_1, such that every sufficiently large positive integer is the sum of k_1 or fewer odd squares.

18. Prove that if a positive integer is the sum of three squares of rational numbers, then it is the sum of three squares of integers. (The converse is obvious.)

19. For positive real numbers x, let $N(x)$ denote the number of positive integers $n \leq x$ that are *not* the sum of three squares, that is, for which $r_3(n) = 0$. Prove that for all $x \geq 1$,

$$\frac{x}{6} - \frac{7}{8}\frac{\log x}{\log 4} - 1 < N(x) < \frac{x}{6} + \frac{\log x}{8 \log 4}.$$

Conclude that $\lim_{x \to \infty} N(x)/x = 1/6$. In other words, on average, five-sixth of the positive integers can be written as the sum of three squares.

20. Extend Lemma 2.1 as follows: Let n be a positive integer. If $\sigma(n)$ is odd, then n is a square or twice a square. If $\sigma(n) \equiv 2 \pmod 4$, then there exist positive integers t and e, and a prime $p \equiv 1 \pmod 4$, such that p does not divide t, $e \equiv 1 \pmod 4$ and either $n = p^e t^2$ or $n = 2p^e t^2$.

21. Prove that if p is prime, then

$$\sum_{i=1}^{p-1} \sigma(i)\sigma(p-i) = \frac{(p^2 - 1)(5p - 6)}{12}.$$

22. Prove that if p and q are prime numbers with $p^2 + 1 = 2q$, then $p^2 + 1 = 2q$ is not the sum of the squares of two primes. (Note that 1 is not a prime.)

23. Prove Bouniakowsky's theorem: If $q \equiv 7 \pmod{16}$ is prime, then there are odd positive integers x, y, e, and a prime $p \equiv 5 \pmod 8$, such that p does not divide y, $e \equiv 1 \pmod 4$ and $q = 2x^2 + p^e y^2$. Show moreover that the number of such representations of q is odd.

24. Let $\psi(n) = \sum_{d|n,\, d > \sqrt{n}} d - \sum_{d|n,\, 0 < d < \sqrt{n}} d$. With the same $f(x)$ we used in Section 2.9 to prove Theorem 2.7, show that the right side of Equation (2.24) is $2\sigma(n) + 2\psi(n)$ when n is odd, and $8\sigma(n) + 4\sigma(2^{e-2}m) + 4\psi(2^{e-2}m)$ when $n = 2^e m$ with m odd and $e \geq 2$. Show that the right side of Equation (2.25) is $2(-1)^{(n-1)/2}(\sigma(n) - \psi(n))$ when n is odd.

25. Extend the table of $T(n)$ and $r_3(n)$ at the end of Section 2.9 to $n = 50$.

26. Show that $T(n)$ is a multiple of 4 for integers $n \geq 2$.

27. Show by direct construction that the class number $H(k)$ in Theorem 2.8 is positive when $k \equiv 3 \pmod 8$.

28. Verify the final step of the proof of Theorem 2.9. Perhaps by computer calculation, show that for each odd k between 1 and 105 for which there is only one odd s between $\sqrt{3k-2}-1$ and $\sqrt{4(k+2)}$, there exist nonnegative integers t, u, v and w, such that either

$$k+1 = t^2 + u^2 + v^2 + w^2 \qquad \text{and} \qquad s - 1 = t + u + v + w$$

or

$$k+1 = t^2 + u^2 + v^2 + w^2 \qquad \text{and} \qquad s + 1 = t + u + v + w.$$

Chapter 3

Bernoulli Numbers

Some of the material in this chapter is used in later chapters. Specifically, we will use the definitions of the Bernoulli and Euler numbers and polynomials, the functional equations for the Riemann zeta function and the Dirichlet L-functions, their values at negative integers, and the definition and basic properties of the generalized Bernoulli numbers.

The rest of the material in this chapter is beautiful mathematics for you to enjoy. We love this material, we hope you do, too, and we believe we have presented it with a novel slant.

3.1   *Overview*

The Bernoulli numbers have important uses in many parts of number theory and, indeed, throughout mathematics. We will need some of their properties in later chapters. In this chapter, we define Bernoulli numbers and prove some of their important and interesting properties.

We will define the Bernoulli numbers B_n as coefficients of a power series

$$\frac{t}{e^t - 1} = \sum_{n=0}^{\infty} B_n \frac{t^n}{n!}.$$

We will show that they are rational numbers and that $B_n = 0$ when n is > 1 and odd. We define the Bernoulli polynomials $B_n(x)$ in a similar fashion and show that

$$\sum_{i=1}^{m} i^n = \frac{B_{n+1}(m+1) - B_{n+1}}{n+1}.$$

For the case $n = 2$ of squares, the subject of this book, the formula says

$$\sum_{i=1}^{m} i^2 = 1 + 4 + 9 + 16 + 25 + \cdots + m^2 = \frac{m(m+1)(2m+1)}{6}.$$

We prove that the derivative of $B_{n+1}(x)$ is $(n+1)B_n(x)$ and use integration by parts to obtain the Euler-MacLaurin sum formula (Theorem 3.5), which is useful for approximating a sum by an integral. We give some applications of this formula, including Stirling's formula for $n!$ and the functional equation of the Riemann zeta function $\zeta(s)$. We prove formulas for $\zeta(2n)$ such as

$$\sum_{n=1}^{\infty} n^{-2} = 1 + \frac{1}{4} + \frac{1}{9} + \frac{1}{16} + \frac{1}{25} + \cdots = \frac{\pi^2}{6}.$$

For even positive integers n write the rational number $B_n = P_n/Q_n$ in lowest terms, with $Q_n > 0$. We will prove the von Staudt-Clausen theorem,

$$Q_n = \prod_{p \text{ prime}, \, (p-1)|n} p,$$

which completely determines the denominators of the Bernoulli numbers. The numerators P_n are much harder to determine. Before Fermat's Last Theorem was proved, some people used to prove it one exponent at a time. Extra work was required in case a prime exponent p divided a numerator P_n. An odd prime p is called *irregular* if it divides P_n for some even n in the interval $2 \le n \le p-3$. Other odd primes are called *regular*. In 1851, Kummer proved that Fermat's Last Theorem holds when the exponent is a regular prime. Irregular prime exponents needed further investigation.

In Section 3.7 we prove Voronoi's congruence and some of its consequences, such as J. C. Adams' theorem, which give a modicum of information about the numerators P_n. We also prove Kummer's congruence, an essential tool in every proof that there are infinitely many irregular primes.

We assume in this chapter that the reader knows some calculus and the basic facts about complex analysis at the level of Ahlfors [2].

3.2   *Definition of the Bernoulli Numbers*

The function $F(t) = t/(e^t - 1)$ is meromorphic in the complex plane and its singularities nearest to zero are $\pm 2\pi i$. Hence we can expand it in a Taylor series

$$F(t) = \frac{t}{e^t - 1} = \sum_{n=0}^{\infty} B_n \frac{t^n}{n!},$$

valid for $|t| < 2\pi$. The coefficients B_n are called the *Bernoulli numbers*. It is easy to find a recursion formula for the B_n. In the theorem the symbol $\binom{n}{i}$ is the *binomial coefficient* $\binom{n}{i} = n!/(i!(n-i)!)$.

THEOREM 3.1 Recursion Formula for the Bernoulli Numbers
Let n be a positive integer. Then

$$\sum_{i=0}^{n-1} \binom{n}{i} B_i = \begin{cases} 1 & \text{if } n = 1 \\ 0 & \text{if } n \neq 1. \end{cases}$$

PROOF   Compare coefficients of t^n in

$$(e^t - 1) \sum_{i=0}^{\infty} B_i \frac{t^i}{i!} = \left(\sum_{j=1}^{\infty} \frac{t^j}{j!} \right) \left(\sum_{i=0}^{\infty} B_i \frac{t^i}{i!} \right) = t.$$

∎

This recursion formula shows that the Bernoulli numbers are rational numbers and enables us to compute them.

Example 3.1

$$1 \cdot B_0 = 1, \text{ so } B_0 = 1,$$

$$B_0 + 2B_1 = 0, \text{ so } B_1 = -\frac{1}{2},$$

$$B_0 + 3B_1 + 3B_2 = 0, \text{ so } B_2 = \frac{1}{6},$$

$$B_0 + 4B_1 + 6B_2 + 4B_3 = 0, \text{ so } B_3 = 0,$$

$$B_0 + 5B_1 + 10B_2 + 10B_3 + 5B_4 = 0, \text{ so } B_4 = -\frac{1}{30}, \text{ etc.}$$

THEOREM 3.2 Odd-Subscripted Bernoulli Numbers Vanish
For every odd integer $n > 1$, $B_n = 0$.

PROOF   Since $t/(e^t - 1) + t/2$ is an even function, we see that $B_n = (-1)^n B_n$ for $n > 1$.   ∎

Here is a table of the first few nonzero Bernoulli numbers.

n	0	1	2	4	6	8	10	12	14	16	18	20
B_n	1	$-\frac{1}{2}$	$\frac{1}{6}$	$-\frac{1}{30}$	$\frac{1}{42}$	$-\frac{1}{30}$	$\frac{5}{66}$	$-\frac{691}{2370}$	$\frac{7}{6}$	$-\frac{3617}{510}$	$\frac{43867}{798}$	$-\frac{174611}{330}$

Now let x be another complex variable and let

$$F(t, x) = \frac{te^{xt}}{e^t - 1} = \sum_{n=0}^{\infty} B_n(x) \frac{t^n}{n!}, \qquad |t| < 2\pi.$$

Thus, $B_n = B_n(0)$. Comparison of coefficients of t^n in

$$\sum_{n=0}^{\infty} B_n(x) \frac{t^n}{n!} = F(t,x) = F(t)e^{xt} = \sum_{n=0}^{\infty} B_n \frac{t^n}{n!} \sum_{i=0}^{\infty} x^i \frac{t^i}{i!}$$

yields

$$B_n(x) = \sum_{i=0}^{n} \binom{n}{i} B_i x^{n-i},$$

which is a polynomial of degree n (since $B_0 = 1$) with rational coefficients. It is called the n-th **Bernoulli polynomial**. We find using Theorem 3.1,

$$B_n(1) = \sum_{i=0}^{n} \binom{n}{i} B_i = \begin{cases} 1 + B_1 = \frac{1}{2} & \text{if } n = 1 \\ B_n & \text{if } n \neq 1. \end{cases} \tag{3.1}$$

Example 3.2

By the method of Example 3.1, the first few Bernoulli polynomials are

$$B_0(x) = 1, \quad B_1(x) = x - \frac{1}{2}, \quad B_2(x) = x^2 - x + \frac{1}{6},$$

$$B_3(x) = x^3 - \frac{3}{2}x^2 + \frac{1}{2}x = x(x-1)\left(x - \frac{1}{2}\right),$$

$$B_4(x) = x^4 - 2x^3 + x^2 - \frac{1}{30},$$

$$B_5(x) = x^5 - \frac{5}{2}x^4 + \frac{5}{3}x^3 - \frac{1}{6}x.$$

See Abramowitz and Stegun [1] for extensive tables of Bernoulli numbers and the coefficients of Bernoulli polynomials.

Bernoulli polynomials were first introduced (by Bernoulli himself) to evaluate the sum

$$S_n(k) = \sum_{i=1}^{k-1} i^n = \sum_{i=0}^{k-1} i^n.$$

THEOREM 3.3 Sum of the First k n-th Powers
For every nonnegative integer n and positive integer k we have

$$S_n(k) = \sum_{i=1}^{k-1} i^n = \frac{B_{n+1}(k) - B_{n+1}}{n+1} = \frac{1}{n+1} \sum_{j=0}^{n} \binom{n+1}{j} B_j k^{n+1-j}.$$

PROOF   Comparing the coefficient of t^n on both sides of $F(t,x) - F(t, x-1) = te^{t(x-1)}$ gives

$$\frac{B_{n+1}(x) - B_{n+1}(x-1)}{n+1} = (x-1)^n. \tag{3.2}$$

Add this equation for $x = 1, 2, \ldots, k$. The sum on the left side telescopes and we obtain the theorem. ∎

Example 3.3

$$S_1(k) = \sum_{i=1}^{k-1} i = \frac{k^2}{2} - \frac{k}{2}, \qquad S_2(k) = \sum_{i=1}^{k-1} i^2 = \frac{k^3}{3} - \frac{k^2}{2} + \frac{k}{6},$$

$$S_3(k) = \sum_{i=1}^{k-1} i^3 = \frac{k^4}{4} - \frac{k^3}{2} + \frac{k^2}{4}, \qquad S_4(k) = \sum_{i=1}^{k-1} i^4 = \frac{k^5}{5} - \frac{k^4}{2} + \frac{k^3}{3} - \frac{k}{30}.$$

THEOREM 3.4 Derivative of the Bernoulli Polynomial
For every nonnegative integer n, $B'_{n+1}(x) = (n+1)B_n(x)$.

PROOF   Replace x by $x+1$ in Equation (3.2) and differentiate with respect to x to get

$$B'_{n+1}(x+1) - B'_{n+1}(x) = (n+1)nx^{n-1} = (n+1)\left(B_n(x+1) - B_n(x)\right).$$

This shows that $f(x) = B'_{n+1}(x) - (n+1)B_n(x)$ is a periodic function with period 1. But $f(x)$ is a polynomial and therefore constant. Hence,

$$f(x) = f(0) = B'_{n+1}(0) - (n+1)B_n(0) = (n+1)B_n - (n+1)B_n = 0,$$

which proves the theorem. ∎

COROLLARY 3.1
For all real $a \leq b$ and nonnegative integers n,

$$\int_a^b B_n(t)\, dt = \frac{1}{n+1}\left(B_{n+1}(b) - B_{n+1}(a)\right).$$

COROLLARY 3.2
We have $\int_0^1 B_0(t)\, dt = 1$ and $\int_0^1 B_n(t)\, dt = 0$ for all integers $n > 0$.

PROOF   This follows from Corollary 3.1 and Equation (3.1). ∎

3.3   The Euler-MacLaurin Sum Formula

.

The material in this section and the next one comes from Rademacher [87]. Our first theorem is that mentioned in the section title. Let $f^{(k)}(x)$ denote the k-th derivative of $f(x)$.

THEOREM 3.5 Euler-MacLaurin Sum Formula

Let $a < b$ be integers and let n be a positive integer. Let $f(x)$ have n continuous derivatives on the closed interval $[a, b]$. Then

$$\sum_{j=a+1}^{b} f(j) = \int_a^b f(x)\,dx + \sum_{k=1}^{n}(-1)^k \frac{B_k}{k!}\left(f^{(k-1)}(b) - f^{(k-1)}(a)\right) + R_n,$$

where

$$R_n = \frac{(-1)^{n-1}}{n!}\int_a^b B_n(x - \lfloor x \rfloor) f^{(n)}(x)\,dx.$$

PROOF   Write $\int_0^1 f(x)\,dx = \int_0^1 B_0(x)f(x)\,dx$ and integrate by parts n times. Using Equation (3.1) we get

$$\int_0^1 f(x)\,dx = \sum_{k=1}^{n}(-1)^{k-1}\frac{B_k(x)}{k!}f^{(k-1)}(x)\Big|_0^1 + (-1)^n \int_0^1 \frac{B_n(x)}{n!}f^{(n)}(x)\,dx$$

$$= \sum_{k=1}^{n}(-1)^{k-1}\frac{B_k}{k!}\left(f^{(k-1)}(1) - f^{(k-1)}(0)\right) + f(1)$$

$$+ (-1)^n \int_0^1 \frac{B_n(x)}{n!}f^{(n)}(x)\,dx.$$

Replace $f(x)$ by $f(j - 1 + x)$ and find

$$f(j) = \int_0^1 f(j - 1 + x)\,dx + \sum_{k=1}^{n}(-1)^k \frac{B_k}{k!}\left(f^{(k-1)}(j) - f^{(k-1)}(j - 1)\right)$$

$$+ (-1)^{n-1}\int_0^1 \frac{B_n(x)}{n!}f^{(n)}(j - 1 + x)\,dx.$$

The theorem follows when we add this equation from $j = a + 1$ to $j = b$.   ∎

For applications of the Euler-MacLaurin sum formula, we need information about the functions $f_n(t) = B_n(t - \lfloor t \rfloor)$ for positive integers n. They are periodic with period 1 and continuous except for $f_1(t)$, which jumps from $+\frac{1}{2}$ to $-\frac{1}{2}$ at each integer. Since they are of bounded variation on finite intervals, they have Fourier expansions

$$f_n(t) = \frac{a_0^{(n)}}{2} + \sum_{j=1}^{\infty}\left(a_j^{(n)}\cos 2\pi jt + b_j^{(n)}\sin 2\pi jt\right),$$

where

$$a_j^{(n)} = 2\int_0^1 B_n(x)\cos 2\pi jx\,dx$$

$$b_j^{(n)} = 2\int_0^1 B_n(x)\sin 2\pi jx\,dx$$

Using Theorem 3.4 and Equation (3.1), we find

$$a_0^{(n)} = 2 \int_0^1 B_n(x) = 0$$

for every positive integer n, and

$$a_j^{(n)} = 2B_n(x) \frac{\sin 2\pi jx}{2\pi j} \Big|_0^1 - \frac{2}{2\pi j} \int_0^1 B_n'(x) \sin 2\pi jx \, dx$$

$$= -\frac{2n}{2\pi j} \int_0^1 B_{n-1}(x) \sin 2\pi jx \, dx = -\frac{n}{2\pi j} b_j^{(n-1)}$$

for integers $n > 1$ and $a_j^{(1)} = 0$ for positive integers j. For integers $n > 1$ we have

$$b_j^{(n)} = -2B_n(x) \frac{\cos 2\pi jx}{2\pi j} \Big|_0^1 + \frac{2}{2\pi j} \int_0^1 B_n'(x) \cos 2\pi jx \, dx$$

$$= \frac{2n}{2\pi j} \int_0^1 B_{n-1}(x) \cos 2\pi jx \, dx = \frac{n}{2\pi j} a_j^{(n-1)},$$

while for $n = 1$,

$$b_j^{(1)} = -2B_1(x) \frac{\cos 2\pi jx}{2\pi j} \Big|_0^1 + \frac{2}{2\pi j} \int_0^1 \cos 2\pi jx \, dx = -\frac{2}{2\pi j}.$$

We have, for $j > 0$ and $n > 0$,

$$a_0^{(n)} = 0, \quad a_j^{(1)} = 0, \quad b_j^{(1)} = -\frac{2}{2\pi j},$$

$$a_j^{(n)} = -\frac{n}{2\pi j} b_j^{(n-1)} = -\frac{n(n-1)}{(2\pi j)^2} a_j^{(n-2)} \quad (n > 1),$$

$$b_j^{(n)} = \frac{n}{2\pi j} a_j^{(n-1)} = -\frac{n(n-1)}{(2\pi j)^2} b_j^{(n-2)} \quad (n > 1).$$

An easy induction on n gives

$$a_j^{(2m-1)} = 0, \quad a_j^{(2m)} = (-1)^{m-1} \frac{2(2m)!}{(2\pi j)^{2m}}$$

$$b_j^{(2m)} = 0, \quad b_j^{(2m-1)} = (-1)^m \frac{2(2m-1)!}{(2\pi j)^{2m-1}},$$

for all positive integers m and j. Hence we have proved Theorem 3.6.

THEOREM 3.6 Fourier Series for Bernoulli Polynomials
For every positive integer m and for all real t we have

$$B_{2m-1}(t - \lfloor t \rfloor) = (-1)^m 2(2m-1)! \sum_{j=1}^{\infty} \frac{\sin 2\pi jt}{(2\pi j)^{2m-1}},$$

$$B_{2m}(t - \lfloor t \rfloor) = (-1)^{m-1} 2(2m)! \sum_{j=1}^{\infty} \frac{\cos 2\pi jt}{(2\pi j)^{2m}},$$

except for $B_1(t - \lfloor t \rfloor)$ at integer t, where the function jumps from $+\frac{1}{2}$ to $-\frac{1}{2}$ while the Fourier series vanishes.

If we let $t = 0$, the first equation gives another proof of Theorem 3.2: $B_n = 0$ for odd $n > 1$, and the second equation gives

$$B_{2m} = 2(-1)^{m-1} \frac{(2m)!}{(2\pi)^{2m}} \sum_{n=1}^{\infty} \frac{1}{n^{2m}},$$

for positive integers m, which, since the sum clearly positive, proves Theorem 3.7.

THEOREM 3.7 Signs of the Nonzero Bernoulli Numbers
For positive integers m, $(-1)^{m-1} B_{2m} > 0$.

In Section 3.5, we will give a proof of Theorem 3.7 that does not use Fourier series.

For complex numbers s with real part > 1, the Riemann zeta function is defined by

$$\zeta(s) = \sum_{n=1}^{\infty} n^{-s} = \prod_{p \text{ prime}} (1 - p^{-s})^{-1}.$$

Thus, the equation just before Theorem 3.7 may be rewritten as Theorem 3.8.

THEOREM 3.8 Riemann Zeta Function at Even Integers
For every positive integer m,

$$\zeta(2m) = (-1)^{m-1} \frac{(2\pi)^{2m}}{2(2m)!} B_{2m} = \frac{(2\pi)^{2m}}{2(2m)!} |B_{2m}|.$$

Example 3.4

With $m = 1, 2, 3$ and 4 we have

$$\sum_{n=1}^{\infty} \frac{1}{n^2} = \frac{\pi^2}{6}, \quad \sum_{n=1}^{\infty} \frac{1}{n^4} = \frac{\pi^4}{90}, \quad \sum_{n=1}^{\infty} \frac{1}{n^6} = \frac{\pi^6}{945}, \quad \sum_{n=1}^{\infty} \frac{1}{n^8} = \frac{\pi^8}{9450}.$$

It is easy to see that $\zeta(s)$ is monotonically decreasing for real $s > 1$ and $\lim_{s \to \infty} \zeta(s) = 1$. This gives the estimates

$$2 \frac{(2m)!}{(2\pi)^{2m}} < |B_{2m}| \leq 2 \frac{(2m)!}{(2\pi)^{2m}} \zeta(2) = \frac{(2m)!}{12(2\pi)^{2m-2}},$$

valid for positive integers m, which shows that $|B_{2m}|$ increases monotonically after its minimum at $B_6 = 1/42$, and

$$|B_{2m}| \sim 2 \frac{(2m)!}{(2\pi)^{2m}}. \tag{3.3}$$

(The notation $f(x) \sim g(x)$ means $\lim_{x \to \infty} f(x)/g(x) = 1$.) We may obtain estimates for $B_n(t - \lfloor t \rfloor)$, too. From Theorem 3.6, we have for $n \geq 2$,

$$|B_n(t - \lfloor t \rfloor)| \leq \frac{2\,n!}{(2\pi)^n} \sum_{j=1}^{\infty} \frac{1}{j^n} = \frac{2\,n!}{(2\pi)^n} \zeta(n) \tag{3.4}$$

$$\leq \frac{2\,n!}{(2\pi)^n} \zeta(2) = \frac{n!}{12(2\pi)^{n-2}},$$

and for even n we have by Theorem 3.8,

$$|B_{2m}(t - \lfloor t \rfloor)| \leq |B_{2m}|. \tag{3.5}$$

We now apply the Euler-MacLaurin sum formula to derive Stirling's formula for $N!$ and to find a series for computing Euler's constant γ. In the next section we will show how the sum formula may be used to compute $\zeta(s)$ for any complex number $s \neq 1$.

Let $f(x) = \log x$ so that the n-th derivative is $f^{(n)}(x) = (-1)^{n-1} x^{-n}(n-1)!$ for $n > 0$. Apply Theorem 3.5 with this $f(x)$, $n = 2m$, $a = 1$ and $b = N$, a positive integer. Since $B_n = 0$ for odd integers $n > 1$, we get

$$\log(N!) = \sum_{n=2}^{N} \log n = \int_1^N \log x \, dx + \frac{1}{2} \log N$$

$$+ \sum_{j=1}^{m} \frac{B_{2j}}{(2j)!} (2j - 2)! \left(N^{1-2j} - 1 \right) + \int_1^N \frac{B_{2m}(t - \lfloor t \rfloor)}{2m} t^{-2m} \, dt,$$

$$= \left(N + \frac{1}{2} \right) \log N - N + \sum_{j=1}^{m} \frac{B_{2j}}{(2j-1)2j} N^{1-2j} - \Omega_{2m}(N) + K_m,$$

where

$$\Omega_{2m}(N) = \frac{1}{2m} \int_N^{\infty} B_{2m}(t - \lfloor t \rfloor) t^{-2m} \, dt$$

and

$$K_m = \frac{1}{2m} \int_1^{\infty} B_{2m}(t - \lfloor t \rfloor) t^{-2m} \, dt - \sum_{j=1}^{m} \frac{B_{2j}}{(2j-1)2j}.$$

For each m the improper integrals converge absolutely by Equation (3.5). Since $\lim_{N\to\infty} \Omega_{2m}(N) = 0$ for each m, and K_m is independent of N, we have

$$\lim_{N\to\infty} \left(\log N! - \left(N + \frac{1}{2} \right) \log N + N \right) = K_m.$$

Therefore, $K_m = K$ is independent of m as well. We will show in Theorem 3.11 that $K = \log \sqrt{2\pi}$. Use Equation (3.5) to estimate $\Omega_{2m}(N)$:

$$|\Omega_{2m}(N)| \le \frac{|B_{2m}|}{2m} \int_N^\infty t^{-2m}\, dt = \frac{|B_{2m}| N^{1-2m}}{2m(2m-1)}.$$

Thus, in our formula for $\log N!$ the error committed by stopping after m terms and ignoring $\Omega_{2m}(N)$ is not greater than the last term (the one with $j = m$). Hence we get Theorem 3.9.

THEOREM 3.9 Stirling's Formula

$$N! \sim \sqrt{2\pi}\, e^{-N} N^{N+1/2} = \sqrt{2\pi N} \left(\frac{N}{e} \right)^N. \qquad (3.6)$$

This estimate for $N!$ is excellent, even for moderately small N. With a little more effort one can show that successive estimates are on opposite sides of $N!$ so that

$$\sqrt{2\pi} N^{N+1/2} \exp(-N) \le N! \le \sqrt{2\pi} N^{N+1/2} \exp\left(-N + \frac{1}{12N} \right).$$

If we combine the estimates (3.3) and (3.6), we find

$$|B_N| \sim 2\sqrt{2\pi N} \left(\frac{N}{2\pi e} \right)^N,$$

for even positive integers N. This estimate is correct within 1% for even integers $N \ge 10$. It implies

$$\lim_{m\to\infty} \frac{|B_{2m}|}{2m} = \infty \qquad (3.7)$$

and shows that the even-subscripted Bernoulli numbers grow in magnitude more swiftly than any polynomial.

We now show that the constant $K = \log \sqrt{2\pi}$ in Stirling's formula. We will derive this fact from Theorem 3.10.

THEOREM 3.10 Wallis' Product Formula

$$\lim_{n\to\infty} \frac{2\cdot 2}{1\cdot 3} \cdot \frac{4\cdot 4}{3\cdot 5} \cdot \frac{6\cdot 6}{5\cdot 7} \cdots \frac{(2n)\cdot(2n)}{(2n-1)\cdot(2n+1)} = \frac{\pi}{2}. \qquad (3.8)$$

and

$$\lim_{n\to\infty} \frac{1 \cdot 3 \cdot 5 \cdots (2n-1)}{2 \cdot 4 \cdot 6 \quad \cdots (2n)} \sqrt{n} = \frac{1}{\sqrt{\pi}}. \tag{3.9}$$

PROOF   In the proof we will use the notation $m!!$ to denote the product of every second integer from m down to 1 or 2. That is, $(2n)!! = (2n)(2n-2)\cdots 4 \cdot 2$ and $(2n-1)!! = (2n-1)(2n-3)\cdots 3 \cdot 1$. With this notation, (3.8) and (3.9) become

$$\lim_{n\to\infty} \left\{ \frac{(2n)!!}{(2n-1)!!} \right\}^2 \frac{1}{2n+1} = \frac{\pi}{2} \quad \text{and} \quad \lim_{n\to\infty} \frac{(2n-1)!!}{(2n)!!} \sqrt{n} = \frac{1}{\sqrt{\pi}}.$$

For nonnegative integers m let $I_m = \int_0^{\pi/2} \sin^m x \, dx$. Integrating by parts we find for $m \geq 2$,

$$I_m = -(\sin^{m-1} x)\cos x \Big|_0^{\pi/2} + (m-1)\int_0^{\pi/2} (\sin^{m-2} x)\cos^2 x \, dx$$

$$= (m-1)\int_0^{\pi/2} \sin^{m-2} x (1 - \sin^2 x) \, dx$$

$$= (m-1)(I_{m-2} - I_m) = (m-1)I_{m-2} - (m-1)I_m,$$

or $I_m = \frac{m-1}{m} I_{m-2}$. Now $I_0 = \int_0^{\pi/2} 1 \, dx = \pi/2$ and $I_1 = \int_0^{\pi/2} \sin x \, dx = 1$. A simple induction argument yields

$$I_m = \begin{cases} \dfrac{(m-1)!!}{m!!} \dfrac{\pi}{2} & \text{if } m \text{ is even} \\ \dfrac{(m-1)!!}{m!!} & \text{if } m \text{ is odd.} \end{cases}$$

For every real x in $0 \leq x \leq \pi/2$ and for every positive integer n we have

$$0 \leq \sin^{2n+1} x \leq \sin^{2n} x \leq \sin^{2n-1} x.$$

Integrating from 0 to $\pi/2$, we have $0 \leq I_{2n+1} \leq I_{2n} \leq I_{2n-1}$. Using the formula for I_m we see

$$\frac{(2n)!!}{(2n+1)!!} \leq \frac{(2n-1)!!}{(2n)!!} \frac{\pi}{2} \leq \frac{(2n-2)!!}{(2n-1)!!},$$

or

$$\frac{1}{2n+1} \left\{ \frac{(2n)!!}{(2n-1)!!} \right\}^2 \leq \frac{\pi}{2} \leq \frac{1}{2n} \left\{ \frac{(2n)!!}{(2n-1)!!} \right\}^2.$$

Let A denote the left side and B denote the right side of this inequality, so that it says $A \leq \pi/2 \leq B$. Now

$$B - A = \left(\frac{1}{2n} - \frac{1}{2n+1} \right) \left\{ \frac{(2n)!!}{(2n-1)!!} \right\}^2$$

$$= \frac{1}{(2n)(2n+1)} \left\{ \frac{(2n)!!}{(2n-1)!!} \right\}^2$$

$$\leq \frac{1}{2n} \frac{\pi}{2} = \frac{\pi}{4n},$$

since $A \leq \pi/2$. Therefore, $\lim_{n \to \infty} (B - A) = 0$ and Formula (3.8) follows. If we take the reciprocal of the square root of Formula (3.8), we find

$$\lim_{n \to \infty} \frac{1 \cdot 3 \cdot 5 \cdots (2n-1)}{2 \cdot 4 \cdot 6 \cdots (2n)} \sqrt{n + \frac{1}{2}} = \frac{1}{\sqrt{\pi}},$$

which is easily seen to be equivalent to Formula (3.9). ∎

THEOREM 3.11 Constant For Stirling's Formula
The constant in Stirling's formula is $K = \log \sqrt{2\pi}$.

PROOF   This constant is defined by

$$K = \lim_{N \to \infty} \left(\log N! - \left(N + \frac{1}{2} \right) \log N + N \right).$$

We already know that the limit exists. Writing c for e^K, we have

$$c = e^K = \lim_{N \to \infty} \frac{N! e^N}{N^{(N+1/2)}}.$$

or $N! \sim c e^{-N} N^{(N+1/2)}$.

If we multiply numerator and denominator in (3.9) by $2 \cdot 4 \cdot 6 \cdots (2n)$, we find

$$\lim_{n \to \infty} \frac{(2n)!}{(2^n n!)^2} \sqrt{n} = \frac{1}{\sqrt{\pi}}.$$

Apply $N! \sim c e^{-N} N^{(N+1/2)}$ with $N = 2n$ and with $N = n$ to obtain

$$\lim_{n \to \infty} \frac{c e^{-2n} (2n)^{(2n+1/2)} \sqrt{n}}{\left(2^n c e^{-n} n^{(n+1/2)} \right)^2} = \frac{1}{\sqrt{\pi}}.$$

This formula simplifies to $\lim_{n \to \infty} \sqrt{2}/c = 1/\sqrt{\pi}$, or $c = \sqrt{2\pi}$, that is, $K = \log \sqrt{2\pi}$. ∎

The limit

$$\gamma = \lim_{N \to \infty} \left(\sum_{n=1}^{N} \frac{1}{n} - \log N \right)$$

exists by the integral test and is called Euler's constant. Its definition provides a poor method for computing it because the expression in parentheses is only within a constant times N^{-1} of γ. Thus, approximately 1,000,000,000 terms in the sum would be required to compute γ to 9 decimal places. Furthermore, if this calculation were performed by standard floating point arithmetic on a computer, the round-off error would be enormous and in the last step two nearly equal large numbers would be subtracted, resulting in the catastrophic loss of significant figures. The Euler-MacLaurin sum formula provides a better way to compute γ.

Apply Theorem 3.5 with $f(x) = 1/x$, $n = m$, $a = 1$ and $b = N$. The details are like those of the derivation of Stirling's formula. We find

$$\sum_{n=1}^{N} \frac{1}{n} = \log N + \gamma + \frac{1}{2N} - \sum_{j=1}^{m} \frac{B_{2j}}{2j} N^{-2j} + \Psi_{2m}(N),$$

where

$$|\Psi_{2m}(N)| \le \frac{|B_{2m}|}{2m} N^{-2j}.$$

Example 3.5

With $N = 6$ and $m = 3$ we have

$$2.45 = 1 + \frac{1}{2} + \frac{1}{3} + \frac{1}{4} + \frac{1}{5} + \frac{1}{6} = \log 6 + \gamma + \frac{1}{12} - \frac{B_2}{2} 6^{-2} - \frac{B_4}{4} 6^{-4} - \frac{B_6}{6} 6^{-6} + \Psi_6(6).$$

We have $B_2 = \frac{1}{6}$, $B_4 = -\frac{1}{30}$ and $B_6 = \frac{1}{42}$, so with an error not greater than

$$|\Psi_6(6)| \le \left| \frac{B_6}{6} \right| 6^{-6} = 7^{-1} 6^{-8} = 0.000000085$$

we find

$$\gamma = 2.45 - \log 6 - \frac{1}{12} + 2^{-1} 6^{-3} - 20^{-1} 6^{-5} + 7^{-1} 6^{-8} \approx 0.577215667.$$

With larger values of m and N one obtains the more precise estimate $\gamma \approx 0.5772156649015328606$. It is not known whether γ is rational or irrational.

3.4   The Riemann Zeta Function

We have seen in Theorem 3.8 that $\zeta(n) = (2\pi)^n |B_n| / 2n!$ for even positive integers n. No analogous formula is known for odd n. When n is an even positive integer, $\zeta(n) = \pi^n$ times a rational number, and therefore irrational. Apéry (see [114]) proved that $\zeta(3)$ is irrational. But it is unknown whether $\pi^{-3}\zeta(3)$ is rational. The status (rational or irrational) of $\zeta(n)$ and $\pi^{-n}\zeta(n)$, for odd $n > 4$, is also unknown. However, setting $t = 1/4$ in Theorem 3.6 gives

$$\sum_{n=0}^{\infty} \frac{(-1)^n}{(2n+1)^{2m-1}} = (-1)^m \frac{(2\pi)^{2m-1}}{2(2m-1)!} B_{2m-1}\left(\frac{1}{4}\right), \qquad (3.10)$$

which equals π^{2m-1} times a rational number. (See also Exercise 12 at the end of this chapter.)

Example 3.6

With $B_1(t) = t - 1/2$ and $B_3(t) = t(t-1)(t-1/2)$ we find

$$\sum_{n=0}^{\infty} \frac{(-1)^n}{2n+1} = 1 - \frac{1}{3} + \frac{1}{5} - \cdots = \frac{\pi}{4} \quad \text{and} \quad \sum_{n=0}^{\infty} \frac{(-1)^n}{(2n+1)^3} = 1 - \frac{1}{3^3} + \frac{1}{5^3} - \cdots = \frac{\pi^3}{32}.$$

We can evaluate a few more sums of this type. For $s > 1$ we have

$$\lambda(s) \stackrel{\text{def}}{=} \sum_{n=0}^{\infty} \frac{1}{(2n+1)^s} = \sum_{k=1}^{\infty} \frac{1}{k^s} - \sum_{k=1}^{\infty} \frac{1}{(2k)^s} = (1 - 2^{-s})\zeta(s)$$

and

$$\eta(s) \stackrel{\text{def}}{=} \sum_{n=1}^{\infty} \frac{(-1)^{n-1}}{n^s} = \sum_{k=1}^{\infty} \frac{1}{k^s} - 2\sum_{k=1}^{\infty} \frac{1}{(2k)^s} = (1 - 2^{1-s})\zeta(s).$$

Example 3.7

By L'Hôpital's rule (and $\zeta'(1) = 1$), $\eta(1) = \log 2$. Theorem 3.8 allows us to evaluate $\lambda(s)$ and $\eta(s)$ for even positive integers s. Thus,

$$\lambda(2) = 1 + \frac{1}{3^2} + \frac{1}{5^2} + \cdots = \frac{\pi^2}{8}, \quad \lambda(4) = 1 + \frac{1}{3^4} + \frac{1}{5^4} + \cdots = \frac{\pi^4}{96},$$

$$\eta(2) = 1 - \frac{1}{2^2} + \frac{1}{3^2} - \cdots = \frac{\pi^2}{12} \quad \text{and} \quad \eta(4) = 1 - \frac{1}{2^4} + \frac{1}{3^4} - \cdots = \frac{7\pi^4}{720}.$$

Let $\Re(s)$ denote the real part of the complex number s. The Riemann ζ-function has been defined for $\Re(s) > 1$. We will extend this domain by analytic continuation and, at the same time, give a convenient formula for computing $\zeta(s)$ for any complex number $s \neq 1$. Apply the Euler-MacLaurin sum formula with $f(x) = x^{-s}$, $a = 1$ and $b = N$. Let n be a positive integer. We have $f^{(n)}(x) = (-1)^n s(s+1)(s+2)\cdots(s+n-1)x^{-s-n}$ and $\zeta(s) = 1 + \lim_{N\to\infty}\sum_{j=2}^{N} f(j)$, so that

$$\zeta(s) - 1 = \lim_{N\to\infty}\left\{\int_1^N x^{-s}dx - \sum_{j=1}^n \frac{B_j}{j!}s(s+1)\cdots(s+j-2)(N^{-s+1-j}-1)\right.$$
$$\left. - \frac{1}{n!}s(s+1)\cdots(s+n-1)\int_1^N B_n(t-\lfloor t\rfloor)t^{-s-n}dt\right\}$$
$$= \lim_{N\to\infty}\left\{\frac{N^{1-s}-1}{1-s} + \frac{1}{2}(N^{-s}-1)\right.$$
$$- \sum_{j=2}^n \frac{B_j}{j!}s(s+1)\cdots(s+j-2)(N^{-s-j+1}-1)$$
$$\left. - \frac{1}{n!}s(s+1)\cdots(s+n-1)\int_1^N B_n(t-\lfloor t\rfloor)t^{-s-n}dt\right\}.$$

For $\Re(s) > 1$ this limit can be performed termwise to give

$$\zeta(s) = \frac{1}{s-1} + \frac{1}{2} + \sum_{j=2}^n \frac{B_j}{j!}s(s+1)\cdots(s+j-2) \qquad (3.11)$$
$$- \frac{1}{n!}s(s+1)\cdots(s+n-1)\int_1^\infty B_n(t-\lfloor t\rfloor)t^{-s-n}dt.$$

But the right side converges in $\Re(s) > 1 - n$, except at $s = 1$. Hence $\zeta(s)$ is continued analytically to a meromorphic function in the whole complex plane with its only singularity a simple pole with residue 1 at $s = 1$.

Formula (3.11) provides an excellent way to compute $\zeta(s)$ for any complex number $s \neq 1$. Take n to be a positive integer with $n > 1 - \Re(s)$. For any such n, $B_n(t - \lfloor t\rfloor)$ is a polynomial in t for t between consecutive integers. Thus we can compute

$$\int_1^N B_n(t-\lfloor t\rfloor)t^{-s-n}dt$$

explicitly for any reasonable positive integers N and n. The remainder may be estimated by (3.4).

With $s = 0$ in (3.11) we have $\zeta(0) = -1/2$. We can also use (3.11) to find $\zeta(-k)$ for positive integers k. When $n = k+1$ and $s = -k$, Equation (3.11) becomes

$$\zeta(-k) = \frac{1}{-k-1} + \frac{1}{2} + \sum_{j=2}^{k+1} \frac{B_j}{j!}(-k)(-k+1)\cdots(-k+j-2)$$

$$= \frac{-1}{k+1}\left(1 + \left(-\frac{1}{2}\right)(k+1) + \sum_{j=2}^{k+1}\binom{k+1}{j}B_j\right)$$

$$= \frac{-1}{k+1}\sum_{j=0}^{k+1}\binom{k+1}{j}B_j = -\frac{B_{k+1}}{k+1},$$

by Theorem 3.1. Thus $\zeta(-2m) = 0$ and $\zeta(1-2m) = -B_{2m}/2m$ for positive integers m. The zeros at the negative even integers are called the "trivial zeros" of the Riemann ζ-function.

See H. M. Edwards, [26], pp. 114–118, for more examples of the use of Equation (3.11).

We will now sketch a proof of the functional equation for the ζ-function. This equation connects $\zeta(s)$ and $\zeta(1-s)$. For more details of the proof, see Rademacher [87], p. 83. Let $n = 3$ in Equation (3.11). Then

$$\zeta(s) = \frac{1}{s-1} + \frac{1}{2} + \frac{B_2}{2}s - \frac{s(s+1)(s+2)}{6}\int_1^\infty B_3(t - \lfloor t \rfloor)t^{-s-3}dt$$

for $\Re(s) > -2$. The improper integral

$$\int_0^1 B_3(t - \lfloor t \rfloor)t^{-s-3}dt = \int_0^1 t(t-1)\left(t - \frac{1}{2}\right)t^{-s-3}dt$$

converges for $\Re(s) < -1$. Integrating by parts three times and using Theorem 3.4, we find that the value of this integral is

$$-\frac{B_2}{2}s - \frac{1}{2} - \frac{1}{s-1}.$$

Hence,

$$\zeta(s) = -\frac{s(s+1)(s+2)}{6}\int_0^\infty B_3(t - \lfloor t \rfloor)t^{-s-3}dt$$

for $-2 < \Re(s) < -1$. Replace $B_3(t - \lfloor t \rfloor)$ by its Fourier series (Theorem 3.6). One can justify interchanging the integral and sum and we have

$$\zeta(s) = -\frac{s(s+1)(s+2)}{6}\sum_{n=1}^\infty 12\int_0^\infty \frac{\sin 2\pi nt}{(2\pi n)^3}t^{-s-3}dt$$

$$= -2s(s+1)(s+2)\sum_{n=1}^\infty \frac{1}{(2\pi n)^{1-s}}\int_0^\infty y^{-s-3}\sin y\,dy,$$

where we have set $y = 2\pi nt$. Hence

$$\zeta(s) = -2^s \pi^{s-1} s(s+1)(s+2) \int_0^\infty y^{-s-3} \sin y \, dy \cdot \zeta(1-s).$$

With a little more work, done in the next section, one can rewrite the equation as

$$\zeta(s) = 2^s \pi^{s-1} \Gamma(1-s) \sin \frac{\pi s}{2} \zeta(1-s), \qquad (3.12)$$

where $\Gamma(s) = \int_0^\infty e^{-t} t^{s-1} dt$ is the Gamma function. Equation (3.12) is Riemann's functional equation for $\zeta(s)$, and holds for $-2 < \Re(s) < -1$. But both sides are meromorphic functions, so it is valid for all complex numbers s for which both sides are defined. Using properties of the Γ-function, it may be written as

$$\Lambda(s) \stackrel{\text{def}}{=} \pi^{-s/2} \Gamma(s/2) \zeta(s) = \pi^{-(1-s)/2} \Gamma((1-s)/2) \zeta(1-s) = \Lambda(1-s). \ (3.13)$$

This form of the functional equation shows that the function $\Lambda(s)$ on the left side is symmetric about the *critical line* $\Re(s) = 1/2$. It is known that $\zeta(s)$ does not vanish for $\Re(s) \geq 1$, that it has only the trivial zeros for $\Re(s) \leq 0$ (these correspond to poles of $\Gamma(s/2)$), and that it has infinitely many zeros in the *critical strip* $0 < \Re(s) < 1$. The famous Riemann hypothesis is the conjecture that every zero of $\zeta(s)$ in the critical strip lies on the critical line. The functional equation (3.12) shows that if s is a zero in the critical strip, then so is $1 - s$. Since by (3.11) $\zeta(s)$ is real for real s, if s is a zero, then so is its complex conjugate. Thus the zeros on the critical line occur in complex conjugate pairs, and (if the Riemann hypothesis is false) those in the critical strip off the critical line occur in quadruples as the vertices of rectangles in the complex plane.

REMARK 3.1   The zeros of the ζ-function are connected with the distribution of prime numbers. Let $\pi(x)$ denote the number of primes $\leq x$ and define $\mathrm{li}(x) = \int_0^x (1/\log t) dt$. The prime number theorem states that $\pi(x) \sim \mathrm{li}(x)$. If Θ is the least upper bound of the real parts of the zeros of $\zeta(s)$, so that $1/2 \leq \Theta \leq 1$, then $|\pi(x) - \mathrm{li}(x)| < cx^\Theta \log x$, where c is a certain positive constant. The Riemann hypothesis is the statement $\Theta = 1/2$.   ∎

3.4.1   *Functional Equation for the Zeta Function*

In this section we prove the following fundamental result due to Riemann.

THEOREM 3.12 Functional Equation for the Riemann Zeta Function
The completed zeta function

$$\Lambda(s) \stackrel{\text{def}}{=} \pi^{-s/2} \Gamma(s/2) \zeta(s),$$

as a function of the complex variable s, defined originally for $\Re(s) > 1$, has a meromorphic continuation to the whole s plane which is regular everywhere except at $s = 1$ and 0, where it has simple poles with residue 1, and it satisfies the relation

$$\Lambda(s) = \Lambda(1 - s). \tag{3.14}$$

Riemann himself gave three conceptually distinct proofs of this theorem.

1. One was based on the theta transformation formula

$$\vartheta(z) \overset{\text{def}}{=} \sum_{n=-\infty}^{\infty} q^{n^2}, \qquad q = e^{i\pi z},$$

namely,

$$\vartheta\left(-\frac{1}{z}\right) = \sqrt{\frac{z}{i}}\vartheta(z);$$

the latter being a consequence of the Poisson summation formula.

2. A second one was based on Cauchy's residue theorem.

3. A third one was based on Riemann's asymptotic analysis of the complex integral

$$\int_L \frac{e^{iw^2/4\pi + aw}}{e^w - 1}\,dw,$$

where L is a straight line inclined at an angle $\pi/4$ to the real axis, and intersecting the imaginary axis between 0 and $2\pi i$.

Riemann used the third proof to obtain the approximate location of a few zeros of $\zeta(s)$ on the line $\Re(s) = \frac{1}{2}$. The same line of ideas was subsequently developed by Siegel into a powerful technique (the Riemann-Siegel formula) capable of yielding the true location of the complex zeros of $\zeta(s)$ high up on the critical line $\Re(s) = \frac{1}{2}$.

The argument we present here is that of the second proof. Its main merit is that it quickly provides the relation between the values $\zeta(1 - 2k)$, for positive integers k, and the Bernoulli numbers B_{2k}. It has the added feature that it generalizes to give the functional equation for the Dirichlet L-functions $L(s, \chi)$, as well as to the relation between the values of $L(s, \chi)$ at the negative integers and the generalized Bernoulli numbers, which we will treat in the next section.

PROOF   We begin with the well-known integral formula for the Gamma function,

$$\Gamma(z) \overset{\text{def}}{=} \int_0^\infty e^{-x} x^s \frac{dx}{x} = \int_0^\infty e^{-ny}(ny)^s \frac{dy}{y}, \qquad (x \mapsto ny),$$

valid for $\Re(s) > 0$, and which we can write in the form

$$\Gamma(s) \cdot \frac{1}{n^s} = \int_0^\infty e^{-ny} y^s \frac{dy}{y} . \tag{3.15}$$

Using the absolute convergence of $\zeta(s)$ for $\Re(s) > 1$ and the formula for the sum of a geometric series, we obtain

$$\sum_{n=1}^\infty \Gamma(s) \cdot \frac{1}{n^s} = \sum_{n=1}^\infty \int_0^\infty e^{-ny} y^{s-1} \, dy = \int_0^\infty \frac{y^{s-1}}{e^y - 1} \, dy .$$

Note that the inversion of the order of summation and integration is valid because for $s = \sigma + it$ we have

$$\sum_{n=1}^\infty \int_0^\infty \left| e^{-ny} y^{s-1} \right| dy = \Gamma(\sigma)\zeta(\sigma) ,$$

and the expression on the right side exists for $\sigma > 1$.

We now consider the path integral

$$F(s) \stackrel{\text{def}}{=} \frac{1}{2\pi i} \int_{\mathcal{C}_c} \frac{z^{s-1}}{e^{-z} - 1} \, dz$$

taken along a path $\mathcal{C}_c = \mathcal{C}_c^- \cup \mathcal{C}_c^0 \cup \mathcal{C}_c^+$ consisting of

$$\begin{aligned}
\mathcal{C}_c^+ &: z = x - \tfrac{\sqrt{-3}}{2} c, \text{ for } x = -\infty \text{ to } -\tfrac{1}{2}, \\
\mathcal{C}_c^0 &: z = c e^{i\vartheta}, \quad\quad\ \text{ for } \vartheta = -\tfrac{2\pi}{3} \text{ to } \tfrac{2\pi}{3}, \\
\mathcal{C}_c^- &: z = x + \tfrac{\sqrt{-3}}{2} c, \text{ for } x = -\tfrac{1}{2} \text{ to } -\infty,
\end{aligned}$$

where c is a real number satisfying $0 < c < 2\pi$. The path $\mathcal{C}_c$ is oriented so that the circular arc $\mathcal{C}_c^0$ about the origin is traversed in the counterclockwise direction. The profile of the path has the shape of a lollipop with an infinitely long stick.

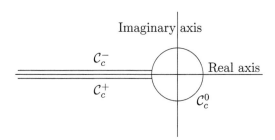

Figure 3.1 The lollipop, $\mathcal{C}_c$.

The complex power $z^s = e^{s \log z}$ is to be understood with respect to the branch of $\log z$, which is continuous in the z plane with the negative axis removed and with $\log z$ real-valued on the positive real axis. With this convention we have for $s = \sigma + it$ and $z = re^{i\vartheta}$,

$$\log z = \log r + i\vartheta, \qquad -\pi \leq \vartheta \leq \pi, \quad r > 0,$$
$$s \log z = \sigma \log r - t\vartheta + i\{\sigma\vartheta + t \log r\},$$

and hence

$$|z^s| = r^\sigma \cdot e^{-t\vartheta}.$$

We note that if the path integral $F(s)$ is convergent, which we show below, then Cauchy's residue theorem implies that $F(s)$ is independent of the path C_c chosen and is in fact a function of s alone.

By decreasing the distance between C_c^+ and C_c^- down to zero while keeping them connected to C_c^0, we arrive at a new path $C = C^- \cup C_c^0 \cup C^+$, in which both C^- and C^+ can be identified with the negative real axis, but with opposite orientation. Using the new path C and employing the inequality

$$\left| \frac{z^{s-1}}{e^{-z} - 1} \right| = \frac{x^{\sigma-1} e^{\pm \pi t}}{e^x - 1} = \frac{x^{\Delta-1} e^{\Delta \pi}}{e^x - 1} < e^{-x/2},$$

valid for $s = \sigma + it$ inside the disk $|s| < \Delta$ and for $x > x_0 = x_0(\Delta)$, we conclude that the convergence is uniform inside $|s| < \Delta$.

Using the parametrizations

$$C^+ : z = xe^{-i\pi}, \text{ for } x = -\infty \text{ to } x = c$$
$$C_c^0 : z = ce^{i\vartheta}, \quad \text{ for } \vartheta = -\tfrac{2\pi}{3} \text{ to } \tfrac{2\pi}{3},$$
$$C^- : z = xe^{i\pi}, \quad \text{ for } x = c \text{ to } x = -\infty,$$

and setting

$$g(z) = \frac{1}{e^{-z} - 1},$$

we get

$$2\pi i F(s) = - \int_c^\infty x^{s-1} e^{-s\pi i} g(-x) \, dx$$
$$+ \int_{-\pi}^{\pi} c^s e^{s\vartheta i} g(ce^{i\vartheta}) i \, d\vartheta$$
$$+ \int_c^\infty x^{s-1} e^{s\pi i} g(-x) \, dx,$$

or equivalently,

$$\pi F(s) = \sin s\pi \int_c^\infty x^{s-1} g(-x) \, dx + \frac{c^s}{2} \int_{-\pi}^{\pi} e^{s\vartheta i} g(ce^{i\vartheta}) \, d\vartheta.$$

Let $F_1(s, c)$ and $F_2(s, c)$ denote the first and second terms in this sum, so that $\pi F(s) = F_1(s, c) + F_2(s, c)$.

Since $zg(z)$ is regular in $|z| < 2\pi$, we have $|zg(z)| < A_1$ for $|z| < \pi$ and a constant $A_1 > 0$. Therefore, if $c < \pi$, then

$$|F_2(s,c)| \le \frac{c^\sigma}{2} \int_{-\pi}^{\pi} e^{-t\vartheta} \frac{A_1}{c} \, d\vartheta \le \pi A_1 e^{|t|\pi} r^{\sigma-1}. \qquad (3.16)$$

If we suppose that $\sigma > 1$, then (3.16) shows clearly that $F_2(s,c) \to 0$ as $c \to 0$ with s fixed.

Hence we have

$$\pi F(s) = \lim_{c \to 0} F_1(s,c) = \sin s\pi \int_0^\infty \frac{x^{s-1}}{e^x - 1} \, dx = \sin s\pi \, \Gamma(s)\zeta(s),$$

or equivalently, using the Gamma formula $\pi/\sin s\pi = \Gamma(s)\Gamma(1-s)$,

$$\zeta(s) = \frac{\pi}{\sin s\pi \, \Gamma(s)} F(s) = \Gamma(1-s)F(s).$$

This formula gives a representation of $\zeta(s)$ as a product of an integral function, $F(s)$, and a meromorphic function, $\Gamma(1-s)$, whose poles are located at the points $s = 1, 2, 3, \ldots$.

Now, if s is an integer > 1, the expression $z^{s-1}/(e^{-z} - 1)$ has the structure of a single-valued integrand regular at $z = 0$. For $s = 1$, $g(z)$ has a simple pole with residue -1 and hence $F(1) = -1$ and $F(2) = F(3) = \cdots = 0$. Using the identity

$$(1-s)\zeta(s) = (1-s)\Gamma(1-s)F(s) = \Gamma(2-s)F(s),$$

we see that $\zeta(s)$ has a simple pole at $s = 1$ with residue 1.

To obtain the functional equation, we now suppose $\sigma < 0$, and consider the modified integral

$$F_N(s) = \frac{1}{2\pi i} \int_{\mathcal{C}(N)} \frac{z^{s-1}}{e^{-z} - 1} \, dz,$$

where $\mathcal{C}(N)$ consists of a circle of radius $R = (2N+1)\pi$, with N a positive integer, together with a path corresponding to the segment of $\mathcal{C} = \mathcal{C}^- \cup \mathcal{C}_c^0 \cup \mathcal{C}^+$ which lies inside the circle of radius $R = (2N+1)\pi$. The full path is oriented so that the orientation of $\mathcal{C}$ is preserved.

We note that on the circle $z = Re^{i\vartheta}$, $\vartheta = \pi$ to $\vartheta = -\pi$, we have

$$\left| \frac{z^{s-1}}{e^{-z} - 1} \right| = R^{\sigma-1} e^{t\vartheta} \left| \frac{1}{e^{-z} - 1} \right| < R^{\sigma-1} e^{|t|\pi} \cdot A_2,$$

for some positive constant A_2.

Note that the poles of $1/(e^{-z} - 1)$ are located at the points $2\pi i n$ for $n = 0$, $\pm 1, \pm 2, \ldots$. Removing small disks of radius $\delta > 0$ centered at the points $2\pi i n$, $n \in \mathbb{Z}$, will leave a region with boundary $\mathcal{C}(N)$ in which $1/(e^{-z} - 1)$

is bounded. Thus we can bound the total contribution to $F_N(s)$ along the boundary of $\mathcal{C}(N)$ by an expression which is

$$\leq \frac{R^{\sigma-1}e^{|t|\pi}}{2\pi}(2\pi R) \cdot A_2 \leq R^\sigma e^{|t|\pi} A_2,$$

which for $\sigma < 0$ approaches 0 as $N \to \infty$, and hence

$$\lim_{N\to\infty} F_N(s) = F(s).$$

The evaluation of $F_N(s)$ by Cauchy's residue theorem gives

$$F_N(s) = \sum_{n=1}^{N} \left\{(2\pi in)^{s-1} + (-2\pi in)^{s-1}\right\}$$

$$= 2(2\pi)^{s-1} \sin\frac{\pi s}{2} \sum_{n=1}^{N} n^{s-1}.$$

We thus obtain

$$F(s) = \lim_{N\to\infty} F_N(s) = 2(2\pi)^{s-1} \sin\frac{s\pi}{2}\,\zeta(1-s),$$

an expression which is valid for

$$\Re(1 - s) = 1 - \sigma > 1.$$

Hence,

$$\zeta(s) = (2\pi)^s \frac{\sin\frac{2\pi s}{2}}{\sin\pi s} \cdot \frac{\zeta(1-s)}{\Gamma(s)}.$$

The duplication formula for $\Gamma(s)$ gives the relation

$$(2\pi)^s \frac{\sin\frac{\pi s}{2}}{\sin\pi s\,\Gamma(s)} = \frac{\pi^{(1-s)/2}\Gamma\left(\frac{1-s}{2}\right)}{\pi^{-s/2}\Gamma\left(\frac{s}{2}\right)} = \pi^{s-1/2}\frac{\Gamma\left(\frac{1-s}{2}\right)}{\Gamma\left(\frac{s}{2}\right)}.$$

This then yields the functional equation for $\zeta(s)$ in its standard form

$$\pi^{-s/2}\Gamma\left(\frac{s}{2}\right)\zeta(s) = \pi^{-(1-s)/2}\Gamma\left(\frac{1-s}{2}\right)\zeta(1-s),$$

and the proof is complete.   ∎

3.4.2 Functional Equation for the Dirichlet L-functions

DEFINITION 3.1   A *Dirichlet character* (or simply a *character*) χ to the modulus n is a complex-valued function defined on the integers

$$\chi : \mathbb{Z} \longrightarrow \mathbb{C}$$

with the properties

1. $\chi(a)$ depends only on the residue class of $a \bmod n$.

2. $\chi(ab) = \chi(a)\chi(b)$ for any $a, b \in \mathbb{Z}$.

3. $\chi(a) \neq 0$ if a is relatively prime to n.

Let χ' be a Dirichlet character to the modulus m and let m be a divisor of a positive integer n. We extend the definition of χ' to a Dirichlet character to the modulus n by prescribing that

$$\chi(a) = \begin{cases} \chi'(a) & \text{if } \gcd(a, n) = 1, \\ 0 & \text{if } \gcd(a, n) > 1. \end{cases}$$

The character χ is said to be *induced from* χ'. A Dirichlet character χ to the modulus n is called *primitive* if it is not induced from any character to a modulus m with $m < n$. In this case the modulus n is called the *conductor* of χ and is denoted by $f = f_\chi$.

If χ_1 and χ_2 are primitive characters of respective conductors f_1 and f_2, then there is a unique character χ, called the *product* of χ_1 and χ_2, such that for any a with $\gcd(a, f_1 f_2) = 1$ we have

$$\chi(a) = \chi_1(a)\chi_2(a).$$

Such a character is also primitive of conductor some divisor of $f_1 f_2$. Under this multiplication, the primitive characters form an abelian group whose identity element is the *principal character* χ_0 with $\chi_0(a) = 1$ for all a. The inverse of χ is its complex conjugate $\overline{\chi}$. By identifying a primitive character χ to the modulus f with a character of the multiplicative group $(\mathbb{Z}/f\mathbb{Z})^\times$ under the natural map $\mathbb{Z} \to \mathbb{Z}/f\mathbb{Z}$, we get a periodic function of period f, that is, $\chi(a + fk) = \chi(a)$ for all integers a and k.

DEFINITION 3.2   If χ is a Dirichlet character of conductor f, we define its L-function by

$$L(s, \chi) = \sum_{n=1}^{\infty} \frac{\chi(n)}{n^s}, \quad \Re(s) > 1.$$

Since χ is a multiplicative function, the unique factorization property of the integers can be used to show that, for a large positive integer N and $\Re(s) > 1$, we have

$$\prod_{p \leq N} \left\{ 1 + \sum_{v=1}^{\infty} \frac{\chi(p^v)}{p^{vs}} \right\} - \sum_{n=1}^{N} \frac{\chi(n)}{n^s} = {\sum_{n>N}}' \frac{\chi(n)}{n^s},$$

the latter being the tail of an absolutely convergent series. Letting $N \to \infty$ we obtain the Euler product representation

$$L(s, \chi) = \prod_{p} \frac{1}{1 - \chi(p)p^{-s}}, \quad \Re(s) > 1.$$

No factor on the right side vanishes, since $|\chi(p)p^{-s}| = p^{-\Re(s)} < \frac{1}{2}$ for $\Re(s) > 1$. As the product is absolutely convergent we have

$$L(s, \chi) \neq 0$$

for $\Re(s) > 1$.

We now give several examples of Dirichlet L-functions, all of them connected to the problem of representing an integer as a sum of squares.

Example 3.8

The principal character χ_0 is of course associated to the Riemann zeta function,

$$L(s, \chi_0) = \sum_{n=1}^{\infty} \frac{1}{n^s} = \zeta(s).$$

Example 3.9

The Gauss character χ_g is defined by

$$\chi_g(a) = \begin{cases} (-1)^{(a-1)/2} & \text{if } a \text{ is odd,} \\ 0 & \text{if } a \text{ is even.} \end{cases}$$

It is not difficult to verify that the conductor of χ_g is 4. This character has already made its appearance in Chapter 2 in the discussion of the Liouville identities. It will appear again in the discussion of the counting formula for the number of representations of an integer as a sum of squares. It is also intimately connected with the arithmetic in the ring of Gaussian integers

$$\mathbb{Z}[i] \overset{\text{def}}{=} \{a + bi \, : \, a, b \in \mathbb{Z}\} \text{ where } i = \sqrt{-1}.$$

Note that $\chi_g(-1) = -1$. The associated Dirichlet L-function is given by

$$L(s, \chi_g) = \sum_{n=1}^{\infty} \frac{\chi_g(n)}{n^s} = \sum_{m=0}^{\infty} \frac{(-1)^m}{(2m+1)^s}.$$

Leibnitz was the first to prove that

$$L(1, \chi_g) = \sum_{m=0}^{\infty} \frac{(-1)^m}{(2m+1)} = \frac{\pi}{4}.$$

We shall see in the next section that there is a vast generalization of the Leibnitz evaluation which applies to all Dirichlet L-functions. This development will lead to the generalized Bernoulli numbers.

Example 3.10

The Eisenstein character χ_e is defined by

$$\chi_e(a) = \begin{cases} 1 & \text{if } a \equiv 1 \pmod 3, \\ -1 & \text{if } a \equiv 2 \pmod 3, \\ 0 & \text{if } a \equiv 0 \pmod 3. \end{cases}$$

This character has conductor $f_{\chi_e} = 3$ and satisfies $\chi_e(-a) = -\chi_e(a)$. The associated Dirichlet L-function is given by

$$L(s, \chi_e) = \sum_{n=1}^{\infty} \frac{\chi_e(n)}{n^s}, \quad \Re(s) > 1.$$

This L-function is closely connected to the arithmetic of the ring of Eisenstein integers

$$\mathbb{Z}[\rho] \overset{\text{def}}{=} \{a + b\rho : a, b \in \mathbb{Z}\},$$

where ρ is a cube root of 1.

Example 3.11

The product character $\chi_{ge} \overset{\text{def}}{=} \chi_g \cdot \chi_e$ is an even function of conductor $f_{\chi_{ge}} = 12$. The associated Dirichlet L-function,

$$L(s, \chi_{ge}) = \sum_{n=1}^{\infty} \frac{\chi_{ge}(n)}{n^s}, \quad \Re(s) > 1,$$

completes the set of four L-functions related to the characters of the group $(\mathbb{Z}/12\mathbb{Z})^{\times}$.

In the following, if χ is an even character, that is, if $\chi(-a) = \chi(a)$, then we put $\delta = \delta_\chi = 0$. Otherwise, if χ is odd, that is, if $\chi(-a) = -\chi(a)$, we put $\delta = \delta_\chi = 1$. Thus, $\chi(-a) = (-1)^\delta \chi(a)$ in all cases.

THEOREM 3.13 Functional Equation for L-functions
Let χ be a primitive Dirichlet character to the modulus f_χ. As a function of the complex variable s, the function

$$L(s, \chi) = \sum_{n=1}^{\infty} \frac{\chi(n)}{n^s}$$

has a meromorphic continuation to the whole s plane which is entire unless $\chi = \chi_0$ is the principal character, in which case it has a simple pole at $s = 1$ with residue 1. There it satisfies the functional equation

$$\Lambda(s, \chi) \overset{\text{def}}{=} \pi^{-(s+\delta)/2} \Gamma\left(\frac{s+\delta}{2}\right) L(s, \chi)$$

$$= W(\chi) \cdot f_\chi^{(1/2)-s} \cdot \Lambda(1 - s, \chi^{-1}),$$

where the constant $W(\chi)$, called the root number, is given by

$$W(\chi) = i^{-\delta} \frac{1}{\sqrt{f_\chi}} \sum_{a=1}^{f} \chi(a) e^{2\pi i a / f},$$

and has absolute value 1.

It is a pleasant exercise to verify that $W(\chi) = 1$ in the four examples given above. We shall verify this equality in the case $\chi = \chi_g$. In this case, $\delta = 1$ and $f = 4$. The values of χ are $\chi(1) = 1$ and $\chi(3) = -1$, and we have

$$W(\chi_g) = i^{-1} \frac{1}{\sqrt{4}} \left\{ e^{2\pi i/4} - e^{2\pi i 3/4} \right\} = i^{-1} \cdot \frac{1}{2} \left\{ i - i^3 \right\} = 1.$$

Since our proof of Theorem 3.13 is similar to that of Theorem 3.12, we merely describe the modifications to the earlier proof needed to handle the more general context.

PROOF   Begin with the integral formula (3.15). Multiply both sides by $\chi(n)$ and sum over $n = 1, 2, 3, \dots$. From the periodicity of χ, we obtain a new formula

$$\Gamma(s) L(s, \chi) = \int_0^\infty g_\chi(y) y^{s-1} \, dy,$$

with

$$g_\chi(y) = \sum_{a=1}^{f} \chi(a) \cdot \frac{e^{-ay}}{1 - e^{-fy}}, \quad \text{where } y > 0.$$

Letting

$$f_\chi(z) = \sum_{a=1}^{f} \frac{\chi(a) z e^{az}}{e^{fz} - 1},$$

we obtain $f_\chi(-y) = y g_\chi(y)$. Both f_χ and g_χ are meromorphic functions of z with poles at the points $2\pi i m / f$ for $m \in \mathbb{Z}$.

Using the same conventions as in the proof of Theorem 3.12 concerning the branch of $\log z$ under consideration and the corresponding value for z^s, we are lead to consider the path integral

$$F_\chi(s) = \int_{C_c} f_\chi(z) z^{s-1} \frac{dz}{z},$$

where the path C_c is the same one as in the proof of Theorem 3.12, but with the proviso that $0 < c < 2\pi/f$. Exactly the same deformation and convergence arguments as before lead to the integral representations

$$2\pi i F_\chi(s) = - \int_c^\infty x^{s-1} e^{-s\pi i} g_\chi(-x)\, dx$$
$$+ \int_{-\pi}^\pi c^s e^{s\vartheta i} g_\chi(ce^{i\vartheta}) i\, d\vartheta$$
$$+ \int_c^\infty x^{s-1} e^{s\pi i} g_\chi(-x)\, dx,$$

or equivalently,

$$\pi F_\chi(s) = \sin s\pi \int_c^\infty x^{s-1} g_\chi(-x)\, dx + \frac{c^s}{2} \int_{-\pi}^\pi e^{s\vartheta i} g_\chi(ce^{i\vartheta})\, d\vartheta. \tag{3.17}$$

Since $zg_\chi(z)$ is regular in $|z| < 2\pi/f$ and since $|zg_\chi(z)| < A_1(\chi)$, a constant depending only on χ, for $|z| < 2\pi/f$, the same limiting procedure as was used in the case of $\zeta(s)$ applied to the second term on the right side of (3.17) yields the integral representation

$$\pi F_\chi(s) = - \sin s\pi \int_0^\infty f_\chi(y) y^{s-1} \frac{dy}{y} = - \sin s\pi \Gamma(s) L(s,\chi).$$

From this we obtain, using that $\pi/\sin s\pi = \Gamma(s)\Gamma(1-s)$,

$$L(s,\chi) = -\Gamma(1-s)\cdot F_\chi(s).$$

If s is a positive integer, $s > 1$ and χ is not the principal character χ_0, so that $\sum_{a=1}^f \chi(a) = 0$, we see that the expression

$$g_\chi(z) = \frac{\chi(a) e^{-az} z^{s-1}}{1 - e^{-fz}}$$

is single-valued and regular at $z = 0$ and therefore $F_\chi(1) = F_\chi(2) = F_\chi(3) = \cdots = 0$. This proves that when χ is not the principal character, $L(s,\chi)$ has a holomorphic extension to the whole s plane.

To obtain the functional equation, we suppose that $\Re(s) = \sigma < 0$ and consider the integral

$$F_{\chi,N} = \frac{1}{2\pi i} \int_{C_\chi(N)} g_\chi(z) z^{s-1}\, dz,$$

where $C_\chi(N)$ consists of a circle of radius

$$R_f = \frac{2\pi}{f}\left(N + \frac{1}{2}\right).$$

The same considerations as in the proof of Theorem 3.12 give

$$\lim_{N\to\infty} F_{\chi,N}(s) = F_\chi(s).$$

We assume here that $N \to \infty$ through a sequence of positive integers.

At this point we encounter the most interesting part of the argument as we deal with the case $\Re(s) = \sigma < 0$.

Using Cauchy's residue theorem, we see that the contribution to $F_\chi(s)$ coming from the residues of $g_\chi(z)z^{s-1}$ at the poles at $z = 2\pi im/f$ with $m \geq 1$ is

$$\frac{1}{f}\sum_{a=1}^{f} \chi(a)e^{-a(2\pi im/f)}\left(\frac{2\pi im}{f}\right)^{s-1}$$

$$= \frac{\chi(m)^{-1}}{f}\sum_{a=1}^{f}\chi(am)e^{-2\pi iam/f}\left(\frac{2\pi}{f}\right)^{s-1}\cdot m^{s-1}e^{\pi i(s-1)/2}$$

$$= \frac{\chi(m)^{-1}}{f}G_\chi \cdot \left(\frac{2\pi}{f}\right)^{s-1}\cdot m^{s-1}e^{\pi i(s-1)/2},$$

where

$$G_\chi = \sum_{a=1}^{f}\chi(a)e^{-2\pi ia/f}.$$

The residue at the point $2\pi im/f$ with $m \leq -1$ is

$$\frac{\chi(m)^{-1}}{f}G_\chi \cdot \chi(-1)\left(\frac{2\pi}{f}\right)^{s-1}\cdot m^{s-1}e^{-\pi i(s-1)/2}.$$

Adding the contributions of both types of residues we obtain

$$F_\chi(s) = \frac{1}{f}G_\chi \cdot \left(\frac{2\pi}{f}\right)^{s-1}\left\{e^{(s-1)\pi i/2} + \chi(-1)e^{-(s-1)\pi i/2}\right\}\sum_{m=1}^{\infty}\chi(m)^{-1}\cdot m^{s-1}$$

$$= \frac{1}{fi}G_\chi \cdot \left(\frac{2\pi}{f}\right)^{s-1}\left\{e^{s\pi i/2} - \chi(-1)e^{-s\pi i/2}\right\}\sum_{m=1}^{\infty}\chi(m)^{-1}\cdot m^{s-1}$$

$$= \frac{1}{2\pi i}G_\chi \cdot \left(\frac{2\pi}{f}\right)^{s}\left\{e^{s\pi i/2} - \chi(-1)e^{-s\pi i/2}\right\}\sum_{m=1}^{\infty}\chi(m)^{-1}\cdot m^{s-1}.$$

Combining this expression with

$$F_\chi(s) = \frac{\sin s\pi}{\pi}\Gamma(s)L(s,\chi),$$

we obtain

$$L(s,\chi) = \frac{G_\chi}{2\pi i}\left(\frac{2\pi}{f}\right)^{s}\cdot \frac{\pi}{\Gamma(s)\sin s\pi}\cdot \left\{e^{s\pi i/2} - \chi(-1)e^{-s\pi i/2}\right\}L(1-s,\chi^{-1}).$$

This in turn can be rewritten in the form

$$L(s,\chi) = i^{-\delta} G_\chi \left(\frac{2\pi}{f}\right)^s \cdot \frac{1}{2} \cdot \frac{L(1-s,\chi^{-1})}{\Gamma(s)\cos\left(\frac{\pi}{2}(s-\delta)\right)},$$

where we have used that $\chi(-1) = i^{-\delta} \cdot i^{-\delta}$. If we recall that

$$W(\chi) = \frac{i^{-\delta} \cdot G_\chi}{\sqrt{f}}$$

and use the Gamma identity

$$\frac{(2\pi)^s}{2\Gamma(s)\cos\pi(s-\delta)/2} = \frac{\pi^{-(1-s-\delta)/2}\Gamma\left(\frac{1-s-\delta}{2}\right)}{\pi^{-(s+\delta)/2}\Gamma\left(\frac{s+\delta}{2}\right)},$$

a result that can be derived from $\pi/\sin s\pi = \Gamma(s)\Gamma(1-s)$ and the duplication formula

$$2^{2s-1}\Gamma(s)\Gamma\left(s+\frac{1}{2}\right) = \pi^{1/2}\Gamma(2s),$$

we arrive at the standard form of the functional equation: If

$$\Lambda(s,\chi) = \pi^{-(s+\delta)/2}\Gamma\left(\frac{s+\delta}{2}\right)L(s,\chi),$$

then

$$\Lambda(s,\chi) = W(\chi) \cdot f_\chi^{(1/2)-s} \cdot \Lambda(1-s,\chi^{-1}).$$

This completes the proof of the functional equation for $L(s,\chi)$. ∎

As a final remark we note that for an integer $n \geq 1$ and $s = 1-n$ we obtain

$$\frac{L(1-n,\chi)}{\Gamma(n)} = \frac{-1}{2\pi i}\int_{C_\chi} f_\chi(z)z^{-n} \cdot \frac{dz}{z} = \frac{-1}{2\pi i}\int_{C_c^0} f_\chi(z)z^{-n-1}\,dz$$

$$= - \text{ the residue of } f_\chi(z)z^{-n-1} \text{ at } z = 0.$$

This formula will be of fundamental importance in the next section, where we discuss the values of $L(1-n,\chi)$.

3.4.3 Generalized Bernoulli Numbers

Let χ be a Dirichlet character of conductor $f = f_\chi$. With the notation of the previous section, and for an integer $n \geq 0$, we define the generalized Bernoulli number $B_{n,\chi}$ as the coefficient of z^n in the Taylor expansion of $f_\chi(z)$ about $z = 0$, that is,

$$\sum_{a=1}^{f} \frac{\chi(a)ze^{az}}{e^{fz}-1} = \sum_{n=0}^{\infty} B_{n,\chi} \cdot \frac{z^n}{n!}.$$

THEOREM 3.14 Value of $L(1 - n, \chi)$

For a primitive Dirichlet character χ and for any integer $n \geq 1$, we have

$$L(1 - n, \chi) = -\frac{B_{n,\chi}}{n}.$$

PROOF   We need only observe that

$$f_\chi(z)z^{-n-1} = \frac{B_{n,\chi}}{n!} \cdot \frac{1}{z} + \sum_{\substack{m=0 \\ m \neq n}}^{\infty} B_{m,\chi} \cdot \frac{z^m}{m}.$$

Thus we have that the residue of $f_\chi(z)z^{-n-1}$ at $z = 0$ is $B_{n,\chi}/n!$. Since $\Gamma(n) = (n-1)!$, the formula at the end of the previous section gives the result claimed. ∎

It is a pleasant exercise to prove that with $\chi(-1) = (-1)^\delta$, where $\delta \in \{0, 1\}$, we have

$$B_{n,\chi} = 0 \text{ if } \chi \neq \chi_0 \text{ and } n \equiv \delta \pmod 2.$$

Examples of evaluations of $L(s, \chi)$ using generalized Bernoulli numbers will appear in Chapter 6, where the singular series is first shown to be the quotient of two Dirichlet L-functions, and the latter are then expressed as Bernoulli numbers.

3.5   *Signs of Bernoulli Numbers Alternate*

In this section we prove an identity involving the sum

$$S_n(k) = \sum_{i=1}^{k-1} i^n$$

and use it to derive a recursion formula for B_{2m}. This formula leads to a simple proof of Theorem 3.7 and the fact that $|B_{2n}|$ increases for $n \geq 4$. This section is based on Uspensky and Heaslet [113], pp. 254 ff.

We will add the numbers in the following $k-1$ by $k-1$ table in two different ways.

1^n	2^n	3^n	$\cdots$	$1^n(k-1)^n$
2^n	4^n	6^n	$\cdots$	$2^n(k-1)^n$
3^n	6^n	9^n	$\cdots$	$3^n(k-1)^n$
$\vdots$	$\vdots$	$\vdots$	$\ddots$	$\vdots$
$(k-1)^n$	$2^n(k-1)^n$	$3^n(k-1)^n$	$\cdots$	$(k-1)^{2n}$

The sum of the numbers in row i is $i^n S_n(k)$. Adding all the rows gives the grand total $(S_n(k))^2$. The sum of the numbers in the i-th L-shaped region is

$$2i^n \left(1^n + 2^n + \cdots + (i-1)^n\right) + i^{2n} = 2i^n S_n(i) + i^{2n}.$$

Therefore,

$$
\begin{aligned}
(S_n(k))^2 &= \sum_{i=1}^{k-1} \left(2i^n S_n(i) + i^{2n}\right) \\
&= \sum_{i=1}^{k-1} \left(\frac{2i^n}{n+1} \left(\sum_{j=0}^{n} \binom{n+1}{j} B_j i^{n+1-j} \right) + i^{2n} \right) \\
&= \sum_{i=1}^{k-1} \left(\frac{2i^{2n+1}}{n+1} + 2B_1 i^{2n} + i^{2n} + \frac{1}{n+1} \sum_{j=1}^{n-1} \binom{n+1}{j+1} B_{j+1} i^{2n-j} \right) \\
&= \frac{2}{n+1} S_{2n+1}(k) + 2 \sum_{j=1}^{n-1} \binom{n}{j} \frac{B_{j+1}}{j+1} S_{2n-j}(k),
\end{aligned}
$$

where we have used Theorem 3.3 and $B_1 = -1/2$. Both sides are polynomials of degree $2n+2$ in k. Since they agree at every positive integer k, they must be the same polynomial. Equate the coefficients of k^2 in each polynomial. The coefficient of k^2 in $(S_n(k))^2$ is

$$\left(\frac{1}{n+1} \binom{n+1}{n} B_n \right)^2 = B_n^2.$$

The coefficient of k^2 in $\frac{2}{n+1} S_{2n+1}(k)$ is

$$\frac{2}{n+1} \frac{1}{2n+2} \binom{2n+2}{2n} B_{2n} = \frac{2n+1}{n+1} B_{2n}.$$

Since the coefficient of k^2 in $S_{2n-j}(k)$ is

$$\frac{1}{2n-j+1}\binom{2n-j+1}{2n-j-1}B_{2n-j-1} = \frac{2n-j}{2}B_{2n-j-1},$$

we have

$$\frac{2n+1}{n+1}B_{2n} + \sum_{j=1}^{n-1}\binom{n}{j}\frac{2n-j}{j+1}B_{j+1}B_{2n-j-1} = B_n^2.$$

Separate the term for $j = n - 1$ from the sum to get

$$\frac{2n+1}{n+1}B_{2n} + \sum_{j=1}^{n-2}\binom{n}{j}\frac{2n-j}{j+1}B_{j+1}B_{2n-j-1} = -nB_n^2.$$

By Theorem 3.2, $B_{j+1} = 0$ when j is an even positive integer, so we can assume that $j = 2s - 1$ is odd. Solving the last equation for B_{2n} we obtain

$$B_{2n} = -\frac{n+1}{2n+1}\left(nB_n^2 + \sum_{s=1}^{(n-2)/2}\frac{2n-2s+1}{2s}\binom{n}{2s-1}B_{2s}B_{2n-2s}\right)$$

for even $n \geq 2$, and

$$B_{2n} = -\frac{n+1}{2n+1}\left(\sum_{s=1}^{(n-1)/2}\frac{2n-2s+1}{2s}\binom{n}{2s-1}B_{2s}B_{2n-2s}\right)$$

for odd $n \geq 3$.

We know that $B_2 = \frac{1}{6} > 0$ and $B_4 = -\frac{1}{30} < 0$. Use complete induction on n. Suppose that $(-1)^{s-1}B_{2s} > 0$ for $1 \leq s < n$. Then the two equations displayed above show that $B_{2n} > 0$ for odd n and $B_{2n} < 0$ for even n. Hence $(-1)^{n-1}B_{2n} > 0$ for every positive integer n. The equations also show that

$$|B_{2n}| \geq \frac{2n-1}{2n+1}\frac{n(n+1)}{12}|B_{2n-2}|$$

for $n \geq 2$, so that $|B_{2n}| > |B_{2n-2}|$ for $n \geq 4$, as was proved in Section 3.3.

3.6   The von Staudt-Clausen Theorem

This theorem and its corollaries describe the denominator and the fractional parts of the Bernoulli numbers, which are rational numbers. The theorem is most easily stated in terms of m-integers. We will state it also without this notation, but the notation will be used in the next few sections.

DEFINITION 3.3   Let m be a positive integer. A rational number r is called an m-integer if m is relatively prime to the denominator of r in lowest terms.

Every integer is an m-integer for every positive integer m. If a rational number r is a p-integer for every prime p, then r is an integer because no prime can divide its denominator. The set of m-integers forms a ring, a subring of the rationals $\mathbb{Q}$. If r and s are two rational numbers and m is a positive integer, we write

$$r \equiv s \pmod{m}$$

if m divides the numerator of the rational number $r - s$ in lowest terms. This relation is an equivalence relation and it generalizes congruence for integers.

For even positive integers m, define integers P_m and Q_m by $B_m = P_m/Q_m$ in lowest terms. This means that P_m and Q_m are relatively prime and $Q_m > 0$. Also let $P_0 = Q_0 = 1$, $P_1 = -1$ and $Q_1 = 2$. Finally, let $P_m = 0$ and $Q_m = 1$ for all odd integers $m > 1$. Thus, $B_m = P_m/Q_m$ for all nonnegative integers m. We will use this notation frequently for the rest of this chapter.

THEOREM 3.15 The von Staudt-Clausen Theorem

Let p be prime and m an even positive integer. If $(p-1) \nmid m$, then B_m is a p-integer (that is, $p \nmid Q_m$). If $(p-1) | m$, then pB_m is a p-integer (that is, if $p^k | Q_m$, then $k \leq 1$), and $pB_m \equiv -1 \pmod{p}$.

PROOF We use induction on m. The primes p for which $(p-1)|2$ are 2 and 3, and these are exactly the primes that divide $Q_2 = 6$. Also, $2B_2 = \frac{1}{3} \equiv -1 \pmod{2}$ and $3B_2 = \frac{1}{2} \equiv -1 \pmod{3}$, so the theorem holds for $m = 2$.

By Theorem 3.3 we have

$$S_m(p) = \frac{1}{m+1} \sum_{j=0}^{m} \binom{m+1}{j} B_j p^{m+1-j},$$

which we write as

$$pB_m = S_m(p) - \sum_{j=0}^{m-1} \frac{1}{m+1} \binom{m+1}{j} p^{m-j} pB_j. \tag{3.18}$$

By induction, pB_j is a p-integer for every even $j < m$. Consider a term

$$\frac{1}{m+1} \binom{m+1}{j} p^{m-j} \tag{3.19}$$

with $0 \leq j < m$. If $p = 2$, then since $m+1$ is odd, (3.19) is a 2-integer and so $\equiv 0 \pmod{2}$. If $p \neq 2$, write (3.19) in the form

$$\frac{1}{m+1} \binom{m+1}{m+1-j} p^{m-j} = \frac{m(m-1)\cdots(j+1)}{(m-j+1)!} p^{m-j}.$$

Let $r = m - j + 1$ and let p^k be the highest power of p that divides $r!$. It is well known that

$$k = \sum_{n=1}^{\infty} \left\lfloor \frac{r}{p^n} \right\rfloor < \sum_{n=1}^{\infty} \frac{r}{p^n} = \frac{r}{p-1} \le \frac{r}{2} \le r - 1 = m - j.$$

Therefore, $p^{m-j}/(m-j+1)!$ and also (3.19) are p-integers and are $\equiv 0 \pmod{p}$. Thus, (3.18) becomes

$$pB_m \equiv S_m(p) \pmod{p}. \tag{3.20}$$

We will be almost finished when we prove

$$S_m(p) \equiv \begin{cases} -1 \pmod{p} & \text{if } (p-1)|m \\ 0 \pmod{p} & \text{if } (p-1) \nmid m. \end{cases} \tag{3.21}$$

If $(p-1)|m$, then $n^m \equiv 1 \pmod{p}$ for $1 \le n \le p-1$ by Fermat's little theorem, and we have

$$S_m(p) = \sum_{n=1}^{p-1} n^m \equiv \sum_{n=1}^{p-1} 1 = p - 1 \equiv -1 \pmod{p}.$$

If $(p-1) \nmid m$, let g be a primitive root modulo p. Then

$$S_m(p) \equiv \sum_{r=0}^{p-2} g^{rm} = \frac{g^{(p-1)m} - 1}{g^m - 1} \equiv 0 \pmod{p}$$

because $g^m \not\equiv 1 \pmod{p}$, while $g^{p-1} = 1 \pmod{p}$. This proves (3.21).

The congruences (3.20) and (3.21) show that if $(p-1) \nmid m$, then $pB_m \equiv 0 \pmod{p}$, so that B_m is a p-integer, and if $(p-1)|m$, then pB_m is a p-integer satisfying $pB_m \equiv -1 \pmod{p}$. ∎

COROLLARY 3.3
If m is an even positive integer, the denominator Q_m of B_m is square-free. In fact, $Q_m = \prod_{(p-1)|m, \, p \text{ prime}} p$.

COROLLARY 3.4
If m is an even positive integer, then there is an integer I_m, such that

$$B_m = I_m - \sum_{(p-1)|m, \, p \text{ prime}} \frac{1}{p}.$$

PROOF   The number I_m defined by this equation is clearly a rational number. If $(p-1) \nmid m$, then $p \nmid Q_m$ and so I_m is a p-integer for that prime p.

If $(p-1)|m$, then $B_m + \frac{1}{p} = (pB_m + 1)/p$ is a p-integer for that prime p. It follows that I_m is a p-integer for every prime p. Hence, I_m is an integer. ∎

REMARK 3.2   The sum in Corollary 3.4 is less than a constant times $\log\log m$, so for large even m, the integer I_m is about as large as B_m. Corollary 3.4 determines the fractional part $B_m - \lfloor B_m \rfloor$ of B_m for even m. It is easy for large m to compute $\zeta(m)$ to many decimal places. If we use this value and Theorem 3.8 to compute B_m with an error less than $\frac{1}{2}$, then we can find B_m exactly because we know the fractional part from Corollary 3.4. This method has been used to produce some tables of Bernoulli numbers. ∎

COROLLARY 3.5
If p is an odd prime and m is an even integer satisfying $1 < m \le p-1$, then

$$pB_m \equiv S_m(p) \pmod{p^2}.$$

PROOF   In the sum in Equation (3.18), when $j > 0$, $(p-1) \nmid j$ because $j < m \le p-1$. Therefore each B_j for $j < m$ is a p-integer. We have saved an extra factor of p in each term of the sum in addition to the one in (3.19), so the sum is $\equiv 0 \pmod{p^2}$. ∎

3.7   *Congruences of Voronoi and Kummer*

We follow Uspensky and Heaslet [113] in our proof of the congruences in the title. In the course of the proof we shall derive several other interesting congruences. The first of these is similar to (3.20), but the argument is more delicate because we allow a composite modulus n in place of the prime p. Since we will use the von Staudt-Clausen theorem in the proof, we do not have an alternate proof of (3.20). When we speak of the numerator or denominator of a rational number, we always mean that the number is in lowest terms.

Let m be an even positive integer and n a positive integer. Begin with Theorem 3.3:

$$
\begin{aligned}
S_m(n) &= \frac{1}{m+1} \sum_{j=0}^{m} \binom{m+1}{j} B_j n^{m+1-j} \\
&= \sum_{j=0}^{m} \binom{m}{m-j} B_j \frac{n^{m+1-j}}{m+1-j} \\
&= nB_m + \sum_{i=1}^{m} \binom{m}{i} B_{m-i} \frac{n^{i+1}}{i+1}.
\end{aligned}
$$

Multiply by the denominator Q_m to get:

$$Q_m S_m(n) = nP_m + n^2 \sum_{i=1}^{m} \binom{m}{i} B_{m-i} Q_m \frac{n^{i-1}}{i+1}. \tag{3.22}$$

We will prove that

$$Q_m S_m(n) \equiv nP_m \pmod{n^2} \tag{3.23}$$

by showing that each term in the sum on the right side of (3.22) is an n-integer. Since $B_{m-1} = 0$ unless $m = 2$, the term with $i = 1$ either vanishes or (when $m = 2$) equals $\binom{2}{1} B_1 Q_2 / 2 = -3$, an integer. For $i \geq 2$ we ignore the binomial coefficient, which is always an integer, and prove simply that

$$B_{m-i} Q_m \frac{n^{i-1}}{i+1} = \frac{P_{m-i} Q_m n^{i-1}}{(i+1) Q_{m-i}} \tag{3.24}$$

is an n-integer. When $i = 2$, Expression (3.24) becomes

$$P_{m-2} \cdot \frac{Q_m}{3} \cdot \frac{n}{Q_{m-2}},$$

which is an n-integer because $3 | Q_m$ and Q_{m-2} is square-free.

Now assume that $i \geq 3$. Let p be a prime divisor of n and write $i+1 = p^a q$, where $p \nmid q$ and a is a nonnegative integer. Since Q_{m-i} is square-free, p divides $(i+1)Q_{m-i}$ to the power $a+1$ at most, and p divides $Q_m n^{i-1}$ at least to the power $i - 1$, and at least to the power i if $p = 2$ or 3. It suffices to show that $i - 1 = p^a q - 2 \geq a + 1$ for $p \geq 5$ and $i = p^a q - 1 \geq a + 1$ for $p = 2$ and 3. If $a > 0$ and $p \geq 5$, then $p^a > a + 3$. If $a = 0$, then $q = i + 1 > 3$. Also, $3^a \geq a + 2$ for $a \geq 1$. Finally, if $p = 2$, then $2^a \geq a + 2$ for $a \geq 2$, and when $a = 1$, we must have $q \geq 3$ since $p^a q = 2q = i + 1 > 3$ and $p \nmid q$, so that $i = 2q - 1 > a + 1$. This completes the proof of (3.23).

It is easy to check that (3.23) holds also for $m = 1$ but not for odd $m > 1$. (Try $n = 2$.)

We need one more congruence, (3.25) below, before we obtain that of Voronoi. Let m, a and n be positive integers with $\gcd(a, n) = 1$. For nonnegative integers s, write $q_s = \lfloor sa/n \rfloor$ and $r_s = sa - nq_s$. Then

$$a^m s^m = (nq_s + r_s)^m \equiv r_s^m + mnq_s r_s^{m-1} \pmod{n^2}.$$

Also $r_s \equiv sa \pmod{n}$, so

$$a^m s^m \equiv r_s^m + mnq_s a^{m-1} s^{m-1} \pmod{n^2}.$$

Sum this congruence from $s = 0$ to $s = n - 1$. Noting that r_s runs through a complete set of residue classes modulo n, we have

$$a^m S_m(n) \equiv S_m(n) + mna^{m-1} \sum_{s=1}^{n-1} s^{m-1} q_s \pmod{n^2}$$

or

$$(a^m - 1)S_m(n) \equiv mna^{m-1} \sum_{s=1}^{n-1} s^{m-1} \left\lfloor \frac{sa}{n} \right\rfloor \pmod{n^2}. \qquad (3.25)$$

THEOREM 3.16 Voronoi's Congruence

Let m, a and n be positive integers with m even and a coprime to n. Then

$$(a^m - 1)P_m \equiv ma^{m-1}Q_m \sum_{s=1}^{n-1} s^{m-1} \left\lfloor \frac{sa}{n} \right\rfloor \pmod{n}.$$

PROOF   Combine (3.23) and (3.25) and divide by n:

$$nP_m(a^m - 1) \equiv Q_m(a^m - 1)S_m(n) \equiv mna^{m-1}Q_m \sum_{s=1}^{n-1} s^{m-1} \left\lfloor \frac{sa}{n} \right\rfloor \pmod{n^2}.$$

∎

Voronoi's congruence holds also when $m = 1$.

REMARK 3.3   Closer examination of the proofs of (3.23) and (3.25) when $n = p$ is prime, $(p-1) \nmid m$, and p^e is the highest power of p that divides m, shows that in this situation the congruences hold modulo p^{e+2}, so that we may divide by m in Voronoi's theorem to obtain

$$(a^m - 1)\frac{B_m}{m} \equiv a^{m-1} \sum_{s=1}^{p-1} s^{m-1} \left\lfloor \frac{sa}{p} \right\rfloor \pmod{p}. \qquad (3.26)$$

It is this version of Voronoi's theorem that is used in Corollaries 3.8 and 3.9.
∎

The first Corollary is called J. C. Adams' theorem.

COROLLARY 3.6

If p is prime, m is even, p^e divides m and p does not divide Q_m, that is, $p-1$ does not divide m, then p^e divides P_m.

Note that p cannot be 2 or 3 because m is even.

PROOF   In Voronoi's theorem take $n = p^e$ and let a be a primitive root modulo p. We obtain $(a^m - 1)P_m \equiv 0 \pmod{p^e}$. But $(p-1) \nmid m$, so $p \nmid (a^m - 1)$. Therefore, $p^e \mid P_m$. ∎

REMARK 3.4   J. C. Adams' theorem tells us that the primes in the denominator of B_m/m are exactly those in Q_m, the denominator of B_m. It enables us to write $m = m_1 m_2$, where m_1 and m_2 are relatively prime, m_1 divides P_m and every prime that divides m_2 also divides Q_m. We will use this fact in the next section.   ∎

COROLLARY 3.7

If a is an integer and k is a positive integer, then

$$\frac{a^{k+1}\left(a^{2k}-1\right)B_{2k}}{2k}$$

is an integer.

PROOF   The statement is clear if $a = 0$, 1 or -1. Write $2kQ_{2k} = \ell m$, where $\gcd(m, a) = 1$ and every prime divisor of ℓ also divides a. Voronoi's theorem gives $(a^{2k} - 1)P_{2k} \equiv 0 \pmod{m}$, that is, $(a^{2k} - 1)P_{2k}/m$ is an integer. On the other hand, let p^e exactly divide ℓ. Then $p^3 \nmid 2Q_{2k}$ because Q_{2k} is square-free. Therefore, $p^{e-2} | k$. (If $e = 1$, replace p^{e-2} by 1.) Hence, $k \geq p^{e-2}$. Also $p | a$, so $p^{k+1} | a^{k+1}$ and $p^{p^{e-2}+1} | a^{k+1}$. But $p^{e-2} + 1 \geq e$, so a^{k+1}/ℓ is a p-integer. Since p is an arbitrary prime we see that

$$\frac{a^{k+1}}{\ell} \cdot \frac{\left(a^{2k}-1\right)P_{2k}}{m} = \frac{a^{k+1}\left(a^{2k}-1\right)B_{2k}}{2k}$$

is an integer.   ∎

We return briefly to analysis for an application of Corollary 3.7. For $|z| < 2\pi$ we have

$$\frac{z}{2}\cdot\frac{e^{z/2}+e^{-z/2}}{e^{z/2}-e^{-z/2}} = \frac{z}{2} + \frac{z}{e^z - 1} = \frac{z}{2} + \sum_{n=0}^{\infty} B_n \frac{z^n}{n!} = \sum_{n=0}^{\infty} B_{2n}\frac{z^{2n}}{(2n)!}.$$

Setting $z = 2ix$ we find for $0 < |x| < \pi$,

$$\cot x = i\,\frac{e^{ix}+e^{-ix}}{e^{ix}-e^{-ix}} = \frac{1}{x}\frac{z}{2}\frac{e^{z/2}+e^{-z/2}}{e^{z/2}-e^{-z/2}}$$

$$= \frac{1}{x}\sum_{n=0}^{\infty}B_{2n}\frac{z^{2n}}{(2n)!} = \sum_{n=0}^{\infty}(-1)^n 2^{2n}B_{2n}\frac{x^{2n-1}}{(2n)!}.$$

Similarly, for $0 < |x| < \pi$,

$$
\begin{aligned}
\tan x &= \frac{1}{i} \frac{e^{ix} - e^{-ix}}{e^{ix} + e^{-ix}} = -\frac{1}{x} \frac{z}{2} \frac{e^{z/2} - e^{-z/2}}{e^{z/2} + e^{-z/2}} \\
&= \frac{1}{x} \left(\frac{2z}{e^{2z} - 1} - \frac{z}{e^z - 1} + \frac{z}{2} \right) \\
&= \frac{1}{x} \left(\sum_{n=0}^{\infty} B_n \frac{(2z)^n}{n!} - \sum_{n=0}^{\infty} B_n \frac{z^n}{n!} - B_1 z \right) \\
&= \frac{1}{x} \sum_{n=0}^{\infty} (2^{2n} - 1) B_{2n} \frac{z^{2n}}{(2n)!} \\
&= \sum_{n=1}^{\infty} T_n \frac{x^{2n-1}}{(2n-1)!},
\end{aligned}
$$

where $T_n = (-1)^{n-1} 2^{2n} (2^{2n} - 1) B_{2n}/2n$. The "tangent coefficients" T_n are positive by Theorem 3.7 and integers by Corollary 3.7 with $a = 2$.

The next corollary is due to Vandiver [115].

COROLLARY 3.8

If p is an odd prime, k and a are positive integers, $a \geq 2$, $(p-1) \nmid 2k$ and $p \nmid a$, then

$$
(1 - a^{2k}) \frac{B_{2k}}{2k} \equiv a^{2k-1} \sum_{j=1}^{a-1} \sum_{s=1}^{\lfloor jp/a \rfloor} s^{2k-1} \pmod{p}.
$$

PROOF   By J. C. Adams' theorem (Corollary 3.6), p cannot divide the denominator of $B_{2k}/2k$, so the left side is meaningful. Apply Voronoi's theorem in the form of Congruence (3.26). For $(j-1)p < as < jp$ we have $\lfloor sa/p \rfloor = j - 1$. Thus,

$$
(a^{2k} - 1) \frac{B_{2k}}{2k} \equiv a^{2k-1} \sum_{j=1}^{a} (j-1) \sum_{(j-1)p < as < jp} s^{2k-1} \pmod{p}.
$$

Since $(p-1) \nmid 2k$ we have $(a-1) \sum_{s=1}^{p-1} s^{2k-1} \equiv 0 \pmod{p}$ by Theorem 3.3 so that

$$
(1 - a^{2k}) \frac{B_{2k}}{2k} \equiv a^{2k-1} \sum_{j=1}^{a-1} (a-j) \sum_{(j-1)p < as < jp} s^{2k-1} \pmod{p}
$$

$$
\equiv a^{2k-1} \sum_{j=1}^{a-1} \sum_{s=1}^{\lfloor jp/a \rfloor} s^{2k-1} \pmod{p}.
$$

∎

The next corollary is also due to Vandiver [115].

COROLLARY 3.9

If $(p-1) \nmid 2k$, then

$$\left(3^{p-2k} + 4^{p-2k} - 6^{p-2k} - 1\right) \frac{B_{2k}}{4k} \equiv \sum_{p/6<s<p/4} s^{2k-1} \pmod{p}.$$

PROOF   Let $a = 3$, 4 and 6 in Corollary 3.8. Since p cannot be 2 or 3 we have $a^{p-1} \equiv 1 \pmod{p}$ by Fermat's little theorem. We obtain

$$a^{p-2k}\left(1 - a^{2k}\right) \frac{B_{2k}}{2k} \equiv \sum_{j=1}^{a-1} \sum_{s=1}^{\lfloor jp/a \rfloor} s^{2k-1} \pmod{p}.$$

Write

$$f(a) = a^{p-2k}\left(1 - a^{2k}\right) \frac{B_{2k}}{2k} \equiv \left(a^{p-2k} - a\right) \frac{B_{2k}}{2k} \pmod{p}$$

and $S_n = \sum s^{2k-1}$, where the sum is over all integers s between $np/12$ and $(n+1)p/12$. With $a = 3$, 4 and 6, we find these congruence modulo p:

$$f(3) \equiv 2(S_0 + S_1 + S_2 + S_3) + (S_4 + S_5 + S_6 + S_7)$$
$$f(4) \equiv 3(S_0 + S_1 + S_2) + 2(S_3 + S_4 + S_5) + (S_6 + S_7 + S_8)$$
$$f(6) \equiv 5(S_0 + S_1) + 4(S_2 + S_3) + 3(S_4 + S_5) + 2(S_6 + S_7) + (S_8 + S_9).$$

Adding the first two congruences and subtracting the third one yields

$$\left(3^{p-2k} + 4^{p-2k} - 6^{p-2k} - 1\right) \frac{B_{2k}}{2k} \equiv S_2 - S_9 \pmod{p}.$$

But

$$S_2 - S_9 \equiv \sum_{p/6<s<p/4} s^{2k-1} - \sum_{3p/4<s<5p/6} s^{2k-1} \pmod{p}$$

$$\equiv \sum_{p/6<s<p/4} \left(s^{2k-1} - (p-s)^{2k-1}\right) \pmod{p}$$

$$\equiv 2 \sum_{p/6<s<p/4} s^{2k-1} \pmod{p},$$

and the corollary follows.   ∎

REMARK 3.5   One use of this corollary is in finding irregular primes, that is, primes p that divide P_{2k} for some $2 \le 2k \le p-3$. For each $2k$ in this interval the sum on the right side is evaluated modulo p. If it does not vanish (as usually happens), then p does not divide P_{2k}. Otherwise, one computes the coefficient of $B_{2k}/4k$ on the left side. If this factor is not a multiple of p, then p is irregular and $p|P_{2k}$. Another test (a slower one) must be used if p divides the coefficient. Some such alternative tests are provided by variations on Corollary 3.9 derived from Corollary 3.8 by choosing different values of a. See Exercise 20 for some examples. See also page 268 of [113]. Also see Buhler et al. [13] for even faster ways to find irregular primes. ∎

COROLLARY 3.10
If $p \equiv 3 \pmod 4$ is prime, then

$$\left(2 - \left(\frac{2}{p}\right)\right) \frac{B_{(p+1)/2}}{(p+1)/2} \equiv - \sum_{s=1}^{(p-1)/2} \left(\frac{s}{p}\right) \pmod{p}, \qquad (3.27)$$

where (s/p) is the Legendre symbol.

PROOF Let $m = (p+1)/2$, $a = 2$ and $n = p$ in Voronoi's congruence. Using Euler's criterion, we have

$$\left(2^{(p+1)/2} - 1\right) \frac{B_{(p+1)/2}}{(p+1)/2} \equiv 2^{(p-1)/2} \sum_{s=1}^{p-1} s^{(p-1)/2} \left\lfloor \frac{2s}{p} \right\rfloor \pmod{p},$$

$$\left(2 \left(\frac{2}{p}\right) - 1\right) \frac{B_{(p+1)/2}}{(p+1)/2} \equiv \left(\frac{2}{p}\right) \sum_{s=(p+1)/2}^{p-1} \left(\frac{s}{p}\right) \pmod{p},$$

$$\left(2 - \left(\frac{2}{p}\right)\right) \frac{B_{(p+1)/2}}{(p+1)/2} \equiv - \sum_{s=1}^{(p-1)/2} \left(\frac{s}{p}\right) \pmod{p}$$

because $\sum_{s=1}^{p-1} \left(\frac{s}{p}\right) = 0$. ∎

REMARK 3.6 The sum on the right side of (3.27) is not divisible by p since it is the sum of an odd number of terms ± 1 and has absolute value less than p. Hence, p does not divide $P_{(p+1)/2}$ when $p \equiv 3 \pmod 4$. This is the only known case of $p \nmid P_{2k}$ with $2 \le 2k \le p - 3$, $k = k(p)$, that holds for infinitely many primes p. ∎

THEOREM 3.17 Kummer's Congruence
Let p be prime and n and k be positive integers, such that $(p-1) \nmid 2k$ (which implies $p > 3$) and $2k \ge n + 1$. Then

$$\sum_{s=0}^{n} (-1)^s \binom{n}{s} \frac{B_{2k+s(p-1)}}{2k + s(p-1)} \equiv 0 \pmod{p^n}.$$

Notation: For positive even m we write $B_m/m = N_m/D_m$, where N_m and D_m are relatively prime integers with $D_m > 0$. We will use this notation in the next section, too. According to J. C. Adams' theorem (Corollary 3.6), a prime $p|D_m$ if and only if $p|Q_m$.

PROOF   Since $(p-1) \nmid 2k + s(p-1)$, we have $p \nmid Q_{2k+s(p-1)}$ and $p \nmid D_{2k+s(p-1)}$ by the theorems of von Staudt-Clausen and J. C. Adams. Let p^e be the highest power of p that divides any of the $n+1$ numbers

$$2k, 2k + (p-1), 2k + 2(p-1), \ldots, 2k + n(p-1).$$

Let g be a primitive root modulo p and $a = g^{p^{n-1}}$. Then $\gcd(a, p^{n+e}) = 1$ and Voronoi's theorem gives (writing $m = 2k + s(p-1)$)

$$(a^m - 1)\, P_m \equiv m Q_m a^{m-1} \sum_{t=1}^{p^{n+e}-1} t^{m-1} \left\lfloor \frac{at}{p^{n+e}} \right\rfloor \pmod{p^{n+e}}.$$

By our choice of e we can write

$$(a^m - 1)\, \frac{B_m}{m} \equiv a^{m-1} \sum_{t=1}^{p^{n+e}-1} t^{m-1} \left\lfloor \frac{at}{p^{n+e}} \right\rfloor \pmod{p^n}$$

for $s = 0, 1, \ldots, n$. But $\phi(p^n) = p^{n-1}(p-1)$ and $\gcd(a^s, p^n) = 1$, so $a^{s(p-1)} = g^{sp^{n-1}(p-1)} \equiv 1 \pmod{p^n}$. Thus

$$\left(a^{2k} - 1\right) \frac{B_{2k+s(p-1)}}{2k + s(p-1)} \equiv a^{2k-1} \sum_{t=1}^{p^{n+e}-1} \left\lfloor \frac{ta}{p^{n+e}} \right\rfloor t^{2k+s(p-1)-1} \pmod{p^n}.$$

Multiply by $(-1)^s \binom{n}{s}$ and add from $s = 0$ to $s = n$. We get

$$\left(a^{2k} - 1\right) \sum_{s=0}^{n} (-1)^s \binom{n}{s} \frac{B_{2k+s(p-1)}}{2k + s(p-1)}$$

$$\equiv a^{2k-1} \sum_{t=1}^{p^{n+e}-1} \left\lfloor \frac{ta}{p^{n+e}} \right\rfloor t^{2k-1} \sum_{s=0}^{n} (-1)^s \binom{n}{s} t^{(p-1)s}$$

$$\equiv a^{2k-1} \sum_{t=1}^{p^{n+e}-1} \left\lfloor \frac{ta}{p^{n+e}} \right\rfloor t^{2k-1} \left(1 - t^{(p-1)}\right)^n \pmod{p^n}.$$

By hypothesis, $2k - 1 \geq n$, so if $p|t$, then $t^{2k-1} \equiv 0 \pmod{p^n}$. If $p \nmid t$, then $t^{p-1} \equiv 1 \pmod{p}$ and $\left(1 - t^{p-1}\right)^n \equiv 0 \pmod{p^n}$. In either case the right side is divisible by p^n. The theorem follows when we note that $a^{2k} - 1 \not\equiv 0 \pmod{p}$ since $\gcd(a,p) = 1$ and $p - 1 \nmid 2k$. ∎

The case $n = 1$ of Theorem 3.17 says that if $p - 1 \nmid 2k$, then

$$\frac{B_{2k}}{2k} \equiv \frac{B_{2k+p-1}}{2k+p-1} \quad (\text{mod } p), \tag{3.28}$$

which shows that B_m/m modulo p is periodic with period $p - 1$ for $m \geq 2$ not divisible by $p - 1$. This fact is used in almost every proof that there are infinitely many irregular primes.

REMARK 3.7   The close examination of the terms (3.19) and (3.24) in the proofs of the von Staudt-Clausen theorem and Voronoi's congruence may be avoided by the use of p-adic numbers. Wells Johnson [61] gave p-adic proofs of Theorems 3.2, 3.15, 3.16 and 3.17 as well as most of the corollaries of Theorems 3.15 and 3.16. ∎

3.8   *Irregular Primes*

Irregular primes were defined and their significance explained in Section 3.1. See Borevich and Shafarevich [11] for a different treatment of much of the material in this section. In 1915, Jensen [60] proved that there are infinitely many irregular primes $\ell \equiv 3 \pmod 4$. We will prove his theorem, but first we give a very short proof due to Carlitz that there are infinitely many irregular primes. Then we will present H. Montgomery's proof that for each $t > 2$ there are infinitely many irregular primes not of the form $mt + 1$.

DEFINITION 3.4   *An odd prime ℓ is a proper divisor of a Bernoulli number B_{2k} if ℓ divides N_{2k} (the numerator of $B_{2k}/2k$ in lowest terms).*

LEMMA 3.1
An odd prime is irregular if and only if it is a proper divisor of some Bernoulli number.

PROOF   If ℓ is irregular, then there is an even index $2 \leq 2k \leq \ell - 3$, such that $\ell | P_{2k}$. Then $\ell \nmid 2k$, so $\ell | N_{2k}$ and ℓ is a proper divisor of B_{2k}.
Suppose $\ell | N_{2k}$. By the von Staudt-Clausen theorem, $(\ell - 1) \nmid 2k$. Let k' be the least nonnegative residue of k modulo $(\ell - 1)/2$, so that $2k \equiv 2k' \not\equiv 0 \pmod{\ell - 1}$ and $2 \leq 2k' \leq \ell - 3$. By Kummer's congruence (3.28),

$$\frac{B_{2k'}}{2k'} \equiv \frac{B_{2k}}{2k} \equiv 0 \quad (\text{mod } \ell).$$

Therefore, $\ell | N_{2k'}$, $\ell | P_{2k'}$ and ℓ is irregular. ∎

THEOREM 3.18 Infinitely Many Irregular Primes
There are infinitely many irregular primes.

PROOF   (Carlitz [15]) Let $p_1, \ldots, p_s$ be distinct odd primes. We will construct an irregular prime ℓ different from all the p_i's. Let $m = (p_1 - 1)(p_2 - 1) \cdots (p_s - 1)$. By (3.7), we can choose a positive integer t, such that

$$\frac{|B_{tm}|}{tm} = \frac{|N_{tm}|}{D_{tm}} > 1.$$

In particular, $N_{tm} \neq \pm 1$ and D_{tm} is even, so there is an odd prime ℓ dividing N_{tm}. By Lemma 3.1, ℓ is irregular. Also, for $1 \leq i \leq s$, we have $\ell \neq p_i$ because $(p_i - 1)|tm$ and hence $p_i|D_{tm}$. ∎

Now we prove Jensen's theorem, which he proved before Carlitz gave the short proof of Theorem 3.18.

THEOREM 3.19 Jensen [60]   Irregular Primes $\equiv 3$ (mod 4)
There are infinitely many irregular primes $\equiv 3$ (mod 4).

PROOF   Let $p_1, \ldots, p_s$ be distinct primes larger than 3. By Dirichlet's theorem, there is a prime $q \equiv 1 \pmod{12 \prod_{i=1}^s p_i(p_i - 1)}$. As always, $6|Q_{2q}$. If $p > 3$ is prime and $p|Q_{2q}$, then $(p - 1)|2q$, so $p - 1 = 2q$ and $p = 2q + 1 \equiv 3 \pmod{12}$ since $q \equiv 1 \pmod 6$. Because this is impossible for $p > 3$, we must have $Q_{2q} = 6$. (Note that this shows that $Q_{2m} = 6$ for infinitely many m.) Also, $2q \equiv 2 \pmod{p_i - 1}$, so by (3.28) we have

$$\frac{B_{2q}}{2q} \equiv \frac{B_2}{2} = \frac{1}{12} \pmod{p_i}.$$

Therefore, for each i, $p_i \nmid P_{2q}$. By (3.23) with $m = 2q$ and $n = 4$ we have

$$4P_{2q} \equiv Q_{2q}\left(1^{2q} + 2^{2q} + 3^{2q}\right) \equiv 6(1 + 0 + 1) = 12 \pmod{16}.$$

Hence $P_{2q} \equiv 3 \pmod 4$ and there must be a prime $\ell \equiv 3 \pmod 4$, such that $\ell|P_{2q}$. Clearly, $\ell \nmid q$, and $\ell \neq p_i$ because $p_i|Q_{2q}$, so $\ell|N_{2q}$ and ℓ is irregular.
∎

LEMMA 3.2
If p is an odd prime and $k \equiv 1 \pmod{p(p-1)}$, then

$$S_{2k}(p) \equiv \frac{p}{6} \pmod{p^2}.$$

PROOF   By Euler's theorem and Example 3.3, we have

$$S_{2k}(p) = \sum_{i=1}^{p-1} i^{2k} \equiv \sum_{i=1}^{p-1} i^2 = \frac{p^3}{3} - \frac{p^2}{2} + \frac{p}{6} \equiv \frac{p}{6} \pmod{p^2}.$$

∎

THEOREM 3.20 Montgomery ([72])   Irregular Primes Not $mp+1$
If p is an odd prime, then there are infinitely many irregular primes not of the form $mp + 1$.

PROOF   Let $p_1, \ldots, p_s$ be distinct primes greater than 3. We will find an irregular prime $p_{s+1} \not\equiv 1 \pmod{p}$ and distinct from $p_1, \ldots, p_s$. Let ℓ' be a prime, such that

$$\ell' \equiv -1 \pmod{12p(p-1)p_1(p_1-1)\cdots p_s(p_s-1)}. \qquad (3.29)$$

Then $\ell'|Q_{2\mu'}$, where $\mu' = (\ell'-1)/2$, since $(\ell'-1)|2\mu'$. Let $\ell \geq 5$ be the least prime (other than 2 or 3) that divides $Q_{2\mu'}$ and such that $\ell \not\equiv 1 \pmod{p}$. With $\mu = (\ell-1)/2$, we have $\ell|Q_{2\mu}$, $\mu|\mu'$ and $Q_{2\mu}|Q_{2\mu'}$ by the von Staudt-Clausen theorem. Hence the only prime factors of $Q_{2\mu}$ that are not of the form $mp+1$ are 2, 3 and ℓ. Note that μ is odd.

Let $d_1, d_2, \ldots, d_r$ be all the distinct divisors of μ and let $\ell_1, \ell_2, \ldots, \ell_r$ be distinct primes larger than $M = 6\ell p(p-1)p_1(p_1-1)\cdots p_s(p_s-1)$. Consider the simultaneous congruences

$$\eta \equiv \frac{1}{\mu} \pmod{M}, \quad \eta \equiv \frac{-1}{2d_i} \pmod{\ell_i^2} \quad (i = 1, 2, \ldots, r). \qquad (3.30)$$

Now $\gcd(\mu, M) = 1$ because $\mu|\mu'$ and $\mu' \equiv -1 \pmod{M/\ell}$ by Congruence (3.29). Also, $\gcd(2d_i, \ell_i) = 1$ because $2d_i < \ell < \ell_i$ and ℓ_i is prime. Thus the congruences make sense. The moduli in Congruences (3.30) are pairwise relatively prime, so by the Chinese remainder theorem there is a solution η to (3.30), which is unique modulo the product of the moduli. Clearly η is relatively prime to each modulus, so by Dirichlet's theorem we may choose η to be a prime satisfying (3.30).

Let $q = \mu\eta$. We claim that $Q_{2q} = Q_{2\mu}$. The divisors of $2q$ are

$$\begin{array}{cccc}
d_1, & d_2, & \ldots, & d_r, \\
2d_1, & 2d_2, & \ldots, & 2d_r, \\
d_1\eta, & d_2\eta, & \ldots, & d_r\eta, \\
2d_1\eta, & 2d_2\eta, & \ldots, & 2d_r\eta.
\end{array} \qquad (3.31)$$

When 1 is added to each member of (3.31), the primes in the first two rows are precisely those dividing $Q_{2\mu}$. The $d_i\eta + 1$ are even and $2d_i\eta + 1 \equiv 0 \pmod{\ell_i^2}$ by (3.30). Therefore $Q_{2q} = Q_{2\mu}$, and hence $Q_{2q}/6 \equiv \ell \pmod{p}$.

By Congruence (3.23) and Lemma 3.2 we have

$$pP_{2q} \equiv Q_{2q} S_{2q}(p) \equiv Q_{2q} \frac{p}{6} \pmod{p^2},$$

so that

$$P_{2q} \equiv \frac{Q_{2q}}{6} \equiv \ell \pmod{p}. \tag{3.32}$$

According to Remark 3.4, J. C. Adams' theorem allows us to write $q = q_1 q_2$, where $\gcd(q_1, q_2) = 1$, $q_1 | P_{2q}$ and every prime that divides q_2 also divides Q_{2q}. The only primes in Q_{2q} not of the form $mp + 1$ are 2, 3 and ℓ, but $q = \mu \eta \equiv 1 \pmod{6\ell}$ by (3.30), so 2, 3 and ℓ do not divide q and *a fortiori* do not divide q_2. Hence all prime divisors of q_2 are of the form $mp+1$. Therefore $q_2 \equiv 1 \pmod{p}$. But $q \equiv 1 \pmod{p}$ by (3.30). Therefore $q_1 \equiv 1 \pmod{p}$ and $N_{2q} = P_{2q}/q_1 \equiv \ell \pmod{p}$ by (3.32). Also by (3.32), $\ell \not\equiv 0 \pmod{p}$ since $\gcd(P_{2q}, Q_{2q}) = 1$. By Theorem 3.7, P_{2q} is a positive integer. Hence the positive integer N_{2q} is not congruent to 0 or 1 modulo p and contains a prime factor p_{s+1} not of the form $mp+1$. By Lemma 3.1, p_{s+1} is irregular. Suppose $p_{s+1} = p_i$ for some $1 \leq i \leq s$. By (3.30), $2q = 2\mu\eta \equiv 2 \not\equiv 0 \pmod{p_i - 1}$ and by (3.28)

$$\frac{B_{2q}}{2q} \equiv \frac{B_2}{2} = \frac{1}{12} \not\equiv 0 \pmod{p_i}.$$

Therefore $p_i \nmid N_{2q}$ and $p_{s+1} \neq p_i$. ∎

Montgomery [72] also proved the following result.

COROLLARY 3.11

If $t > 2$ is an integer, then there are infinitely many irregular primes not of the form $mt + 1$.

PROOF   The result follows from Theorem 3.20 if t has an odd prime factor and from Theorem 3.19 if t is a power of 2. ∎

Example 3.12

For $t = 3$ and $t = 6$ we have $\phi(t) = 2$ and Corollary 3.11 tells us that there are infinitely many irregular primes of each of the forms $3m + 2$ and $6m + 5$. Of course, there are also infinitely many irregular primes of the form $4m + 3$ by Theorem 3.19.

REMARK 3.8   Using methods like those of Montgomery, T. Metsänkylä [71] has proved that if $t > 2$ and H is a proper subgroup of the multiplicative group of reduced residue classes modulo t, then there are infinitely many irregular primes whose residue class is not in H. ∎

3.9 Fractional Parts of Bernoulli Numbers

The *fractional part* of a real number x is $x - \lfloor x \rfloor$, that is, the distance to the largest integer $\leq x$. In this section we investigate the distribution of the fractional parts of the Bernoulli numbers. First of all, $B_0 = 1$ and $B_{2k+1} = 0$ for $k \geq 1$ by Theorem 3.2, so these Bernoulli numbers are integers and have fractional part 0. We will ignore these Bernoulli numbers and also $B_1 = -1/2$, the only Bernoulli number with fractional part $1/2$.

One might guess that the other Bernoulli numbers, B_{2k} for $k \geq 1$, are more or less random numbers and thus their fractional parts should have a uniform distribution between 0 and 1. Nothing could be further from the truth. Corollary 3.4 to the von Staudt-Clausen theorem tells us what we need to know about the fractional parts of B_{2k}. It says that

$$B_{2k} = I_{2k} - \sum_{(p-1)|2k,\ p \text{ prime}} \frac{1}{p},$$

where I_{2k} is an integer. Hence the fractional part of B_{2k} is that same as the fractional part of $-\sum_{(p-1)|2k,\ p \text{ prime}} 1/p$. Also, the Bernoulli numbers B_{2j} and B_{2k} have the same fractional part if and only if the primes p for which $p - 1|2j$ are exactly the same as the primes p for which $p - 1|2k$.

For real $x > 0$ and $0 \leq z \leq 1$, let $F_x(z) = x^{-1} \cdot$ (the number of positive integers $k \leq x$ for which the fractional part of $B_{2k} < z$). The graph in Figure 3.2 shows $F_x(z)$ for $x = 100000$ and $0 \leq z \leq 1$. The graph would be a straight line from $(0,0)$ to $(1,1)$ if the fractional parts were uniformly distributed in the unit interval. The graph looks very much like that of $F_x(z)$ for $x = 1000$ and $x = 10000$. Is there a limiting distribution function $F(z) = \lim_{x\to\infty} F_x(z)$ for $0 \leq z \leq 1$?

DEFINITION 3.5   The *asymptotic density* of a set S of positive integers is $d(S) = \lim_{x\to\infty} S(x)/x$, where $S(x)$ is the number of integers $\leq x$ in S, provided the limit exists.

Example 3.13

The even integers have asymptotic density $1/2$, as do the odd integers. The prime numbers have asymptotic density 0. Exercise 19 of Chapter 2 shows that the asymptotic density of the set of integers representable as a sum of three squares is $5/6$.

For positive integers j let d_{2j} be the asymptotic density (if the limit exists) of the set of positive integers k for which B_{2k} has the same fractional part as B_{2j}. It turns out that the limits do exist and that d_{2j} is approximately the height of the vertical jump of the graph at $z =$ the fractional part of B_{2j}. For example, the graph shows that about one-sixth of the Bernoulli numbers B_{2j}, $j \geq 1$, have fractional part $1/6$.

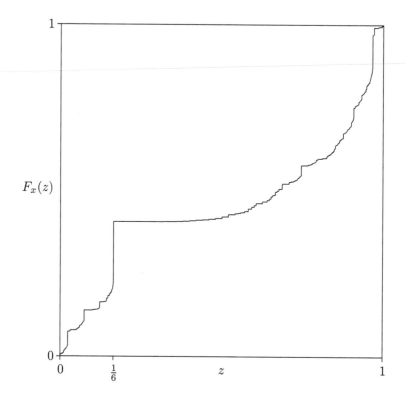

Figure 3.2 Distribution of the fractional parts of the Bernoulli numbers.

The graph appears to be horizontal in places, for example from $1/6$ to about 0.3. This observation suggests that some intervals contain the fractional part of no Bernoulli number. We will prove now that this is not so. We will show that the fractional parts of the Bernoulli numbers are dense in the unit interval $(0, 1)$.

For positive integers j, let $S_{2j} = \sum_{p-1|2j} 1/p$. By Corollary 3.3, the denominator of B_{2j} is greater than 1 when j is a positive integer. Is follows from this fact and the remarks above that the fractional part of B_{2j} is never 0, and it equals $1-$ the fractional part of S_{2j} when j is a positive integer. Hence it suffices to show that the fractional parts of the sums S_{2j}, $j \geq 1$, are dense in $(0, 1)$. This we will show in Theorem 3.21 below. The proof of this theorem requires three lemmas, which are interesting in their own right.

The first lemma counts the number of primes p, for which $p - 1$ is square-free, lying in an arithmetic progression. In it, $\mu(n)$ is the Möbius function. It is multiplicative, $\mu(1) = 1$, $\mu(p) = -1$ if p is prime, and $\mu(k) = 0$ if k is divisible by a square > 1.

LEMMA 3.3

For all $A > 1$, if k and b are positive integers with $\gcd(b,k) = \gcd(b-1,k) = 1$ and $k \le (\log x)^A$, then

$$\sum_{\substack{p \le x \\ p \equiv b \ (\mathrm{mod}\ k)}} \mu^2(p-1) = c(k)\,\mathrm{li}(x) + O\left(\frac{x}{(\log x)^A}\right),$$

where

$$c(k) = \frac{1}{\phi(k)} \prod_{p \nmid k} \left(1 - \frac{1}{p(p-1)}\right). \tag{3.33}$$

PROOF Note that $\mu^2(a) = \sum_{d^2|a} \mu(a)$. To see this, observe that we may always write $a = st^2$ with s square-free. Thus a is square-free if and only if $t = 1$, so we have $\mu^2(a) = \sum_{d|t} \mu(d)$, which is 1 if $t = 1$ and 0 if $t > 1$. Since $d|t$ if and only if $d^2|a$, the identity follows.

Using this identity with $a = p - 1$, we have

$$\sum_{\substack{p \le x \\ p \equiv b \ (\mathrm{mod}\ k)}} \mu^2(p-1) = \sum_{\substack{p \le x \\ p \equiv b \ (\mathrm{mod}\ k)}} \sum_{d^2|(p-1)} \mu(d) = \sum_{d \le \sqrt{x}} \mu(d) \sum_{\substack{p \le x,\, p \equiv b \ (\mathrm{mod}\ k) \\ p \equiv 1 \ (\mathrm{mod}\ d^2)}} 1.$$

Since $\gcd(b-1,k) = 1$, the conditions $p \equiv b \ (\mathrm{mod}\ k)$ and $p \equiv 1 \ (\mathrm{mod}\ d^2)$ are inconsistent if $\gcd(d,k) > 1$. Hence,

$$\sum_{d \le \sqrt{x}} \mu(d) \sum_{\substack{p \le x,\, p \equiv b \ (\mathrm{mod}\ k) \\ p \equiv 1 \ (\mathrm{mod}\ d^2)}} 1 = \sum_{\substack{d \le \sqrt{x} \\ \gcd(d,k)=1}} \mu(d)\, \pi(x; d^2 k, c),$$

where c depends on d, b and k. (Specifically, c is obtained by applying the Chinese remainder theorem to the congruences $c \equiv 1 \ (\mathrm{mod}\ d^2)$ and $c \equiv b \ (\mathrm{mod}\ k)$.) Write the above as $\Sigma_1 + \Sigma_2$, where $d \le (\log x)^A$ in Σ_1 and $d > (\log x)^A$ in Σ_2. The Siegel-Walfisz theorem says that for all $A > 1$, $\pi(x, m, e) = \mathrm{li}(x)/\phi(m) + O(x/(\log x)^A)$ uniformly for $1 \le m \le (\log x)^A$. By this theorem, the first sum is

$$\Sigma_1 = \sum_{\substack{d \le (\log x)^A \\ \gcd(d,k)=1}} \mu(d) \left(\frac{\mathrm{li}(x)}{\phi(d^2 k)} + O\left(\frac{x}{(\log x)^{2A}}\right)\right)$$

$$= \frac{\mathrm{li}(x)}{\phi(k)} \sum_{\substack{d \le (\log x)^A \\ \gcd(d,k)=1}} \frac{\mu(d)}{\phi(d^2)} + O\left(\frac{x}{(\log x)^A}\right).$$

The last sum may be written as an Euler product

$$\sum_{\substack{d \le (\log x)^A \\ \gcd(d,k)=1}} \frac{\mu(d)}{\phi(d^2)} = \prod_{p \nmid k} \left(1 - \frac{1}{p(p-1)}\right) + O\left(\frac{x}{(\log x)^A}\right).$$

To estimate Σ_2 we note that the number of primes $p \leq x$, such that $p \equiv c \pmod{d^2 k}$, does not exceed the number of integers in this arithmetic progression. Thus,

$$\Sigma_2 \ll \sum_{(\log x)^A < d \leq \sqrt{x}} \left(\frac{x}{d^2 k} + 1\right) \ll \frac{x}{k(\log x)^A} + \sqrt{x}.$$

The lemma follows when these estimates are combined.   ∎

The next lemma gives the principle of "partial summation," also called Abel's summation formula. The lemma and its proof are from Hardy and Wright [52] (Theorem 421 in Section 22.5).

LEMMA 3.4
Suppose $c_1, c_2, \ldots$ is a sequence of real numbers. For real $x \geq 1$, let $C(x) = \sum_{1 \leq i \leq x} c_i$. Suppose the real-valued function $f(t)$ has a continuous derivative for $t \geq 1$. Then for $x \geq 1$,

$$\sum_{1 \leq n \leq x} c_n f(n) = C(x) f(x) - \int_1^x C(t) f'(t)\, dt.$$

PROOF   Writing $N = \lfloor x \rfloor$, we see that

$$\sum_{1 \leq n \leq x} c_n f(n) = \sum_{n=1}^N c_n f(n)$$

$$= C(1) f(1) + \sum_{n=2}^N (C(n) - C(n-1)) f(n)$$

$$= \sum_{n=1}^{N-1} C(n)(f(n) - f(n+1)) + C(N) f(N).$$

Since $C(t) = C(n)$ when $n \leq t < n+1$, we have

$$C(n)(f(n) - f(n+1)) = -\int_n^{n+1} C(t) f'(t)\, dt.$$

If we sum this equation for $n = 1$ to $n = N - 1$, we get

$$\sum_{1 \leq n \leq x} c_n f(n) = C(N) f(N) - \int_1^N C(t) f'(t)\, dt.$$

Finally,

$$C(N) f(N) = C(x) f(x) - \int_N^x C(t) f'(t)\, dt,$$

and the proof is complete. ∎

LEMMA 3.5
For all integers $k > 2$ and b satisfying $1 \le b < k$, $\gcd(k, b) = 1$ and $\gcd(k, b - 1) = 1$, we have

$$\sum_{\substack{p \le x \\ p \equiv b \,(\mathrm{mod}\ k)}} \mu^2(p-1)\frac{1}{p} = c(k) \log \log x + K + O\left(\frac{1}{\log x}\right), \qquad (3.34)$$

where $c(k)$ is defined in Equation (3.33) and K is a constant depending on k.

PROOF   Lemma 3.5 follows from Lemma 3.3 by partial summation. Let k and b be fixed. We will apply Lemma 3.4 with $f(x) = 1/x$ and c_i the sequence defined by $c_i = 1$ if i is prime, $i \equiv b \pmod{k}$ and $i - 1$ is square-free, and $c_i = 0$ if i does not satisfy these three conditions. Write $C(x) = \sum_{1 \le i \le x} c_i$. Since $c_1 = 0$, we have $C(x) = \sum_{2 \le i \le x} c_i$. Note that Lemma 3.3 estimates $C(x)$. Lemma 3.4 gives

$$\sum_{\substack{p \le x \\ p \equiv b \,(\mathrm{mod}\ k)}} \mu^2(p-1)\frac{1}{p} = \sum_{1 \le n \le x} c_n f(n) = \frac{C(x)}{x} + \int_2^x \frac{C(t)}{t^2} dt.$$

Apply Lemma 3.3 with $A > 2$ to get

$$\frac{C(x)}{x} = \frac{c(k)\,\mathrm{li}(x)}{x} + O\left(\frac{1}{(\log x)^A}\right) = O\left(\frac{1}{\log x}\right).$$

Again by Lemma 3.3, the integral is

$$\int_2^x \frac{C(t)}{t^2} dt = \int_2^x \frac{c(k)\,\mathrm{li}(t)}{t^2} dt + O\left(\int_2^x \frac{dt}{t^2 (\log t)^A}\right).$$

The O term here is $O((\log x)^{-A+1})$, which is $O((\log x)^{-1})$. If we write $\mathrm{li}(t) = t/\log t + O(t/(\log t)^2)$, we find that the first integral is

$$\int_2^x \frac{c(k)\,\mathrm{li}(t)}{t^2} dt = c(k) \int_2^x \frac{dt}{t \log t} + c(k)\, O\left(\int_2^x \frac{dt}{t(\log t)^2}\right).$$

The first integral on the right side gives $c(k) \log \log x + K_1$. The integral in the O term may be evaluated as

$$\left(\int_2^\infty - \int_x^\infty\right) \frac{dt}{t(\log t)^2} = \frac{1}{\log 2} + O\left(\frac{1}{\log x}\right).$$

The lemma follows when all the estimates are combined. ∎

THEOREM 3.21 The S_{2j} Are Dense

For all real $\alpha \geq 5/6$ and $\varepsilon > 0$, there are infinitely many positive integers j for which $|S_{2j} - \alpha| < \varepsilon$.

PROOF   Let p_n denote the n-th prime. Let r be an integer to be chosen later. For positive integers s, let $A_s = 2p_r p_{r+1} \cdots p_{r+s}$. If p is prime, $p - 1$ is square-free and $p \equiv -1 \pmod{p_2 p_3 \cdots p_{r-1}}$, then $(p-1)|A_s$ for all sufficiently large s.

Now apply Lemma 3.5 with $k = p_2 p_3 \cdots p_{r-1}$ and $b = k - 1$. Since k is odd, it is easy to check that $\gcd(k, b) = \gcd(k, b-1) = 1$. Lemma 3.5 concludes that $\sum 1/p$ diverges, where p runs over all primes $p \equiv -1 \pmod{p_2 p_3 \cdots p_{r-1}}$ with $p - 1$ square-free. Therefore we can choose s large enough so that $S(A_s) > \alpha$. To prove the theorem we will remove the factors p_{r+s}, p_{r+s-1}, etc., from A_s, one by one, until $S(A_s)$ is close to α. It is enough to show that $S(A_s) - S(A_{s-1}) < \varepsilon$, provided p_r is large enough.

Let $d_1, \ldots, d_k$ be all of the positive divisors of A_{s-1}. Let $q = p_{r+s}$. Then $d_1, \ldots, d_k, qd_1, \ldots, qd_k$ are all of the positive divisors of A_s. Using σ to denote the sum of divisors function, we find

$$S(A_s) - S(A_{s-1}) = \sum_{\substack{p-1|A_s, \text{ but} \\ p-1 \nmid A_{s-1}}} \frac{1}{p} = \sum_{\substack{p-1=qd_i \\ \text{for some } i}} \frac{1}{p} = \sum_{\substack{i=1 \\ 1+qd_i \text{ is prime}}}^{k} \frac{1}{1+qd_i}$$

$$\leq \frac{1}{q} \sum_{i=1}^{k} \frac{1}{d_i} = \frac{1}{q} \sum_{i=1}^{k} \frac{d_i}{A_{s-1}} = \frac{\sigma(A_{s-1})}{qA_{s-1}}.$$

By Theorem 323 of [52], $\sigma(A_{s-1})/A_{s-1} < c_1 \log \log A_{s-1}$ for some positive constant c_1. By Theorem 414 of [52],

$$\log A_{s-1} = \log(2p_r p_{r+1} \cdots p_{r+s-1}) < c_2(p_{r+s-1} - p_r)$$

for some positive constant c_2. Combining these estimates yields

$$S(A_s) - S(A_{s-1}) < \frac{c_3 \log q}{q} \leq \frac{c_3 \log p_r}{p_r}$$

for a positive constant c_3. If we choose r large enough so that $c_3 \log p_r / p_r < \varepsilon$, then $S(A_s) - S(A_{s-1}) < \varepsilon$, which proves the theorem.   ∎

COROLLARY 3.12

The fractional parts of the Bernoulli numbers (with even subscripts) are dense in the unit interval $(0, 1)$.

Erdős and Wagstaff [29] proved the following theorem, which explains many properties of the graph above.

THEOREM 3.22 Fractional Parts of B_{2j}

For every positive integer j, the asymptotic density d_{2j} of the set of all k for which B_{2k} has the same fractional part as B_{2j} exists and is positive. The distribution function $F(z) = \lim_{x \to \infty} F_x(z)$ (for $0 \le z \le 1$) exists and is a jump function, that is, F increases only by jumping. The convergence is uniform and the sum of the heights of the jumps of F is 1.

3.10 Exercises

The first thirteen exercises deal with the Euler numbers E_n and Euler polynomials $E_n(x)$, which are defined by

$$\frac{2e^{xt}}{e^t + 1} = \sum_{n=0}^{\infty} E_n(x)\frac{t^n}{n!} \quad \text{and} \quad E_n = 2^n E_n\left(\frac{1}{2}\right).$$

These numbers and polynomials enjoy many properties analogous to the Bernoulli numbers and Bernoulli polynomials, as the exercises show.

1. Compute E_0, E_1, E_2, E_3 and E_4. Prove that the Euler numbers are all integers and that $E_n = 0$ when n is odd.

2. Prove that $(-1)^n E_{2n} > 0$ for every nonnegative integer n.

3. For all real numbers x and nonnegative integers n, prove

$$E_n(x) = \sum_{k=0}^{n} \binom{n}{k} \frac{E_k}{2^k} \left(x - \frac{1}{2}\right)^{n-k}.$$

4. For all real numbers x and nonnegative integers n, prove $E_n(x + 1) + E_n(x) = 2x^n$.

5. For all real numbers x and positive integers n, prove $E_n'(x) = nE_{n-1}(x)$.

6. For all nonnegative integers m and n, define the alternating sum

$$A_n(m) = \sum_{k=1}^{m} (-1)^{m-k} k^n = m^n - (m-1)^n + \cdots + (-1)^{m-1} 1^n.$$

Prove the formula

$$A_n(m) = \frac{1}{2}(E_n(m+1) + (-1)^m E_n(0))$$

for all nonnegative integers m and n.

7. Prove these Fourier series expansions for the Euler polynomials. Show that for every positive integer m and for all real noninteger t we have

$$E_{2m-1}(t - \lfloor t \rfloor) = (-1)^m 4(2m-1)! \sum_{j=0}^{\infty} \frac{\cos(2j+1)\pi t}{((2j+1)\pi)^{2m}},$$

$$E_{2m}(t - \lfloor t \rfloor) = (-1)^m 4(2m)! \sum_{j=0}^{\infty} \frac{\sin(2j+1)\pi t}{((2j+1)\pi)^{2m+1}}.$$

8. Prove that for real x with $|x| < \pi/2$ we have

$$\sec x = \sum_{n=0}^{\infty} (-1)^n E_{2n} \frac{x^{2n}}{(2n)!}.$$

9. Prove that if $n \geq 1$ and $p \geq 3$ is prime, then $E_{2n+(p-1)} \equiv E_{2n} \pmod{p}$. This is an analogue of Kummer's congruence.

10. Prove this analogue of J. C. Adams' theorem. Let p be an odd prime, n a positive integer and e a nonnegative integer. Suppose $(p-1)p^e$ divides n. Then $E_n \equiv 0$ or $2 \pmod{p^{e+1}}$ according as $p \equiv 1$ or $3 \pmod 4$.

11. Define the alternating sum

$$D_n(m) = \sum_{k=1}^{m} (-1)^{m-k}(2k-1)^n = (2m-1)^n - (2m-3)^n + \cdots + (-1)^{m-1}1^n$$

for integers $m \geq 1$ and $n \geq 0$. Prove that if $m \geq 1$ and $n \geq 0$, then

$$D_n(m) = \sum_{k=0}^{n-1} \binom{n}{k} 2^{n-k-1} E_k m^{n-k} + \frac{1-(-1)^m}{2} E_n.$$

12. Prove that for every nonnegative integer m, we have

$$\sum_{n=0}^{\infty} \frac{(-1)^n}{(2n+1)^{2m+1}} = \frac{\pi^{2m+1}|E_{2m}|}{2^{2m+2}(2m)!}.$$

Compare with Equation (3.10).

13. Prove that for all nonnegative integers n and k we have $E_{2n} \equiv E_{2n+2^k} + 2^k \pmod{2^{k+1}}$. As a corollary, show that the set $\{E_0, E_2, \ldots, E_{2^k-2}\}$ forms a reduced set of residues modulo 2^k for $k \geq 1$.

14. Prove that $B_n(1/2) = (2^{1-n} - 1) B_n$ and

$$B_{2n}(1/3) = B_{2n}(2/3) = \frac{(3^{1-2n} - 1) B_{2n}}{2}$$

for every nonnegative integer n.

15. Prove $\int_0^1 B_n(t)B_m(t)\,dt = (-1)^{m-1}B_{m+n}/\binom{m+n}{m}$ for all positive integers m and n. (The left side is symmetric in m and n. Why is there just a "$(-1)^{m-1}$" and not also a "$(-1)^{n-1}$" on the right side? It is because $B_{m+n} = 0$ when m and n are positive integers of opposite parity.)

16. If $p > 3$ is prime, show that

$$1 + \frac{1}{2} + \frac{1}{3} + \cdots + \frac{1}{p-1} \equiv 0 \pmod{p^2}.$$

17. If p is an odd prime, prove

$$1^{p-2} + 2^{p-2} + 3^{p-2} + \cdots + \left(\frac{p-1}{2}\right)^{p-2} \equiv (2^p - 2)B_{p-1} \pmod{p}.$$

18. If $m > 1$, show that $2B_{2m} \equiv 1 \pmod 4$, and if $m > 2$, then $2B_{2m} \equiv 3 - (-1)^m 2 \pmod 8$.

19. Prove that if m, a and b are positive integers and $\gcd(a,b) = 1$, then $(a^{2m} - 1)(b^{2m} - 1)B_{2m}/2m$ is an integer.

20. Let $p > 3$ be a prime and k be a positive integer, such that $(p-1)\nmid 2k$. Prove that

$$\left(2^{p-2k} + 3^{p-2k} - 4^{p-2k} - 1\right)\frac{B_{2k}}{4k} \equiv \sum_{p/4 < s < p/3} s^{2k-1} \pmod{p},$$

and that if $p > 5$, then

$$\left(4^{p-2k} + 5^{p-2k} - 8^{p-2k} - 1\right)\frac{B_{2k}}{4k} \equiv$$
$$\sum_{p/8 < s < p/5} s^{2k-1} + \sum_{3p/8 < s < 2p/5} s^{2k-1} \pmod{p},$$

21. When p and k are fixed, x and y are real numbers, and a, b, c are integers, let $S(x, y) = \sum_{xp < s < yp} s^{2k-1}$ and $C(a, b, c) = a^{p-2k} + b^{p-2k} - c^{p-2k} - 1$. Then the congruences of Exercise 20 become

$$C(2, 3, 4)B_{2k}/4k \equiv S(1/4, 1/3) \pmod p$$

and

$$C(4, 5, 8)B_{2k}/4k \equiv S(1/8, 1/5) + S(3/8, 2/5) \pmod p.$$

These congruences have about $p/12$ and $p/10$ total terms in their sums, respectively. Prove that

$$C(2, 5, 6)B_{2k}/4k \equiv S(1/6, 1/5) + S(1/3, 2/5) \pmod p.$$

Exploit the fact that the endpoints of the second summation interval are exactly twice those of the first summation interval and show that

$$C(2,5,6)B_{2k}/4k \equiv (1+2^{2k-1})S(1/6,1/5) - 2^{2k-1}S(3/10,1/3) \pmod{p},$$

with only about $p/15$ total terms in the sums. Can you find similar congruences for B_{2k} with fewer terms in the sums?

22. In case p is prime, $12 \le 2k < p - 2$ and $2^{p-2k} \equiv 3^{p-2k} \equiv 5^{p-2k} \equiv 1 \pmod{p}$, the coefficients $C(a,b,c)$ of all four congruences mentioned in the previous exercises vanish, and they cannot be used to compute $B_{2k} \bmod p$. In this situation, show that the coefficient of B_{2k} in the congruence

$$(2^{2k-1} + 3^{2k-1} + 6^{2k-1} - 1)B_{2k}/4k \equiv \sum_{0<s<p/6} (p - 6s)^{2k-1} \pmod{p^2}$$

is not $\equiv 0 \pmod{p}$. Prove the congruence, which is due to E. Lehmer.

23. Prove this result of Carlitz: If m and n are nonnegative integers, then

$$(-1)^m \sum_{r=0}^{m} \binom{m}{r} B_{n+r} = (-1)^n \sum_{s=0}^{n} \binom{n}{s} B_{m+s}.$$

24. Prove this result of Carlitz: If $p^e(p - 1)|2k$ for some $e > 1$, then p^e divides the numerator of $B_{2k} + \frac{1}{p} - 1$.

25. Without using Theorem 3.22, prove that infinitely many Bernoulli numbers have denominator $Q_n = 30$.

26. Show that if p and q are distinct odd primes, then $Q_{p-1} \ne Q_{q-1}$.

27. Let p be an irregular prime and $p|P_m$, where m is even and $2 \le m \le p - 3$. Then $p|N_{m+(p-1)s}$ for every nonnegative integer s. Define a_t by $N_{m+(p-1)t} \equiv pa_t \pmod{p^2}$ and $0 \le a_t < p$. Let $n = m + (p - 1)t$ for some nonnegative integer t. Prove that $p^2|N_n$ if and only if $-a_0 \equiv (a_1 - a_0)t \pmod{p}$. Prove that when $a_0 \ne 0$ and $a_1 \ne a_0$, there are infinitely many even positive integers n, such that $p \nmid n$ and $p^2|P_n$.

28. Let $Q(x)$ denote the number of square-free positive integers $\le x$. It is well known (see Theorem 333 of [52]) that

$$Q(x) = \frac{6x}{\pi^2} + O(\sqrt{x}).$$

Prove that

$$\sum_{\substack{1 \le n \le x \\ n \text{ square-free}}} \frac{1}{n} = \frac{6}{\pi^2} \log x + O(1).$$

Chapter 4

Examples of Modular Forms

4.1 Introduction

The main theme of this chapter is to exhibit the close relation between the higher reciprocity laws of number theory and the classical theory of modular forms. The simplest such relation is perhaps noticed in connection with the quadratic reciprocity law: let $f(x) = ax^2 + bx + c$ be a quadratic polynomial with integer coefficients and negative discriminant $d = b^2 - 4ac$, let $\chi(n) = (d/n)$ be the Jacobi symbol, and let $\mathbb{F}_p$ be the p-element field. Then for a prime p, which is relatively prime to d, we have the equality

$$|\{x \in \mathbb{F}_p \; : \; ax^2 + bx + c = 0\}| = 1 + \chi(p).$$

It should be observed that on the left hand side we have a number that arises from purely diophantine data; as opposed to this, we have on the right hand side a number that can be thought of as arising from purely analytical considerations. In fact, the number $1 + \chi(p)$ is an eigenvalue of the operator T_p acting on the function

$$E(q) = \frac{1}{2}L(0, \chi) + \sum_{n=1}^{\infty} \left(\sum_{d|n} \chi(d) \right) q^n.$$

The Hecke operator T_p is the linear map defined on formal power series, such as

$$E(q) = \sum_{n=0}^{\infty} a(n)q^n,$$

by

$$(E|T_p)(q) \stackrel{\text{def}}{=} \sum_{n=0}^{\infty} a(np)q^n + \chi(p) \sum_{n=0}^{\infty} a(n)q^{pn}.$$

The eigenvalue property of $E(q)$ is then the statement

$$(E|T_p)(q) = (1 + \chi(p))E(q).$$

In this chapter we will discuss four classical examples that are intended to serve as an indication of the basic questions treated in the theory of modular forms and their connection with number theory.

General references for this chapter include Hardy [49], Samuel [95] and Serre [97].

4.2   An Example of Jacobi and Smith

The example we are about to describe lies close to the roots of the theory of modular forms and the higher reciprocity laws of number theory. It came into existence at the end of two decades of work by Gauss, Jacobi, Eisenstein and others on the cubic and biquadratic reciprocity laws. It appeared for the first time explicitly in Smith's *Number Theory Report to the British Mathematical Association* [104] in 1865. Its interest from a historical point of view stems from the fact that it is the first instance of an explicit computation of a numerical example of a cusp form of weight 1 that is an eigenfunction of the Hecke operators and what was later called an Artin *L*-function. The result, which is contained in Section 128 of Smith's report [104], is the following.

THEOREM 4.1
  (a) *We have the following identities:*

$$\phi(q) \overset{\text{def}}{=} q \prod_{n=1}^{\infty} \left(1 - q^{8n}\right)\left(1 - q^{16n}\right)$$

$$= \sum_{m,n \in \mathbb{Z}} (-1)^n q^{(4m+1)^2 + 8n^2}$$

$$= \sum_{m,n \in \mathbb{Z}} (-1)^{m+n} q^{(4m+1)^2 + 16n^2}.$$

  (b) *The coefficient* $a(n)$ *in the power series identity*

$$\phi(q) = \sum_{n=1}^{\infty} a(n)q^n$$

is a multiplicative arithmetic function.

REMARK 4.1   In Theorem 4.4 below, which is an example of a higher reciprocity law, we will evaluate completely the coefficients $a(n)$.   ∎

PROOF   We will prove Part (a) here, and postpone the proof of Part (b) until later.

The well-known Jacobi triple product identity

$$\sum_{m\in\mathbb{Z}} \nu^m q^{m^2} = \prod_{m=1}^{\infty} \left(1 - q^{2m}\right)\left(1 + q^{2m-1}\nu\right)\left(1 + q^{2m-1}\nu^{-1}\right)$$

can be specialized to yield the following four identities:

$$\sum_{m\in\mathbb{Z}} q^{2m^2+m} = \prod_{m=1}^{\infty} \left(1 - q^{4m}\right)\left(1 + q^{2m-1}\right) \tag{4.1}$$

$$\sum_{n\in\mathbb{Z}} (-1)^n q^{n^2} = \prod_{m=1}^{\infty} \left(1 - q^{2m}\right)\left(1 - q^{2m-1}\right)^2 \tag{4.2}$$

$$\sum_{m\in\mathbb{Z}} (-1)^m q^{2m^2+m} = \prod_{m=1}^{\infty} \left(1 - q^{4m}\right)\left(1 - q^{2m-1}\right) \tag{4.3}$$

$$\sum_{n\in\mathbb{Z}} (-1)^n q^{2n^2} = \prod_{m=1}^{\infty} \left(1 - q^{4m}\right)\left(1 - q^{4m-2}\right). \tag{4.4}$$

For instance, to obtain Identity (4.1) replace q by q^2 and put $\nu = q$ in the triple product. Identity (4.3) is obtained from (4.1) by replacing q by $-q$. Multiply Identities (4.1) and (4.2) together and then multiply Identities (4.3) and (4.4) together to get respectively

$$\sum_{m\in\mathbb{Z}} q^{2m^2+m} \sum_{n\in\mathbb{Z}} (-1)^n q^{n^2} = \prod_{n=1}^{\infty} \left(1 - q^n\right)\left(1 - q^{2n}\right)$$

and

$$\sum_{m\in\mathbb{Z}} (-1)^m q^{2m^2+m} \sum_{n\in\mathbb{Z}} (-1)^n q^{2n^2} = \prod_{n=1}^{\infty} \left(1 - q^n\right)\left(1 - q^{2n}\right).$$

Now replace q by q^8 and then multiply both sides of the last two identities by q to obtain

$$q \prod_{n=1}^{\infty} \left(1 - q^{8n}\right)\left(1 - q^{16n}\right) = \sum_{m,n\in\mathbb{Z}} (-1)^n q^{(4m+1)^2+8n^2}$$

$$= \sum_{m,n\in\mathbb{Z}} (-1)^{m+n} q^{(4m+1)^2+16n^2}.$$

This proves Part (a).   ∎

Before we prove the multiplicativity of the coefficients $a(n)$, we examine, as a first approximation to the higher reciprocity laws, the connection between the identities in the statement of Theorem 4.1(a) and Gauss' two criteria for the biquadratic character of 2.

THEOREM 4.2 Gauss' First Criterion
Let p be prime, $p \equiv 1 \pmod 8$ and let $p = (4a + 1)^2 + 8b^2$. Then

$$X^4 - 2 \equiv 0 \pmod p$$

has a solution and hence four solutions if and only if $a \equiv 0 \pmod 2$.

THEOREM 4.3 Gauss' Second Criterion
Let p be prime, $p \equiv 1 \pmod 8$ and let $p = (4\alpha + 1)^2 + 16\beta^2$. Then

$$X^4 - 2 \equiv 0 \pmod p$$

has a solution and hence four solutions if and only if $\beta \equiv 0 \pmod 2$.

PROOF From the representability of p as $p = (4a + 1)^2 + 8b^2$, we get

$$-8b^2 \equiv (4a + 1)^2 \pmod p. \tag{4.5}$$

We now raise both sides to the power $(p - 1)/4$ and observe that, modulo p,

$$(-1)^{(p-1)/4} \equiv 1, \quad 4^{(p-1)/4} = 2^{(p-1)/2} \equiv \left(\frac{2}{p}\right) = 1$$

by Euler's criterion. Also, if $2^k || b$, then, with $b - 2^k b'$, b' odd, we have, modulo p,

$$(b^2)^{(p-1)/4} = b^{(p-1)/2} \equiv \left(\frac{b}{p}\right) \equiv \left(\frac{2}{p}\right)^k \left(\frac{b'}{p}\right) = \left(\frac{p}{b'}\right) = 1,$$

since $p \equiv (4a + 1)^2 \pmod {b'}$. Similarly,

$$(4a+1)^{2 \cdot (p-1)/4} \equiv \left(\frac{4a + 1}{p}\right) = \left(\frac{p}{4a + 1}\right) = \left(\frac{8b^2}{4a + 1}\right) = \left(\frac{2}{4a + 1}\right) = (-1)^a.$$

Collecting these congruences we have

$$2^{(p-1)/4} \equiv (-1)^a \pmod p.$$

Gauss' First Criterion now readily follows from the simple arithmetical fact that the congruence
$$X^n \equiv a \pmod p$$
has $s = \gcd(n, p - 1)$ solutions if $a^{(p-1)/s} \equiv 1 \pmod p$ and has no solutions otherwise.

We derive Gauss' Second Criterion from the First Criterion by a method due to Dirichlet. We observe first that the equality $p = (4a + 1)^2 + 8b^2 = (4\alpha + 1)^2 + 16\beta^2$ implies

$$\{4(a + \beta) + 1\}\{4(a - \beta) + 1\} = (4\alpha + 1)^2 - 8b^2. \tag{4.6}$$

Now if d is a common divisor > 1 of $4\alpha + 1$ and b, then it cannot divide the greatest common divisor of $4(a + \beta) + 1$ and $4(a - \beta) + 1$, for if it did, it would follow, since d is odd, that $d|\beta$; but if this is the case, then $d^2|p$, which is impossible because p is prime. Therefore we can write

$$\gcd((4\alpha + 1)^2, b^2) = \delta^2 \delta_0^2, \quad \gcd(\delta, \delta_0) = 1,$$

with $\delta^2|4(a + \beta) + 1$ and $\delta_0^2|4(a - \beta) + 1$. Now from Equation (4.6) we obtain

$$\left\{ \frac{4(a + \beta) + 1}{\delta^2} \right\} \left\{ \frac{4(a - \beta) + 1}{\delta_0^2} \right\} = x^2 - 8y^2$$

with x odd and $\gcd(x, y) = 1$. From the congruence

$$x^2 - 2(2y)^2 \equiv 0 \pmod{D}, \qquad D = \frac{4(a + \beta) + 1}{\delta^2},$$

we get that

$$1 = \left(\frac{2}{D} \right) = (-1)^{(D^2 - 1)/8},$$

or equivalently, $D^2 \equiv 1 \pmod{16}$. But since $\delta^4 \equiv 1 \pmod{16}$, this reduces to $(4(a + \beta) + 1)^2 \equiv 1 \pmod{16}$ or $8(a + \beta) \equiv 0 \pmod{16}$. Therefore,

$$a + \beta \equiv 0 \pmod 2. \tag{4.7}$$

Thus we get that $\beta \equiv 0 \pmod 2$ if and only if $a \equiv 0 \pmod 2$. This proves Gauss' Second Criterion. ∎

REMARK 4.2   The equality $p = (4a+1)^2 + 8b^2 = (4\alpha+1)^2 + 16\beta^2$ implies that $(4a+1)^2 + 8b^2 \equiv (4\alpha+1)^2 \pmod{16}$, or equivalently, $16a^2 + 8a + 1 + 8b^2 \equiv 16\alpha^2 + 8\alpha + 1 \pmod{16}$, from which we get

$$a + b \equiv \alpha \pmod 2;$$

here we have used the simple fact that $b^2 \equiv b \pmod 2$. This last congruence together with Congruence (4.7) give for a prime $p \equiv 1 \pmod 8$,

$$b \equiv \alpha + \beta \pmod 2, \tag{4.8}$$

where b, α and β are the numbers appearing in Gauss' two criteria. It then appears that the identities in Theorem 4.1(a) are a generalization of the equivalence of Gauss' two criteria for the biquadratic character of 2. ∎

Now we prove Part (b) of Theorem 4.1.

PROOF   We have to show that the function $a(n)$ is multiplicative. We first make a few trivial remarks about the obvious values of $a(n)$.

1. $a(n) = 0$ if $n \not\equiv 1 \pmod{8}$. This is seen from the expansion of the infinite product that defines $a(n)$.

2. $a(n) = 0$ if $n \equiv 1 \pmod{8}$ and $p^{2v+1}||n$, with a prime $p \equiv 3$, 5 or 7 (mod 8). This is seen by observing that a representation of the form $n = (4\alpha + 1)^2 + (4\beta)^2$ implies that the congruence $x^2 \equiv -1 \pmod{p}$ has a solution, which is impossible if $p \equiv 3$ or 7 (mod 8). Similarly, a representation of the form $n = (4a + 1)^2 + 2(2b)^2$ implies that the congruence $x^2 \equiv -2 \pmod{p}$ has a solution, which is impossible if $p \equiv 5 \pmod{8}$.

To evaluate $a(n)$ it suffices to consider integers n of the form

$$n = \prod_{(1)} p^{\alpha(p)} \prod_{(3)} q^{2\alpha(q)} \prod_{(5)} r^{2\alpha(r)} \prod_{(7)} s^{2\alpha(s)}, \tag{4.9}$$

where the products $\prod_{(1)}, \prod_{(3)}, \prod_{(5)}$ and $\prod_{(7)}$ run respectively over the prime divisors of n satisfying $p \equiv 1 \pmod{8}$, $q \equiv 3 \pmod{8}$, $r \equiv 5 \pmod{8}$ and $s \equiv 7 \pmod{8}$. To prove the multiplicativity of $a(n)$ it suffices to consider integers n of the form (4.9) and to show for these that if $n = n' \cdot n''$, with $\gcd(n', n'') = 1$, then $a(n) = a(n')a(n'')$. This we now proceed to do.

Two well-known facts from the theory of binary quadratic forms are ([64], p. 187):

1. (Dirichlet). The number of solutions of $n = x^2 + 8y^2$ with $x \equiv 1 \pmod{4}$ is

$$\sum_{d|n} \left(\frac{-8}{d} \right). \tag{4.10}$$

2. (Euler).

$$(ax^2 + cy^2)(z^2 + acw^2) = aX^2 + cY^2, \tag{4.11}$$

$$\text{where } X = xz - cyw, \ Y = axw + zy.$$

To a representation of n in the form $n = (4a + 1)^2 + 8b^2$ we attach the binary character

$$b(n) \equiv b \pmod{2}.$$

Let us first consider how representations of n' and n'' give rise to a representation of $n = n' \cdot n''$. Suppose

$$n' = (4a' + 1)^2 + 8b'^2 = x'^2 + 8y'^2$$
$$n'' = (4a'' + 1)^2 + 8b''^2 = x''^2 + 8y''^2.$$

From Euler's identity (4.11) we have

$$n = n' \cdot n'' = (x'x'' - 8b'b'')^2 + 8(x'b'' + x''b')^2. \tag{4.12}$$

Since the counting function for the number of representations of n by the form $x^2 + 8y^2$ with $x \equiv 1 \pmod 4$ is multiplicative, we see that all representations of n are obtained by *composition* from all the representations of n' and n''. Now, since $x' \equiv x'' \equiv 1 \pmod 4$, we get from (4.12) that $b(n) = x'b'' + x''b' \equiv b' + b'' \pmod 2$, that is

$$b(n' \cdot n'') \equiv b(n') + b(n'') \pmod 2. \tag{4.13}$$

If we multiply the sums

$$a(n') = \sum_{a', b'} (-1)^{b'}, \quad a(n'') = \sum_{a'', b''} (-1)^{b''},$$

where a', b' (respectively a'', b'') run over all solutions of $n' = (4a'+1)^2 + 8b'^2$ (respectively $n'' = (4a''+1)^2 + 8b''^2$), and use (4.12) and (4.13), we get

$$\sum_{a', b'} (-1)^{b'} \cdot \sum_{a'', b''} (-1)^{b''} = \sum_{a, b} (-1)^b,$$

where the last sum runs over all representations of $n = (4a+1)^2 + 8b^2$. But this last sum is precisely $a(n)$. This proves that $a(n)$ is multiplicative. ∎

REMARK 4.3 The multiplicativity of $a(n)$ is equivalent to the fact, which we will establish later, that the modular form

$$\phi(q) = \sum_{n=1}^{\infty} a(n)q^n$$

is an eigenfunction of the Hecke operators, that is,

$$\phi|T_p = \sum_{n=1}^{\infty} a(pn)q^n + \left(\frac{-8}{p}\right) \sum_{n=1}^{\infty} a(n)q^{pn} = a(p)\phi.$$

∎

The multiplicativity of $a(n)$ reduces the evaluation of $a(n)$ to the following four cases.

1. $n = p^v$, with a prime $p \equiv 1 \pmod 8$,

2. $n = p^{2v}$, with a prime $p \equiv 3 \pmod 8$,

3. $n = p^{2v}$, with a prime $p \equiv 5 \pmod 8$,

4. $n = p^{2v}$, with a prime $p \equiv 7 \pmod 8$.

The last case is the simplest to treat, since here the only representation of p^{2v} by the quadratic form $Q_2 = (4a+1)^2 + 8b^2$ is

$$p^{2v} = ((-1)^v p^v)^2 + 8 \cdot 0^2.$$

Hence the sum

$$a\left(p^{2v}\right) = \sum_{a,b}(-1)^b$$

contains only one term and this is $(-1)^0 = 1$.

Since -1 is not a quadratic residue modulo a prime $p \equiv 3 \pmod 8$, the only representation of p^{2v} by the quadratic form $Q_1 = (4\alpha+1)^2 + 16\beta^2$ is

$$p^{2v} = ((-1)^v p^v)^2 + 16 \cdot 0^2.$$

To find the parity of α we use the binomial theorem to obtain $(-1)^v p^v \equiv 1 + 4v \pmod 8$ and observe that $\alpha \equiv v \pmod 2$. Therefore in Case 2 we have

$$a\left(p^{2v}\right) = (-1)^{\alpha+\beta} = (-1)^v.$$

Now since -2 is not a quadratic residue modulo a prime $p \equiv 5 \pmod 8$, the only representation of p^{2v} by the quadratic form $Q_2 = (4a+1)^2 + 8b^2$ is

$$p^{2v} = ((-1)^v p^v)^2 + 8 \cdot 0^2,$$

and hence the sum $\sum_{a,b}(-1)^b$, $p^{2v} = (4a+1)^2 + 8b^2$, contains only one term which is 1.

There remains the first case. Here p^v is representable by each form

$$\sum_{d|p^v}\left(\frac{-1}{d}\right) = \sum_{d|p^v}\left(\frac{2}{d}\right) = v + 1$$

times. Of these $v+1$ representations two are *primitive*, that is, $p^v = (4a+1)^2 + 8b^2$ with $\gcd(4a+1,b) = 1$, and the remaining $v - 1$ are *derived*, that is, $\gcd(4a+1,b) > 1$. We want to show that $\alpha + \beta \equiv b \pmod 2$ for all representations by $Q_1 = (4\alpha+1)^2 + 16\beta^2$ and $Q_2 = (4a+1)^2 + 8b^2$. To see this, let $p^v = (4\alpha+1)^2 + 16\beta^2$ be a fixed *primitive* representation and let $p^v = (4a+1)^2 + 8b^2$ run through all representations of p^v by the form $Q_2 = (4a+1)^2 + 8b^2$. Now the equality $(4a+1)^2 + 8b^2 = (4\alpha+1)^2 + 16\beta^2$ gives

$$\{4(a+\beta)+1\}\{4(a-\beta)+1\} = (4\alpha+1)^2 - 8b^2. \qquad (4.14)$$

As in an earlier argument, if $d > 1$ divides $\gcd(4\alpha+1, b^2)$, then d does not divide $4(a+\beta)+1$ and $4(a-\beta)+1$. For otherwise, since d is odd, we would have $d|\beta$, which is impossible because of the assumption $\gcd(4\alpha+1,\beta) = 1$.

Therefore, $\gcd((4\alpha + 1)^2, b^2) = \delta^2 \cdot \delta_0^2$, $\gcd(\delta, \delta_0) = 1$, with $\delta^2|4(a + \beta) + 1$, and we have

$$\left\{\frac{4(a + \beta) + 1}{\delta^2}\right\}\left\{\frac{4(a - \beta) + 1}{\delta_0^2}\right\} = x^2 - 8y^2, \qquad \gcd(x, y) = 1,$$

and x is odd. This implies that

$$x^2 - 2(2y)^2 \equiv 0 \pmod{D}, \qquad D = \frac{4(a + \beta) + 1}{\delta^2}.$$

Hence 2 is a quadratic residue modulo D. But $(2/D) = (-1)^{(D^2-1)/8}$, and hence $D^2 \equiv 1 \pmod{16}$. We also have $\delta^4 \equiv 1 \pmod{16}$ and thus $a + \beta \equiv 0 \pmod 2$. On the other hand we always have $\alpha \equiv a+b \pmod 2$. This proves

$$b \equiv \alpha + \beta \pmod 2. \tag{4.15}$$

By a similar argument we can show that if we fix a *primitive* representation $p^v = (4a + 1)^2 + 8b^2$ and let $p^v = (4\alpha + 1)^2 + 16\beta^2$ run through all representations, primitive and derived, then the identity (4.14) would imply that a divisor d of $\gcd(4\alpha + 1, b)$ cannot divide $4(a + \beta) + 1$ and $4(a - \beta) + 1$, unless $d = 1$, for otherwise $d|(4a + 1)$, which is impossible since we assumed $\gcd(4a + 1, b) = 1$. Therefore the relations (4.14) and (4.15) again hold. This shows that $\alpha + \beta \equiv b \pmod 2$ holds for all the representations of p^v by $Q_1 = (4\alpha + 1)^2 + 16\beta^2$ and $Q_2 = (4a + 1)^2 + 8b^2$. We thus have

$$a(p^v) = (-1)^b(v + 1).$$

The parity of b is obtained by observing that the representation $p^v = (4a + 1)^2 + 8b^2$ implies the congruence $(p^v - 1)/8 \equiv a + b \pmod 2$. Also if we let $2^{(p-1)/4} \equiv \varepsilon \pmod p$, then

$$\left(-8b^2\right)^{(p-1)/4} \equiv (4a + 1)^{(p-1)/2} \pmod p,$$
$$2^{(p-1)/4}b^{(p-1)/2} \equiv (4a + 1)^{(p-1)/2} \pmod p,$$
$$\varepsilon\left(\frac{b}{p}\right) \equiv \left(\frac{4a + 1}{p}\right) \pmod p,$$

and by the quadratic reciprocity law this is

$$\varepsilon\left(\frac{p}{b'}\right) \equiv \left(\frac{p}{4a + 1}\right) \pmod p,$$

where $b = 2^k b'$, b' odd. Now raising both sides to the v-th power we get

$$\varepsilon^v\left(\frac{p^v}{b'}\right) \equiv \left(\frac{p^v}{4a + 1}\right) \pmod p$$

and

$$\varepsilon^v \equiv \left(\frac{2}{4a+1}\right) \equiv (-1)^a \pmod{p}. \tag{4.16}$$

Therefore we obtain

$$(-1)^b = \varepsilon^v (-1)^{(p^v-1)/8}$$

and

$$a(p^v) = \varepsilon^v (-1)^{(p^v-1)/8}(v+1), \qquad p \equiv 1 \pmod{8}.$$

This completes the evaluation of $a(n)$.

In summary, we have proved this corollary.

COROLLARY 4.1

(i) $a(n) = 0$ if $n \not\equiv 1 \pmod 8$ or if n is not of the form

$$n = \prod_{(1)} p^{\alpha(p)} \prod_{(3)} q^{2\alpha(q)} \prod_{(5)} r^{2\alpha(r)} \prod_{(7)} s^{2\alpha(s)},$$

where the product $\prod_{(a)}$ runs over the prime divisors of n that are $\equiv a \pmod 8$.
(ii) $a(p^v) = \varepsilon^v (-1)^{(p^v-1)/8}(v+1)$ if $p \equiv 1 \pmod 8$ and $\varepsilon \equiv 2^{(p-1)/4} \pmod p$.
(iii) $a(p^{2v}) = (-1)^v$ if $p \equiv 3 \pmod 8$.
(iv) $a(p^{2v}) = 1$ if $p \equiv 5$ or $7 \pmod 8$.

Our next goal is to exhibit, within the framework of this example, the true connection between the coefficients in the power series expansion

$$q \prod_{n=1}^{\infty} \left(1 - q^{8n}\right) \left(1 - q^{16n}\right) = \sum_{n=1}^{\infty} a(n) q^n$$

and the higher reciprocity laws. This is accomplished in the next theorem, which is stated without using the language of class field theory.

THEOREM 4.4 Arithmetic Congruence Relation

Let p be an odd prime and let $\mathbb{F}_p$ be the p-element field. Let $a(n)$ be the n-th coefficient of the power series expansion

$$q \prod_{n=1}^{\infty} \left(1 - q^{2n}\right) \left(1 - q^{16n}\right) = \sum_{n=1}^{\infty} a(n) q^n.$$

Then

$$|\{x \in \mathbb{F}_p : x^4 - 2x^2 + 2 = 0\}| = 1 + \left(\frac{-1}{p}\right) + a(p). \tag{4.17}$$

REMARK 4.4   We have called the equality (4.17) a *congruence relation* because on one hand we are counting the number of closed points on the scheme

$$\text{Spec}\left(\overline{\mathbb{F}}_p[x]/(x^4 - 2x^2 + 2)\right), \qquad \overline{\mathbb{F}}_p = \text{algebraic closure of } \mathbb{F}_p,$$

left fixed by the action of Frobenius

$$\Phi_p : a \longmapsto a^p,$$

and on the other hand we have an expression involving the eigenvalues of the Hecke operators.   ∎

PROOF   We begin by studying the factorization of the polynomial $f(x) = x^4 - 2x^2 + 2$ in the finite field $\mathbb{F}_p$. Let us first consider the case of a prime $p \equiv 1 \pmod 8$. Let i denote a solution of the congruence $x^2 \equiv -1 \pmod p$ and observe that in $\mathbb{F}_p$

$$x^4 - 2x^2 + 2 = (x^2 - (1 + i))(x^2 - (1 - i)) = f_1(x) \cdot f_{-1}(x).$$

Now, since $(1 + i)(1 - i) = 2$ is a square in $\mathbb{F}_p$, either both $f_1(x)$ and $f_{-1}(x)$ split into two linear factors or both remain irreducible. But the polynomial $x^2 - (1 + i)$ splits into two linear factors over $\mathbb{F}_p$ if and only if $(1 + i)$ is a square in $\mathbb{F}_p$, that is, if and only if

$$(1 + i)^{(p-1)/2} \equiv 1 \pmod p.$$

But recall from Corollary 4.1(ii) that, for a prime $p \equiv 1 \pmod 8$,

$$a(p) \equiv 2(-1)^{(p-1)/8} 2^{(p-1)/4} \pmod p.$$

Now $(1+i)^2 = 2i$ and hence $(-1)^{(p-1)/8} 2^{(p-1)/4} = (2i)^{(p-1)/4} = (1+i)^{(p-1)/2}$. Therefore, for a prime $p \equiv 1 \pmod 8$, $a(p) = 2$ if $x^4 - 2x^2 + 2$ has four solutions, while $a(p) = -2$ if $x^4 - 2x^2 + 2$ splits over $\mathbb{F}_p$ into two irreducible quadratic factors.

To consider the case of a prime $p \equiv 5 \pmod 8$ we observe that again the congruence $x^2 \equiv -1 \pmod p$ has a solution, which we denote by i. Therefore we have the splitting

$$x^4 - 2x^2 + 2 = (x^2 - (1 + i))(x^2 - (1 - i)) = f_1(x) \cdot f_{-1}(x).$$

Now, we have that $(1 + i)(1 - i) = 2$ is not a square in $\mathbb{F}_p$ and hence exactly one of the two numbers $1 + i$, $1 - i$ is a square in $\mathbb{F}_p$, and therefore exactly one of the two polynomials $f_1(x)$, $f_{-1}(x)$ splits. Therefore, if $p \equiv 5 \pmod 8$, then the polynomial $f(x)$ splits over $\mathbb{F}_p$ into the product of two linear factors and an irreducible quadratic polynomial.

If $p \equiv 3 \pmod 8$, then the equation $x^4 - 2x^2 + 2 = 0$ has no solution in $\mathbb{F}_p$, for otherwise the equivalent expression $(x^2 - 1)^2 + 1 = 0$ would imply that -1 is a square in $\mathbb{F}_p$. Therefore the polynomial $x^4 - 2x^2 + 2$ stays irreducible or splits into two irreducible quadratic factors. We claim that this last alternative is impossible. Suppose that we have over $\mathbb{F}_p$

$$x^4 - 2x^2 + 2 = (x^2 + ax + b)(x^2 + cx + d).$$

A comparison of the coefficients on both sides gives $a + c = 0$, $ad + bc = 0$, $d + b + ac = -2$ and $bd = 2$, from which it follows that $a = -c$, $a(d - b) = 0$ and $b + d - a^2 = -2$. If $a = 0$, then $d + b = -2$ and $bd = 2$. Thus, if $a = 0$, then $d + 2/d + 2 = 0$, or equivalently, $d^2 + 2d + 2 = 0$, has solutions in $\mathbb{F}_p$. But the discriminant of this quadratic equation is $\Delta = -4$, which is not a square in $\mathbb{F}_p$. Suppose then that $a \neq 0$. This implies $b = d$ and hence $d^2 = 2$, a contradiction because 2 is not a square in $\mathbb{F}_p$. Therefore $f(x)$ remains irreducible in $\mathbb{F}_p$ when p is a prime $\equiv 3 \pmod 8$.

It remains to consider the case $p \equiv 7 \pmod 8$. Here again $x^4 - 2x^2 + 2 = 0$ has no solutions in $\mathbb{F}_p$ because -1 is not a square. We claim that $x^4 - 2x^2 + 2 = 0$ splits into the product of two irreducible quadratics

$$x^4 - 2x^2 + 2 = (x^2 + ax + b)(x^2 + cx + d).$$

Let d be a root of $x^2 - 2 = 0$ in $\mathbb{F}_p$. We want to produce an $a \neq 0$, such that the above splitting into quadratics is possible. Now $b = d$, $2d - a^2 = -2$, and hence $2d = a^2 - 2$. We square both sides of this last equality and use the fact that $d^2 = 2$ to obtain $(2d)^2 = 8 = a^4 - 4a^2 + 4$, or equivalently,

$$a^4 - 4a^2 - 4 = 0.$$

The solutions of this quadratic are $a^2 = 2(1 \pm d)$. But $(1 + d)(1 - d) = -1$ is not a square in $\mathbb{F}_p$ and therefore one of the two numbers $2(1 + d)$, $2(1 - d)$ is a square in $\mathbb{F}_p$. Hence a can be found. Now the quadratic polynomial $x^2 + ax + b$ splits into the product of two linear factors if and only if its discriminant $a^2 - 4b$ is a square in $\mathbb{F}_p$. But if it splits, so does the polynomial $x^2 - ax + b$. But then $x^4 - 2x^2 + 2 = 0$ would have a root and -1 would be a square in $\mathbb{F}_p$, which is impossible. Therefore, if $p \equiv 7 \pmod 8$, then $x^4 - 2x^2 + 2 = 0$ splits into irreducible quadratic factors over $\mathbb{F}_p$.

Now let us observe that for a prime $p \equiv 1 \pmod 8$, the expression

$$N_p = 1 + \left(\frac{-1}{p}\right) + a(p)$$

is equal to the number of linear factors of $x^4 - 2x^2 + 2$ defined over $\mathbb{F}_p$. For $p \equiv 3 \pmod 8$, $N_p = 0$ and this agrees with the fact that $f(x)$ has no linear factors. If $p \equiv 5 \pmod 8$, then $N_p = 2$ and this is the number of linear factors of $f(x)$. For $p \equiv 7 \pmod 8$, $N_p = 0$ and again $f(x)$ has no linear factors. This completes the proof of the arithmetical congruence relation. ∎

COROLLARY 4.2

Over the finite field $\mathbb{F}_p$ the polynomial $f(x) = x^4 - 2x^2 + 2$ is

(i) the product of four linear factors if $p \equiv 1$ (mod 8) and $a(p) = 2$,

(ii) the product of two irreducible quadratic factors if $p \equiv 1$ (mod 8) and $a(p) = -2$,

(iii) an irreducible quartic if $p \equiv 3$ (mod 8),

(iv) the product of two linear factors and an irreducible quadratic factor if $p \equiv 5$ (mod 8).

(v) the product of two irreducible quadratic factors if $p \equiv 7$ (mod 8).

Example 4.1

If A is the ring of integers of a number field, then we consider a point of Spec (A) of degree e as lying on the intersection of e "sheets." If we think of Spec $\mathbb{Z}[\alpha]$, where $\alpha^4 - 2\alpha^2 + 2 = 0$, as a "covering" of Spec $\mathbb{Z}$, then the following picture shows that part of the "covering" that lies above the primes ≤ 47.

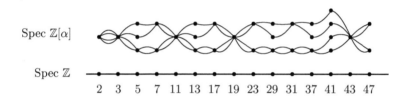

Spec $\mathbb{Z}[\alpha]$

Spec $\mathbb{Z}$

    2 3 5   7 11 13 17 19 23 29 31 37 41 43 47

For example, above the point 11 there is a point in Spec $\mathbb{Z}[\alpha]$ of degree 4; above the point 41 there are four points in Spec $\mathbb{Z}[\alpha]$ each of degree 1. From Theorem 4.4 and Corollary 4.2 we can read easily the values of $a(p)$, in particular, $a(11) = 0$ and $a(41) = 2$.

4.3   An Example of Ramanujan and Mordell

Some of Ramanujan's calculations concerning the coefficients in the formal power series expansion

$$\Delta(q) = q \prod_{n=1}^{\infty} (1 - q^n)^{24} = \sum_{n=1}^{\infty} \tau(n)q^n. \qquad (4.18)$$

lead him to conjecture, among other things, that the arithmetical function $\tau(n)$ is multiplicative. This fact was subsequently established by Mordell in 1917 by a method based on the analytic and automorphic properties of $\Delta(q)$. Mordell's basic idea consisted in using certain linear operators, now called Hecke operators, that had been studied extensively by Hurwitz in his theory of modular correspondences. In retrospect, this is not very surprising and in fact in his work on modular correspondences on the Riemann surface, which admits the simple group of order 168 as a group of automorphisms, Hurwitz had been led to the possibility of establishing the multiplicativity of certain arithmetical functions by exploiting the recurrence relations satisfied by the linear operators associated to the modular correspondences. (See Fricke-Klein [32], Chapter VI, 4, Section 3, especially page 606.) Since the theory of Hecke operators plays such a central role in the modern theory of modular forms, it is perhaps appropriate that we present Mordell's simple proof as an introduction to the Hecke theory, which will be treated more fully in the next chapter. The main result we prove here is the following.

THEOREM 4.5 Properties of Ramanujan's τ-Function
(a) *The Ramanujan function $\tau(n)$ is multiplicative, that is, $\tau(mn) = \tau(m)\tau(n)$ whenever m and n are relatively prime integers.*
(b) *We have formally for each prime p*

$$\sum_{k=0}^{\infty} \tau(p^k)T^k = \frac{1}{1 - \tau(p)T + p^{11}T^2}.$$

The proof of Theorem 4.5 depends on various propositions which we now proceed to establish.

PROPOSITION 4.1
The discriminant function

$$\Delta(z) = q \prod_{n=1}^{\infty} (1 - q^n)^{24} \qquad q = \exp(2\pi i z),$$

with $z = x + iy$, $y > 0$, satisfies the functional equation

$$\Delta\left(\frac{az+b}{cz+d}\right) = (cz+d)^{12}\Delta(z),$$

with a, b, c and $d \in \mathbb{Z}$ and $ad - bc = 1$.

To prove this proposition we observe that it suffices to prove that

$$\Delta(z+1) = \Delta(z) \quad \text{and} \quad \Delta\left(\frac{-1}{z}\right) = z^{12}\Delta(z).$$

This is because of the following slightly more general statement.

LEMMA 4.1
The unimodular group

$$\Gamma = \left\{ \begin{pmatrix} a & b \\ c & d \end{pmatrix} : a, b, c, d \in \mathbb{Z}, \quad ad - bc = 1 \right\}$$

is generated by the elements

$$T = \begin{pmatrix} 1 & 1 \\ 0 & 1 \end{pmatrix} \quad and \quad S = \begin{pmatrix} 0 & -1 \\ 1 & 0 \end{pmatrix}.$$

PROOF We need to show that any element $\sigma = \begin{pmatrix} a & b \\ c & d \end{pmatrix}$ in Γ can be written as a word in T and S. We use induction on $m = |a|$. For $m = 0$ we see that the condition $ad - bc = 1$ implies $-bc = 1$ and hence in this case we may suppose without loss of generality that $b = 1$ and $c = -1$. The claim in this case follows trivially from the matrix identity

$$\begin{pmatrix} 0 & 1 \\ -1 & d \end{pmatrix} = \begin{pmatrix} 0 & -1 \\ 1 & 0 \end{pmatrix} \begin{pmatrix} 1 & -d \\ 0 & 1 \end{pmatrix} \begin{pmatrix} -1 & 0 \\ 0 & -1 \end{pmatrix} = ST^{-d}S^2.$$

Let us suppose then that the claim holds true for all matrices with $0 \le |a| \le m - 1$ and prove it for a matrix with $|a| = m > 0$. We may assume also that $c > 0$. For if $c = 0$, then $a = d = \pm 1$ and clearly either

$$\begin{pmatrix} a & b \\ c & d \end{pmatrix} = \begin{pmatrix} a & b \\ 0 & a \end{pmatrix} = \begin{pmatrix} 1 & b \\ 0 & 1 \end{pmatrix} = T^b$$

or

$$\begin{pmatrix} a & b \\ c & d \end{pmatrix} = \begin{pmatrix} -1 & b \\ 0 & -1 \end{pmatrix} = \begin{pmatrix} 1 & -b \\ 0 & 1 \end{pmatrix} S^2 = T^{-b}S^2.$$

If $c < 0$, then

$$\sigma = \begin{pmatrix} a & b \\ c & d \end{pmatrix} = S^2 \begin{pmatrix} -a & -b \\ -c & -d \end{pmatrix}$$

and clearly $-c > 0$. We may suppose further that $|a| \ge c$. For if $|a| < c$,

$$\sigma = \begin{pmatrix} a & b \\ c & d \end{pmatrix} = \begin{pmatrix} 0 & -1 \\ 1 & 0 \end{pmatrix} \begin{pmatrix} c & d \\ -a & -b \end{pmatrix} = S \begin{pmatrix} c & d \\ -a & -b \end{pmatrix}$$

and then σ can be replaced by a matrix with $|a| \ge c$. Now let h be such that $a = ch + r$ with $|r| < c \le |a|$. Then

$$T^{-h}\sigma = T^{-h} \begin{pmatrix} a & b \\ c & d \end{pmatrix} = \begin{pmatrix} a - ch & b - dh \\ c & d - ch \end{pmatrix} = \begin{pmatrix} r & b - dh \\ c & d - ch \end{pmatrix},$$

and the induction hypothesis applied to $T^{-h}\sigma$ implies it can be written as a word in T and S. This proves the lemma. $\blacksquare$

Note that $\Delta(z + 1) = \Delta(z)$ because $\exp(2\pi i(z + 1)) = \exp(2\pi i z)$. To prove Proposition 4.1, it suffices then to show that $\Delta\left(\frac{-1}{z}\right) = z^{12}\Delta(z)$. Since $\Delta(z) = \eta(z)^{24}$, this will follow from the following transformation formula for $\eta(z)$.

LEMMA 4.2

The Dedekind function

$$\eta(z) = q^{1/24} \prod_{n=1}^{\infty} (1 - q^n), \qquad q = \exp(2\pi i z),$$

satisfies the functional equation

$$\eta\left(\frac{-1}{z}\right) = \left(\frac{z}{i}\right)^{\frac{1}{2}} \eta(z).$$

REMARK 4.5   The proof we give below is a particular application of a general method developed by Rademacher [86] and applied by him to get the full transformation formula of $\eta(z)$ under a unimodular transformation. Rademacher's paper [86], despite its profound originality, seems to have gone unnoticed by many authors who subsequently rediscovered particular cases. This situation is hard to understand in view of the fact that Rademacher's paper, which appeared in Crelle's journal, is followed by an often quoted paper of Hecke. We follow here André Weil's note published in 1962.   ∎

PROOF   The proof is based on Hecke's observation that the functional equation satisfied by a Dirichlet series implies a transformation formula for the corresponding Mellin transform. This idea is applied to Riemann's Euler product (Equation (3.13))

$$\Lambda(s) = \pi^{-s/2}\Gamma(s/2)\zeta(s) = \pi^{-s/2}\Gamma(s/2) \prod_{p \, \text{prime}} \frac{1}{1 - p^{-s}}$$

$$= G_{\mathbb{R}}(s) \sum_{n=1}^{\infty} n^{-s}, \text{ where } G_{\mathbb{R}}(s) = \pi^{-s/2}\Gamma(s/2),$$

which satisfies the functional equation $\Lambda(s) = \Lambda(1 - s)$. These properties in turn imply the following functional equation for

$$\Phi(s) = \Lambda(s)\Lambda(s + 1)$$

$$= G_{\mathbb{R}}(s)G_{\mathbb{R}}(s + 1) \prod_{p \, \text{prime}} \frac{1}{1 - (1 + p^{-1})p^{-s} + p^{-1-2s}}$$

$$= (2\pi)^{-s}\Gamma(s) \sum_{m=1}^{\infty} \sum_{n=1}^{\infty} \frac{m^{-1}}{(mn)^s} \qquad (4.19)$$

$$= \Phi(-s),$$

or more simply, $\Phi(-s) = \Lambda(-s)\Lambda(-s + 1) = \Lambda(1 - (-s))\Lambda(1 - (-s + 1)) = \Lambda(1 + s)\Lambda(s) = \Phi(s)$. Now $\Phi(s)$ is entire except at $s = 0, \pm 1$, where it has the following Laurent expansions:

$$\text{at } s = 0: \qquad \Phi(s) = \frac{1}{2s^2} + \sum_{n=0}^{\infty} A_0(n) s^n$$

$$\text{at } s = 1: \qquad \Phi(s) = \frac{\pi}{12} \cdot \frac{1}{s-1} + \sum_{n=0}^{\infty} A_1(n)(s-1)^n$$

$$\text{at } s = -1: \qquad \Phi(s) = -\frac{\pi}{12} \cdot \frac{1}{s+1} + \sum_{n=0}^{\infty} A_{-1}(n)(s+1)^n.$$

The Dirichlet series in Equation (4.19) has the same coefficients as the q-expansion

$$F(z) = \sum_{m=1}^{\infty} \sum_{n=1}^{\infty} m^{-1} q^{mn}$$

$$= \sum_{n=1}^{\infty} \left(\sum_{m=1}^{\infty} m^{-1} (q^n)^m \right)$$

$$= -\sum_{n=1}^{\infty} \log(1 - q^n).$$

We can thus write

$$F(z) = \frac{\pi i z}{12} - \log \eta(z),$$

where $\eta(z)$ is the Dedekind function. Now $F(z)$ is the Mellin transform of $\Phi(s)$ and in fact we have

$$\Phi(s) = \int_0^{\infty} F(it) t^{s-1} dt,$$

valid for $\Re(s) > 1$, and

$$F(z) = \frac{1}{2\pi i} \int_{\sigma-i\infty}^{\sigma+i\infty} \Phi(s) \left(\frac{z}{i} \right)^{-s} ds,$$

valid for $\sigma > 1$ and $\Im(z) > 0$. In the latter formula the integral is taken along the line $\Re(s) = \sigma$. Hecke's method consists in moving the contour of integration to the line $\Re(s) = -\sigma$ and making use of the bounded behavior of $\Phi(s)$ on vertical strips away from the real axis. In doing this the poles of the integrand as $s = 0, \pm 1$ give rise to residue contributions that are easily calculated. Thus we obtain

$$F(z) = \frac{2\pi i}{24} \left(z + \frac{1}{z} \right) - \frac{1}{2} \log \left(\frac{z}{i} \right) + \frac{1}{2\pi i} \int_{-\sigma-i\infty}^{-\sigma+i\infty} \Phi(s) \left(\frac{z}{i} \right)^{-s} ds.$$

Now, using the functional equation $\Phi(s) = \Phi(-s)$, the last integral is seen to be equal to $F(-1/z)$. Therefore we obtain

$$\log \eta \left(-\frac{1}{z} \right) = \log \eta(z) + \frac{1}{2} \log \left(\frac{z}{i} \right).$$

The exponentiation of this formula leads to the result stated in Lemma 4.2.
∎

This completes the proof of Proposition 4.1.

The upper half plane (in $\mathbb{C}$) is denoted by $\mathcal{H} = \{z = x + iy : y > 0\}$. A fundamental domain for the action of Γ on $\mathcal{H}$ is given by

$$D = \{z \in \mathcal{H} : |\Re(z)| \le \frac{1}{2}, \ |z| \ge 1\}.$$

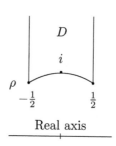

Real axis

The two main properties that a fundamental domain must satisfy are:

1. For every $z \in \mathcal{H}$, there is a $\sigma \in \Gamma$ such that $\sigma(z) \subset D$, and

2. If $z \in D$ and $\sigma \in \Gamma$ are such that $\sigma(z) \in D$ and σ is not the identity, then either $\Re(z) = \pm\frac{1}{2}$ and $\sigma(z) = z \pm 1$ or $|z| = 1$ and $\sigma(z) = -1/z$.

A more precise definition of fundamental domain for arbitrary discrete subgroups of Γ will be given in the next chapter. Here we simply establish that these two conditions are indeed satisfied by D. To verify Property 1 we observe that for $\sigma = \begin{pmatrix} a & b \\ c & d \end{pmatrix} \in \Gamma$ and $z = x + iy \in \mathcal{H}$, we have

$$\Im(\sigma(z)) = \frac{y}{|cz + d|^2} = \frac{y}{(cx + d)^2 + (cy)^2}. \tag{4.20}$$

Since c and d are integers, there are at most a finite number of matrices $\begin{pmatrix} * & * \\ c & d \end{pmatrix} \in \Gamma$ for which $(cx + d)^2 + (cy)^2$ is bounded from above by a given positive number. Hence for each $z \in \mathcal{H}$, a $\sigma \in \Gamma$ exists, which makes $\Im \sigma(z)$ a maximum. Now choose an integer, such that $T^n \sigma(n)$ has real part between $-1/2$ and $1/2$. We claim that the element $z' = T^n \sigma(z)$ belongs to D, that is to say, $|z'| \geq 1$. To see this we observe that if the inequality $|z'| < 1$ holds instead, then the element $-1/z' = ST^n\sigma(z)$ would have an imaginary part strictly larger than $\Im(z') = \Im\sigma(z)$, which is impossible. Hence we have $T^n\sigma(z) \in D$.

To verify Property 2 we let $z \in D$ and let $\sigma = \begin{pmatrix} a & b \\ c & d \end{pmatrix} \in \Gamma$ be such that $\sigma(z) \in D$. We may suppose that $\Im\sigma(z) \geq \Im(z)$, for otherwise we may consider the pair $(\sigma(z), \sigma^{-1})$ instead of (z, σ). By Equation 4.20 we then have $(cx + d)^2 + c^2y^2 \leq 1$, which holds only if $|c| \leq 1$. We consider the cases $c = 0$, 1, -1 separately. For $c = 0$ we have $d = \pm 1$ and $\sigma = \begin{pmatrix} \pm 1 & b \\ 0 & \pm 1 \end{pmatrix} \in \Gamma$, that is, σ is a translation. Since z and $\sigma(z)$ both lie in D, they are at most a unit distance apart and hence $b = 0$ or $b = \pm 1$. Therefore σ is either the identity or $\sigma(z) = z \pm 1$ and $\Re(z) = \pm 1/2$. If $c = 1$ (resp. $c = -1$), then the condition $\Im\sigma(z) \geq \Im(z)$ implies $|z + d| \leq 1$, which forces $d = 0$ unless $z = \rho$ (resp. $z = 1 + \rho$), $\rho^3 = 1$, $\rho \neq 1$, in which case $|\rho| \leq 1$ and $|\rho - 1| \leq 1$ (resp. $|1 - \rho| \leq 1$ and $|(1 - \rho) - 1| \leq 1$). The case $d = 0$ gives $|z| \leq 1$ and hence $|z| = 1$, that is, z lies on the lower boundary of D. In this last case we have $b = -1$ since $ad - bc = 1$.

In the proof of the multiplicativity of the Ramanujan function $\tau(n)$ the need arises for computing the dimension of a certain space of holomorphic functions defined on the upper half plane. It is here that some weak form of the Riemann-Roch theorem plays an important role. In our discussion of $\tau(n)$ an appeal to the maximum modulus principle suffices.

LEMMA 4.3
Suppose that $f(z)$ is a bounded holomorphic function defined on the upper half plane $\mathcal{H}$ and suppose $f(z)$ is Γ-invariant, that is,

$$f(z + 1) = f(z) \quad \text{and} \quad f\left(-\frac{1}{z}\right) = f(z), \quad \text{for all } z \in \mathcal{H}.$$

Then $f(z)$ is a constant.

PROOF (Heilbron). As a function of $q = e^{2\pi iz}$, $g(q) = f(z)$ cannot have an essential singularity at the origin (since otherwise it would assume arbitrarily large values near the origin), and it cannot have a pole at $q = 0$ (since then it would not be bounded). Thus $g(q)$ must be regular at the origin. By subtracting the value $g(0)$ we may assume that $f(z) \to 0$ as $\Im(z) \to \infty$. Now, since for any point $z \in \mathcal{H}$, there is a $\sigma \in \Gamma$ with $\sigma(z) \in D$ and $f(\sigma(z)) = f(z)$, $|f(z)|$ attains its maximum at some point in D, which must line on the boundary ∂D but not at infinity. Suppose $z_0 \in \partial D$ is the point where $|f(z)|$ attains its maximum $|f(z_0)| = M$. If $f(z)$ were not constant, we could find some point z_1 in a neighborhood of z with $|f(z_1)| > M$. But for some $\sigma \in \Gamma$, we have $\sigma(z_1) \in D$. Hence $|f(z_1)| > M \geq |f(\sigma(z_1))| = |f(z_1)|$, which is a contradiction. This proves the lemma. ∎

As we already indicated, Mordell's proof (in Theorem 4.5) of the multiplicativity of $\tau(n)$ uses a family of linear operators closely related to the modular correspondences studied by Hurwitz. Mordell's basic idea is to show that these operators applied to the Ramanujan modular form $\Delta(z)$ leave it invariant up to constant multiples.

Let k be an integer and $\sigma = \begin{pmatrix} a & b \\ c & d \end{pmatrix}$ be a 2×2 matrix with real coefficients and $ad - bc > 0$. If $f(z)$ is a function defined on the upper half plane $\mathcal{H}$, we put

$$f \circ [\sigma]_k(z) = f(\sigma(z))(cz + d)^{-k}(\det \sigma)^{k/2}. \qquad (4.21)$$

Observe that if $\det \sigma = 1$ and $f(\sigma(z)) = (cz+d)^k f(z)$, then $f \circ [\sigma]_k(z) = f(z)$. In particular we have

$$\Delta \circ [\sigma]_{12}(z) = \Delta(z), \qquad \sigma = \begin{pmatrix} a & b \\ c & d \end{pmatrix} \in \Gamma.$$

Now fix a prime p and consider the linear operator T_p:

$$T_p f(z) = p^{(k/2)-1} \left(f \circ \begin{bmatrix} p & 0 \\ 0 & 1 \end{bmatrix}_k (z) + \sum_{v=0}^{p-1} f \circ \begin{bmatrix} 1 & v \\ 0 & p \end{bmatrix}_k (z) \right). \qquad (4.22)$$

This can also be written as

$$T_p f(z) = p^{k-1} f(pz) + \frac{1}{p} \sum_{v=0}^{p-1} f\left(\frac{z+v}{p}\right).$$

If $f(z)$ has a Fourier expansion

$$f(z) = \sum_{n=1}^{\infty} a(n)q^n, \qquad q = \exp(2\pi i z),$$

then

$$T_p f(z) = p^{k-1} f(pz) + \sum_{n=1}^{\infty} a(n) \exp(2\pi i n z/p) \left(\frac{1}{p} \sum_{v=0}^{p-1} \exp(2\pi i n v/p) \right).$$

But

$$\frac{1}{p} \sum_{v=0}^{p-1} \exp(2\pi i n v/p) = \begin{cases} 1 & \text{if } p|n \\ 0 & \text{otherwise,} \end{cases}$$

and hence

$$T_p f(z) = \sum_{n=1}^{\infty} a(np)q^n + p^{k-1} \sum_{n=1}^{\infty} a(n)q^{np}. \tag{4.23}$$

From this it follows easily that $T_p f(z)$ is periodic of period 1, that is,

$$T_p f \circ \begin{bmatrix} 1 & 1 \\ 0 & 1 \end{bmatrix}_k (z) = T_p f(z+1)$$
$$= T_p f(z).$$

Suppose $f \circ [\sigma]_k (z) = f(z)$ for all $\sigma \in \Gamma$, write Equation (4.22) in the form

$$T_p f(z) = p^{k-1} f(pz) + \frac{1}{p} f\left(\frac{z}{p}\right) + p^{k/2-1} \sum_{v=1}^{p-1} f \circ \begin{bmatrix} 1 & v \\ 0 & p \end{bmatrix}_k (z),$$

and consider the effect of the transformation $S : z \longmapsto -1/z$ on the function $T_p f(z)$

$$T_p \circ [S]_k f(z) = p^{k-1} f(pz) \circ [S]_k + \frac{1}{p} f\left(\frac{z}{p}\right) \circ [S]_k + \tag{4.24}$$
$$p^{k/2-1} \sum_{v=1}^{p-1} f \circ \begin{bmatrix} 1 & v \\ 0 & p \end{bmatrix}_k \circ [S](z).$$

Since $f(-1/z) = z^k f(z)$, we see that the first two terms in the right-hand side of Equation (4.24) simply get permuted, that is,

$$p^{k-1} f(pz) \circ [S]_k + \frac{1}{p} f\left(\frac{z}{p}\right) \circ [S]_k = \frac{1}{p} f\left(\frac{z}{p}\right) + p^{k-1} f(pz).$$

The rest of the sum in Equation (4.24) also remains invariant. To see this we observe that

$$f \circ \begin{bmatrix} 1 & v \\ 0 & p \end{bmatrix}_k [S](z) = f \circ \left[\begin{pmatrix} 1 & v \\ 0 & p \end{pmatrix} \cdot S \right]_k (z).$$

But if $v \neq 0$, we can find a $w \neq 0$ with $up - vw = 1$ and

$$\begin{pmatrix} 1 & v \\ 0 & p \end{pmatrix} \cdot S = SA^{-1} \begin{pmatrix} 1 & w \\ 0 & p \end{pmatrix}$$

with some $SA^{-1} \in \Gamma$. Furthermore, as v ranges over the nonzero residues of p, so does w. From the invariance property of $f(z)$ we have

$$f \circ \left[\begin{pmatrix} 1 & v \\ 0 & p \end{pmatrix} \cdot S \right]_k (z) = f \circ [SA^{-1}]_k \circ \begin{bmatrix} 1 & w \\ 0 & p \end{bmatrix}_k (z)$$
$$= f \circ \begin{bmatrix} 1 & w \\ 0 & p \end{bmatrix}_k (z).$$

This then shows that for all $\sigma \in \Gamma$,

$$T_p f \circ [\sigma]_k(z) = T_p f(z).$$

The observations applied to the function $f(z) = \Delta(z)$ with $k = 12$ give that the function $g(z) = T_p \Delta(z)/\Delta(z)$ satisfies

$$g(\sigma(z)) = g(z), \qquad \sigma \in \Gamma,$$

and, since for $z \in \mathcal{H}$,

$$\Delta(z) = e^{2\pi i z} \prod_{n=1}^{\infty} \left(1 - e^{2\pi i n z}\right)^{24}$$

is nonzero and bounded in the finite part of the plane, we can apply Lemma 4.3 to the function

$$g(z) = \frac{\tau(p)q + \tau(2p)q^2 + \cdots}{q + \tau(2)q^2 + \cdots}$$
$$= \tau(p) + (*)q + \cdots .$$

Since $g(z) \to \tau(p)$ as $\Im(z) \to \infty$ we get $g(z) = \tau(p)$; that is to say

$$T_p \Delta(z) = \tau(p)\Delta(z).$$

Using the expression (4.23) we can write this in the form

$$\sum_{n=1}^{\infty} \tau(p)\tau(n)q^n = \sum_{n=1}^{\infty} \tau(pn)q^n + \sum_{n=1}^{\infty} \tau(n)q^{pn}.$$

A comparison of equal powers of q gives for p with $\gcd(p, n) = 1$

$$\tau(pn) = \tau(p)\tau(n)$$

and

$$\tau(p^\lambda n) = \tau(p)\tau(p^{\lambda-1}n) - p^{11}\tau(p^{\lambda-2}n), \quad \lambda \geq 2. \qquad (4.25)$$

To prove that $\tau(n)$ is multiplicative it suffices to show that for $\lambda \geq 0$

$$\tau(p^\lambda n) = \tau(p^\lambda)\tau(n), \quad \gcd(n, p) = 1.$$

Let

$$U_\lambda = \tau(p^\lambda n) - \tau(p^\lambda)\tau(n)$$

and observe that Equation (4.25) gives the difference equation

$$U_\lambda = \tau(p)U_{\lambda-1} - p^{11}U_{\lambda-2}.$$

Now, since $U_0 = 0$ and $U_1 = 0$, we have $U_\lambda = 0$ for all $\lambda \geq 0$. This proves the multiplicativity of $\tau(n)$.

Part (b) of Theorem 4.5 is obtained from the recursion formula (4.25), which implies the formal identity

$$(1 - \tau(p)T + p^{11}T^2)f(T) = 1$$

with

$$f(T) = \sum_{v=0}^{\infty} \tau(p^v)T^v.$$

4.4 An Example of Wilton: $\tau(n)$ Modulo 23

The following example, taken from Wilton [121] and [120], gives a simple and effective method for computing the value of $\tau(n)$ modulo 23, and it also exhibits in a direct way the connection that exists between modular forms and the higher reciprocity laws of number theory. Although many of the congruences satisfied by $\tau(n)$ had been known for some time, it is only in recent times, after the development of ℓ-adic cohomology and ℓ-adic representations, that it has become possible to relate the coefficients of modular forms like $\Delta(z)$ to arithmetic-geometric objects (motives!) and their associated Galois representations. The result we want to prove is the following.

THEOREM 4.6 $\tau(n)$ Modulo 23
Let $b(n)$ be the n-th coefficient in the formal expansion

$$q \prod_{n=1}^{\infty} (1 - q^n)(1 - q^{23n}) = \sum_{n=1}^{\infty} b(n)q^n.$$

Then
(a) $\tau(n) \equiv b(n) \pmod{23}$, and
(b) $|\{x \in \mathbb{F}_p : x^3 - x - 1 = 0\}| = 1 + b(p)$ for prime $p \neq 23$.

REMARK 4.6 It will be seen below that if the Legendre symbol $(-23/p) = -1$, then $x^3 - x - 1 = 0$ has one and only one solution in the finite field $\mathbb{F}_p$. From Part (b) of the theorem we then recover the classical statement that $\tau(p) \equiv 0 \pmod{23}$ for

$$p \equiv 5, 7, 10, 11, 14, 15, 17, 19, 20, 21, 22 \pmod{23}.$$

∎

The proof of Theorem 4.6 will proceed in several steps. First we will evaluate explicitly the coefficients $b(p)$ in terms of the behavior of the prime p in the quadratic extension $\mathbb{Q}(\sqrt{-23})$. We shall then examine the factorization of the polynomial $f(x) = x^3 - x - 1 = 0$. Finally we relate these factorizations to the computation of the Frobenius conjugacy classes in the splitting extension of $f(x)$. The end result will be the identification of the Dirichlet series $\sum_{n=1}^{\infty} b(n)n^{-s}$ with the Artin L-function $L(s, \rho)$ associated with the unique two-dimensional representation of the Galois group of the splitting extension of $f(x)$.

Let us first observe that Theorem 4.6(a), which is equivalent to the formal congruence

$$q \prod_{n=1}^{\infty} (1 - q^n)^{24} \equiv q \prod_{n=1}^{\infty} (1 - q^n)(1 - q^{23n}) \quad (\text{mod } 23),$$

follows easily, by the binomial expansion, from the congruence

$$(1 - x)^{24} \equiv (1 - x)(1 - x^{23}) \quad (\text{mod } 23).$$

The evaluation of $b(p)$ uses Euler's well-known identity and arithmetic properties of the field $\mathbb{Q}(\sqrt{23})$. We first need this lemma, called Euler's identity. This is in essence the general method used by Jacobi to derive formal power series expansions from infinite products.

LEMMA 4.4
   We have

$$\prod_{n=1}^{\infty} (1 - q^n) = \sum_{n=0}^{\infty} a(n)q^n, \tag{4.26}$$

where

$$a(n) = \begin{cases} 1 & \text{if } n = 0 \\ (-1)^m & \text{if } n > 0 \text{ and } n = m(3m \pm 1)/2 \\ 0 & \text{otherwise.} \end{cases}$$

PROOF   Just as in the proof of Theorem 4.1, Jacobi's triple product identity leads to the formal expansion

$$\sum_{n \in \mathbb{Z}} (-1)^n q^{an^2 + bn} = \prod_{n=1}^{\infty} (1 - q^{2an})(1 - q^{2an - a + b})(1 - q^{2an - a - b}).$$

When the values $a = 3/2$ and $b = 1/2$ are substituted into this equality there results Euler's well-known identity

$$\prod_{n=1}^{\infty} (1 - q^n) = \sum_{n \in \mathbb{Z}} (-1)^n q^{n(3n+1)/2}$$

$$= \sum_{n=0}^{\infty} a(n)q^n.$$

This proves the lemma. ∎

REMARK 4.7   Note that Lemma 4.4 is a q-series analogue of Euler's recurrence formula for $\sigma(m)$ in Theorem 2.4. ∎

Now observe that

$$q \prod_{n=1}^{\infty} \left(1 - q^n\right) \left(1 - q^{23n}\right) = q \sum_{k=0}^{\infty} a(k) q^k \sum_{m=0}^{\infty} a(m) q^{23m}$$

$$= \sum_{n=1}^{\infty} b(n) q^n,$$

where

$$b(n) = \sum a(k) a(m),$$

and the sum runs over all solutions of $n = 1 + k + 23m$. Hence we have

$$b(n) = \sum_{0 \le m \le h} a(m) a(n - 1 - 23m) \tag{4.27}$$

$$= a(n-1) + a(1) a(n-24) + \cdots + a(h) a(n - 1 - 23h),$$

where $h = \lfloor (n-1)/23 \rfloor$. Now if the sum (4.27) is nonempty, then from the definition of the $a(n)$, there must exist integers v and w, such that

$$n = 1 + \frac{v}{2}(3v \pm 1) + \frac{w}{2}(3w \pm 1),$$

or equivalently,

$$24n = (6v)^2 \pm 2(6v) + 1 + 23 \left\{ (6w)^2 \pm 2(6w) + 1 \right\}$$

$$= (6v \pm 1)^2 + 23(6w + 1)^2,$$

which implies that the Legendre symbol $\left(\frac{24n}{23} \right) = 1$, that is,

$$n \equiv 1, 2, 3, 4, 6, 8, 9, 12, 13, 16, 18 \pmod{23}.$$

We state this formally using the quadratic reciprocity law as

$$b(p) = 0 \quad \text{if} \quad \left(\frac{-23}{p} \right) = -1. \tag{4.28}$$

For the remaining cases we assume that p is a quadratic residue of 23 and observe that

$$b(p) = \sum a\left(\frac{m}{2}(3m \pm 1) \right) a\left(p - 1 - \frac{23m}{2}(3m \pm 1) \right)$$

$$= \sum (-1)^{m+n}, \tag{4.29}$$

where the sum runs over those values of m and n for which

$$p - 1 - 23\left\{\frac{m}{2}(3m \pm 1)\right\} = \frac{n}{2}(3n \pm 1),$$

that is,

$$24p = (6n \pm 1)^2 + 23(6m \pm 1)^2.$$

To evaluate $b(p)$ we must make a study of the character of the integers u and v that satisfy the equation

$$24p = u^2 + 23v^2. \tag{4.30}$$

This will be done in terms of the decomposition of the prime p in the imaginary quadratic field $\mathbb{Q}(\sqrt{23})$. Recall that if d_k is the discriminant of the quadratic field $k = \mathbb{Q}(\sqrt{23})$ and if

$$1 \quad \text{and} \quad \vartheta = \frac{d_k + \sqrt{d_k}}{2}$$

is an integral basis for the ring of integers of k, where ϑ is a root of the polynomial

$$f(x) = x^2 - d_k x + \frac{d_k^2 - d_k}{4},$$

then a given rational prime p stays prime in k if $(d_k/p) = -1$ and otherwise is the product of two prime ideals

$$(p) = (p, r - \vartheta)(p, s - \vartheta),$$

where r and s are integers that satisfy the congruence

$$x^2 - d_k x + \frac{d_k^2 - d_k}{4} \equiv 0 \pmod{p}.$$

A proof of this result can be found in almost any book on algebraic number theory, for example [59]. We apply this result to the field

$$k = \mathbb{Q}(\alpha) \quad \text{with} \quad \alpha = (-23)^{1/2}$$

and obtain that if p is a quadratic residue of 23, then it splits into the product of two prime ideals

$$(p) = \pi \cdot \pi', \quad \pi, \pi' = \left(p, s - \frac{1}{2} \cdot 23 \mp \frac{1}{2} \cdot \alpha\right),$$

where s is an integer that satisfies the congruence $x^2 + 23x + 138 \equiv 0 \pmod{p}$. The class group $CL(k)$ of $k = \mathbb{Q}(\alpha)$ is of order 3: the principal class which contains the principal ideals and the two classes represented by the ideal factors of 2,

$$\pi_2 = \left(2, \frac{1}{2} + \frac{1}{2}\alpha\right), \quad \pi_2' = \left(2, \frac{1}{2} - \frac{1}{2}\alpha\right).$$

where the sum runs over those values of m and n for which

$$p - 1 - 23 \left\{ \frac{m}{2}(3m \pm 1) \right\} = \frac{n}{2}(3n \pm 1),$$

that is,

$$24p = (6n \pm 1)^2 + 23(6m \pm 1)^2.$$

To evaluate $b(p)$ we must make a study of the character of the integers u and v that satisfy the equation

$$24p = u^2 + 23v^2. \tag{4.30}$$

This will be done in terms of the decomposition of the prime p in the imaginary quadratic field $\mathbb{Q}(\sqrt{23})$. Recall that if d_k is the discriminant of the quadratic field $k = \mathbb{Q}(\sqrt{23})$ and if

$$1 \quad \text{and} \quad \vartheta = \frac{d_k + \sqrt{d_k}}{2}$$

is an integral basis for the ring of integers of k, where ϑ is a root of the polynomial

$$f(x) = x^2 - d_k x + \frac{d_k^2 - d_k}{4},$$

then a given rational prime p stays prime in k if $(d_k/p) = -1$ and otherwise is the product of two prime ideals

$$(p) = (p, r - \vartheta)(p, s - \vartheta),$$

where r and s are integers that satisfy the congruence

$$x^2 - d_k x + \frac{d_k^2 - d_k}{4} \equiv 0 \pmod{p}.$$

A proof of this result can be found in almost any book on algebraic number theory, for example [59]. We apply this result to the field

$$k = \mathbb{Q}(\alpha) \quad \text{with} \quad \alpha = (-23)^{1/2}$$

and obtain that if p is a quadratic residue of 23, then it splits into the product of two prime ideals

$$(p) = \pi \cdot \pi', \quad \pi, \pi' = \left(p, s - \frac{1}{2} \cdot 23 \mp \frac{1}{2} \cdot \alpha \right),$$

where s is an integer that satisfies the congruence $x^2 + 23x + 138 \equiv 0 \pmod{p}$. The class group $CL(k)$ of $k = \mathbb{Q}(\alpha)$ is of order 3: the principal class which contains the principal ideals and the two classes represented by the ideal factors of 2,

$$\pi_2 = \left(2, \frac{1}{2} + \frac{1}{2}\alpha \right), \quad \pi_2' = \left(2, \frac{1}{2} - \frac{1}{2}\alpha \right).$$

This proves the lemma. ∎

REMARK 4.7   Note that Lemma 4.4 is a q-series analogue of Euler's recurrence formula for $\sigma(m)$ in Theorem 2.4. ∎

Now observe that

$$q\prod_{n=1}^{\infty}(1-q^n)\left(1-q^{23n}\right) = q\sum_{k=0}^{\infty}a(k)q^k\sum_{m=0}^{\infty}a(m)q^{23m}$$

$$= \sum_{n=1}^{\infty}b(n)q^n,$$

where

$$b(n) = \sum a(k)a(m),$$

and the sum runs over all solutions of $n = 1 + k + 23m$. Hence we have

$$b(n) = \sum_{0\le m\le h}a(m)a(n-1-23m) \tag{4.27}$$

$$= a(n-1) + a(1)a(n-24) + \cdots + a(h)a(n-1-23h),$$

where $h = \lfloor(n-1)/23\rfloor$. Now if the sum (4.27) is nonempty, then from the definition of the $a(n)$, there must exist integers v and w, such that

$$n = 1 + \frac{v}{2}(3v\pm 1) + \frac{w}{2}(3w\pm 1),$$

or equivalently,

$$24n = (6v)^2 \pm 2(6v) + 1 + 23\left\{(6w)^2 \pm 2(6w) + 1\right\}$$
$$= (6v\pm 1)^2 + 23(6w\pm 1)^2,$$

which implies that the Legendre symbol $\left(\frac{24n}{23}\right) = 1$, that is,

$$n \equiv 1, 2, 3, 4, 6, 8, 9, 12, 13, 16, 18 \pmod{23}.$$

We state this formally using the quadratic reciprocity law as

$$b(p) = 0 \quad\text{if}\quad \left(\frac{-23}{p}\right) = -1. \tag{4.28}$$

For the remaining cases we assume that p is a quadratic residue of 23 and observe that

$$b(p) = \sum a\left(\frac{m}{2}(3m\pm 1)\right)a\left(p-1-\frac{23m}{2}(3m\pm 1)\right)$$
$$= \sum(-1)^{m+n}, \tag{4.29}$$

In fact the class group is isomorphic to the group of ideal classes of binary quadratic forms of discriminant -23, of which a representative system is given by the three binary quadratic forms,

$$x^2 + xy + 6y^2, \quad 2x^2 + xy + 3y^2, \quad 2x^2 - xy + 3y^2.$$

In the following argument we shall also need the prime ideal factors of 3,

$$\pi_3 = \left(3, \frac{1}{2} - \frac{1}{2}\alpha\right), \quad \pi_3' = \left(3, \frac{1}{2} + \frac{1}{2}\alpha\right),$$

where the notation has been chosen so that π_2 and π_3 belong to the same ideal class.

The value of $b(p)$ for a prime p with $(-23/p) = 1$ depends on whether the prime ideal factors of p in $k = \mathbb{Q}(\alpha)$ are in the principal class or not.

Case 1. $(p) = \pi \cdot \pi'$ and π is a principal ideal, that is,

$$(\pi) = \left(\frac{1}{2}a + \frac{1}{2}b\alpha\right), \quad a - b \text{ even}.$$

In this case we have

$$4p = 4\pi \cdot \pi' = a^2 + 23b^2,$$

so that

$$a^2 - b^2 = 4(p - 6b^2),$$

from which it follows that a and b must both be even, for otherwise $a^2 - b^2 \equiv 0 \pmod 8$. We therefore have

$$p = h^2 + 23k^2. \tag{4.31}$$

Also, since the prime ideal decomposition $(p) = \pi \cdot \pi'$ is unique there is only one solution in positive integers of Equation (4.31) and

$$\begin{aligned}
24p &= (1 + \alpha)(1 - \alpha)(h + k\alpha)(h - k\alpha) \\
&= (h - 23k)^2 + 23(h + k)^2 \\
&= (h + 23k)^2 + 23(h - k)^2.
\end{aligned}$$

Hence Equation (4.30) has the two solutions

$$u = h + 23k, \ v = |h - k| \quad \text{and} \quad u = |h - 23k|, \ v = h + k.$$

But for both solutions one of the congruences

$$u + v \equiv 0 \pmod{24}, \quad u - v \equiv 0 \pmod{24}$$

must hold. Hence in the equation $24p = (6n \pm 1)^2 + 23(6m \pm 1)^2$, m and n are either both even or both odd. For otherwise, $6(m \pm n) \not\equiv 0 \pmod{24}$. Therefore the sum in Equation (4.29) contains only two terms and for both of these $m + n$ is even. Hence $b(p) = \sum(-1)^{m+n} = 2$. We have thus shown that if p splits in $\mathbb{Q}(\alpha)$ into two principal prime ideals, then $b(p) = 2$.

Case 2. $(p) = \pi \cdot \pi'$ and π is a nonprincipal ideal. In this case we choose a root r of $x^2 + 23x + 138 \equiv 0 \pmod{p}$ so that the prime ideal factor of p,

$$\pi = \left(p, 2 + \frac{1}{2} \cdot 23 + \frac{1}{2}\alpha \right),$$

is in the same class with π_2 and π_3, and also π' is in the same class with π'_2 and π'_3. Hence $\pi_2\pi_3\pi$ is a principal ideal and $24p$ is expressible in the form $u^2 + 23v^2$ in only one way:

$$\pi_2\pi_3\pi = \left(\frac{1}{2}(3 + 3\alpha), 1 - \alpha \right) \left(p, r + \frac{1}{2}(23 + \alpha) \right) \tag{4.32}$$

$$= \left(p \cdot \frac{1}{2}(3 + 3\alpha), p(1 - \alpha)(1 - \alpha) \left(r + \frac{1}{2}(23 + \alpha) \right), \right.$$

$$\left. \frac{1}{2}(3 + 3\alpha) \left(r + \frac{1}{2}(23 + \alpha) \right) \right)$$

$$= \left(\frac{1}{2}u + \frac{1}{2}v\alpha \right),$$

where u and v are solutions of $24p = u^2 + 23v^2$. It follows that there must exist integers h, k, h' and k', such that

$$p \cdot \frac{1}{2}(3 + 3\alpha) = \left(\frac{1}{2}u + \frac{1}{2}v\alpha \right) \cdot \left(\frac{1}{2}h + \frac{1}{2}k\alpha \right)$$

$$p \cdot (1 - \alpha) = \left(\frac{1}{2}u + \frac{1}{2}v\alpha \right) \cdot \left(\frac{1}{2}h' + \frac{1}{2}k'\alpha \right).$$

The second of these equalities gives

$$p(1 - \alpha)(u - v\alpha) = 6p(h' - k'\alpha),$$

from which we deduce that

$$u - 23v = 6h' \quad \text{and} \quad u + v = 6k'.$$

Similarly,

$$p(1 + \alpha)(u - v\alpha) = 4p(h + k\alpha),$$

from which we deduce that

$$u + 23v = 4h \quad \text{and} \quad u - v = 4k.$$

From the equality $u^2 - v^2 = 24(p - v^2)$ we see that both u and v are relatively prime to 6. Now this implies that either

$$u = 6n + 1, \; v = 6m - 1 \quad \text{or} \quad u = 6n - 1, \; v = 6m + 1$$

and

$$3n - 3m \pm 1 = 2k.$$

But this implies that $m + n$ is odd; now since the prime ideal factorization (4.32) is unique, the sum in Equation (4.29) contains only one term and hence $b(p) = -1$. We collect these results in the following statement.

PROPOSITION 4.2
The coefficients $b(n)$ in the expansion

$$q \prod_{n=1}^{\infty} (1 - q^n) \left(1 - q^{23n}\right) = \sum_{n=1}^{\infty} b(n) q^n$$

have the following values for a prime p:

1. $b(p) = 0$ if $(-23/p) = -1$,

2. $b(p) = 2$ if $(-23/p) = 1$ and p splits in $\mathbb{Q}(\alpha)$ into principal prime ideals,

3. $b(p) = -1$ if $(-23/p) = 1$ and p splits in $\mathbb{Q}(\alpha)$ into nonprincipal prime ideals,

4. $b(23) = -1$.

In order to complete the proof of Theorem 4.6 we need to recall two basic propositions that describe the decomposition of rational primes in algebraic extensions. Here we simply recall their statement and basic definitions.

4.4.1 Factorization in Nonnormal Extensions of $\mathbb{Q}$

The key tool in effectively computing Frobenius conjugacy classes in finite extensions of the rationals is Theorem 4.7 below, which is an adaptation of results in [59], page 124. Let $K/\mathbb{Q}$ be a Galois extension with group G. Let E be a subextension of K with Galois group $H = \mathrm{Gal}(K/E)$. Consider a coset decomposition of G given by

$$G = H\sigma_1 \cup H\sigma_2 \cup \cdots \cup H\sigma_r.$$

The group G acts as a permutation group on the set $S = \{H\sigma_1, \ldots, H\sigma_r\}$ via right multiplication: for $g \in G$ and $H\sigma \in S$ we have $g : H\sigma \longmapsto H\sigma g$. By a cycle of length t for $g \in G$ we mean a sequence

$$H\sigma, H\sigma g, \ldots, H\sigma g^{t-1}$$

of elements in S in which the t cosets are distinct and $H\sigma = H\sigma g^t$. For a fixed $g \in G$, the corresponding permutation has a cycle decomposition

$$S = C_1 \cup C_2 \cup \cdots \cup C_s, \quad \text{length } C_i = t_i, 1 \le i \le s.$$

Let D_K be the discriminant of K and consider a rational prime p with $p \nmid D_K$. Let Q be a prime in K that divides p and let σ_Q be the Frobenius element in G, that is,

$$\sigma_Q(\alpha) \equiv \alpha^{NQ} \pmod{Q}$$

for all $\alpha \in A_K$, the ring of integers in K. To p one attaches the Frobenius conjugacy class in G,

$$\sigma_p = \{\sigma_Q \in G : Q|p\}.$$

THEOREM 4.7 Factorization of Rational Primes in an Extension Field
Let the notation be as above. Let σ be the Frobenius element associated to the prime ideal Q in $K/\mathbb{Q}$. Suppose σ has cycles of length $t_1, \ldots, t_s$ when acting upon the cosets of $H = \mathrm{Gal}(K/E)$ in G. Then the rational prime p is the product of s distinct primes in E having degrees $t_1, \ldots, t_s$ over $\mathbb{Q}$.

REMARK 4.8   For the applications we have in mind, we usually have knowledge of the factorization of p in E, but lack information about the Frobenius conjugacy class σ_p. Often it turns out that if we survey the permutations corresponding to various conjugacy classes in G and match the cycle decompositions with the factorization of p in suitable extensions E, it may be possible to determine effectively the Frobenius conjugacy class σ_p.   ∎

Next we consider the problem of the effective factorization of a prime p in a field extension K of the rationals.

THEOREM 4.8 Kummer-Dedekind Theorem
Let $K = \mathbb{Q}(\alpha)$ be a field extension of degree n and discriminant $D(K)$. Let α be a root of a polynomial $f(x) \in \mathbb{Z}[x]$ which is irreducible and monic. Let $D(f)$ be the discriminant of $f(x)$. If the prime p does not divide $D(f)/D(K)$, then there is a one-to-one correspondence between the factorization of the ideal (p) in K,

$$(p) = Q_1^{e_1} \cdots Q_g^{e_g},$$

and the factorization of $f(x)$ modulo p into irreducible factors,

$$f(x) \equiv f_1(x)^{e_1} \cdots f_g(x)^{e_g} \pmod{p},$$

in the sense that the prime ideal Q_i corresponds to the irreducible factor $f_i(x)$ and

$$\mathrm{degree}\, Q_i = \mathrm{degree}\, f_i(x), \quad 1 \leq i \leq g.$$

REMARK 4.9   If α can be selected so that $D(K) = D(f)$, then the Kummer-Dedekind theorem provides the effective factorization in K of all rational primes. The proof of Theorem 4.8 gives the representation $Q_i = (p, f_i(\alpha))$.   ∎

4.4.2   Table for the Computation of Frobeniuses

The information contained in Theorems 4.7 and 4.8 when applied to cubic extensions of the rationals leads to the following table. Let $f(x) \in \mathbb{Z}[x]$ be an irreducible cubic polynomial with roots, α, β, γ and discriminant $D(f)$. Let $K_f = \mathbb{Q}(\alpha, \beta, \gamma)$ be the splitting field of $f(x)$ and suppose

$$D(f) = ((\alpha - \beta)(\beta - \gamma)(\gamma - \alpha))^2$$

is not a square in $\mathbb{Q}$. Then the quadratic extension $\mathbb{Q}\left(\sqrt{D(f)}\right)$ is contained in K_f and the Galois group of the equation $f(x) = 0$ is the symmetric group S_3 on three letters:

$$\mathrm{Gal}(K_f/\mathbb{Q}) = \{r, s : r^3 = s^2 = 1, \ srs^{-1} = r^2\};$$

the conjugacy classes are

$$C_1 = \{1\}, \quad C_2 = \{r, r^2\}, \quad C_3 = \{s, sr, sr^2\}.$$

Fix a prime p which does not divide the discriminant and let χ_0 be the trivial character, sgn the signature character and χ_ρ the character of the two-dimensional representation. The first three rows in the table below give the character table of S_3.

χ_0	1	1	1
sgn	1	1	-1
χ_ρ	2	-1	0
σ_p	C_1	C_2	C_3
(p)	$Q_1 \cdot Q_2 \cdot Q_3$	Q	$Q_1 \cdot Q_2$
$\deg Q_i$	1, 1, 1	3	1, 2
$f(x) \bmod p$	$(x-\alpha)(x-\beta)(x-\gamma)$	$f(x)$ irreducible	$(x-\alpha)(x^2 + ax + b)$

Going up from the last row to the row of conjugacy classes is accomplished by first applying Theorem 4.8 and then Theorem 4.7. From this table we can read easily the following result.

COROLLARY 4.3
Let the notation be as above. For a prime p that does not divide $D(f)$ we have
$$|\{x \in \mathbb{F}_p : f(x) = 0\}| = 1 + \chi_\rho(\sigma_p).$$

We are now ready to complete the proof of Theorem 4.6. We apply the above table to the example $f(x) = x^3 - x - 1$ with roots α, β, γ and discriminant $D(f) = -23$, which is also the discriminant of the splitting field $K_f = \mathbb{Q}(\alpha, \beta, \gamma)$. The field $k = \mathbb{Q}(\sqrt{-23})$ is contained in K_f and

$\mathrm{Gal}(K_f/k) = \{1, r, r^2\}$. We have already shown that for a prime $p \neq 23$ with $(-23/p) = -1$, we have $b(p) = 0$. Now a well-known theorem from the elementary Galois theory of finite fields, whose proof we outline in the exercises, guarantees in the present case that for a prime p for which -23 is a quadratic nonresidue, the congruence $f(x) \equiv 0 \pmod{p}$ has only one root $x \in \mathbb{F}_p$. Hence for these primes the claim (b) of Theorem 4.6 is true. Now according to the table above the corresponding Frobenius conjugacy class is $\sigma_p = C_3 = \{s, sr, sr^2\}$, and in this case $\chi_\rho(\sigma_p) = 0$. Therefore, we have for these primes $b(p) = \chi_\rho(\sigma_p)$. It remains to consider the case of a prime p with $(-23/p) = 1$, that is, p splits in $\mathbb{Q}(\sqrt{-23})$. We need to recall some elementary facts from class field theory. The class group $CL(k)$ of k is cyclic of order 3:

$$CL(k) = \{A_0, A_1, A_2\},$$

where A_0 is the ideal class of the principal ideals, A_1 is one of the two classes containing nonprincipal ideals, and A_2 is the inverse of A_1. Taking norms from k to $\mathbb{Q}$ we see that $CL(k)$ is isomorphic to the group of binary quadratic classes of discriminant -23, of which a reduced representative system is

$$x^2 + xy + 6y^2, \quad 2x^2 + xy + 3y^2, \quad 2x^2 - xy + 3y^2.$$

The Artin reciprocity law can be made quite explicit in this example by rather elementary arguments (see Cohn [19] p. 103). K_f is the Hilbert class field of $\mathbb{Q}(\sqrt{-23})$ that is, the maximal unramified extension of k, and

$$CL(k) \xrightarrow{\sim} H = \{1, r, r^2\},$$

where $A_0 \Leftrightarrow 1$, $A_1 \Leftrightarrow r$ and $A_2 \Leftrightarrow r^2$.

Now if $(-23/p) = 1$ and $p = \pi \cdot \pi'$ in k, then, since $\deg \pi = 1$, the restriction formula for Frobenius elements

$$(\sigma_\Pi|_H)^{\deg \pi} = \sigma_\pi, \quad \Pi \text{ a prime ideal in } K_f \text{ above } \pi,$$

shows that when π is principal, $\sigma_\pi = 1$, $\sigma_p = \{1\}$, and $\chi_\rho(\sigma_p) = 2$. If π is nonprincipal, then $\sigma_\pi = r$ or r^2 and so $\sigma_p = \{r, r^2\}$ and $\chi_\rho(\sigma_p) = 1$. The above computations together with Proposition 4.2 complete the proof of Theorem 4.6.

REMARK 4.10   The method used above gives an effective algorithm for computing the unramified local factors of the Artin L-function $L(s, \rho)$ associated to the two-dimensional representation of S_3 and of a nonnormal cubic extension.   ∎

4.5   An Example of Hamburger

One of Hecke's earliest and most striking results was his proof of the analytic continuation and functional equation for the Dedekind zeta function of a number field. This result provided Hecke with infinite families of Dirichlet series that satisfy similar functional equations and have Euler product expansions. Because of the many examples that were available and perhaps because of the then recent publication of Hamburger's argument characterizing the Riemann zeta function $\zeta(s)$ as the unique function that satisfies the functional equation

$$\Lambda(s) \stackrel{\text{def}}{=} \pi^{-s/2}\Gamma(s/2) \prod_p \frac{1}{1-p^{-s}} = \Lambda(1-s)$$

and also because of Mordell's argument (see Theorem 4.5, which establishes the Euler product expansion

$$\sum_{n=1}^{\infty} \tau(n)n^{-s} = \prod_p \frac{1}{1 - \tau(p)p^{-s} + p^{11-2s}}),$$

Hecke was lead in the late 1920s to consider the problem of characterizing all functions $\phi(s)$ of the complex variable s, which are of signature $\{N, \gamma, k\}$. Here $N > 0$ and $k > 0$ are real numbers and γ is a complex number. Recall that a function $\phi(s)$ has signature $\{N, \gamma, k\}$ if it has these three properties:

1. We have

$$\Lambda(s) \stackrel{\text{def}}{=} G_{\mathbb{R}}(s)G_{\mathbb{R}}(s+1)\phi(s), \qquad G_{\mathbb{R}}(s) = \pi^{-s/2}\Gamma(s/2),$$
$$= \gamma N^{(1/2)-s}\Lambda(1-s),$$

2. $(s-k)\phi(s)$ is an entire function of finite genus (that is, there is an $M > 0$, such that $\log|\phi(s)| \ll |s|^M$ as $|s| \to \infty$).

3. $\phi(s)$ is a Dirichlet series that converges in a half plane.

Hecke was very successful in solving the problem for signature $\{1, \gamma, k\}$, k integral, and showing, as we will do later, that the vector space of functions $\{1, \gamma, k\}$ is finite dimensional and has a basis consisting of Dirichlet series with Euler product expansions. In fact the dimension of $\{1, \gamma, k\}$ is given by

$$\frac{1}{(1-T^4)(1-T^6)} = \sum_{k=0}^{\infty} \dim_{\mathbb{C}}\{1, \gamma, k\}T^k.$$

When $k = 12$, dim $\{1, \gamma, 12\} = 2$ and a basis is given by

$$\phi_{\Delta}(s) = \sum_{n=1}^{\infty} \tau(n)n^{-s} = \prod_p \frac{1}{1 - \tau(p)p^{-s} + p^{11-2s}}$$

and

$$\phi_E(s) = \sum_{n=1}^{\infty} \sigma_{11}(n)n^{-s} = \prod_p \frac{1}{1 - \sigma_{11}(p)p^{-s} + p^{11-2s}},$$

where $\sigma_{11}(n) = \sum_{d|n} d^{11}$.

The last example we want to consider is Hamburger's theorem, which gives a characterization of Riemann's zeta function $\zeta(s)$. We can think of this theorem as describing completely the modified space of functions $\{1, \gamma, 1\}_0$, where the subscript 0 indicates that in Property 1 above we define $\Lambda(s) = G_{\mathbb{R}}(s)\phi(s)$.

THEOREM 4.9 Hamburger's Theorem
Let $G(s)$ be an entire function of finite order, $p(s)$ a polynomial and $f(s) = G(s)/p(s)$. Let

$$f(s) = \sum_{n=1}^{\infty} a(n)n^{-s}$$

be absolutely convergent for $\sigma = \Re(s) > 1$. Let

$$G_{\mathbb{R}}(s)f(s) = G_{\mathbb{R}}(1-s)g(1-s), \quad G_{\mathbb{R}}(s) = \pi^{-s/2}\Gamma(s/2), \qquad (4.33)$$

where

$$g(1-s) = \sum_{n=1}^{\infty} b(n)n^{-(1-s)},$$

the series being absolutely convergent for $\sigma < -\alpha$, for some constant $\alpha < 0$. Then $f(s) = c\zeta(s)$, where c is a constant.

In the proof of Hamburger's theorem we need an elementary result due to Mellin.

LEMMA 4.5
For any real $a > 0$ we have

$$e^{-x} = \frac{1}{2\pi i} \int_{a-i\infty}^{a+i\infty} \Gamma(s)x^{-s}\,ds.$$

PROOF   The usual integral definition of the Gamma function can be written in the form

$$\Gamma(s) = \int_0^{\infty} e^{-x}x^{s-1}\,dx$$

$$= \int_0^1 e^{-x}x^{s-1}\,dx + E(s), \quad \text{where } E(s) \text{ is an entire function,}$$

$$= \sum_{k=0}^{\infty} \frac{(-1)^k}{k!} \int_0^1 x^{k+s-1}\,dx + E(s)$$

$$= \sum_{k=0}^{\infty} \frac{(-1)^k}{k!} \cdot \frac{1}{s+k} + E(s).$$

This last expression exhibits clearly the location of the poles and the corresponding residues of the function $\Gamma(s)$. To complete the proof of Mellin's formula the line of integration is moved to the left so as to lie half way between two poles of $\Gamma(s)$. One then uses Cauchy's formula to obtain

$$\frac{1}{2\pi i}\int_{a-i\infty}^{a+i\infty}\Gamma(s)x^{-s}\,ds = \sum_{n=0}^{k-1}\frac{(-1)^n}{n!}x^n + \frac{1}{2\pi i}\int_{\frac{1}{2}-k-i\infty}^{\frac{1}{2}-k+i\infty}\Gamma(s)x^{-s}\,ds.$$

The asymptotic estimate

$$\Gamma(\sigma+it) \sim e^{-\pi|t|/2}|t|^{\sigma-1/2}(2\pi)^{1/2}, \qquad (4.34)$$

valid in $\sigma_0 \le \sigma \le \sigma_1$ as $|t| \to \infty$, gives on the line $(1/2) - k + it$

$$|\Gamma(s)x^{-s}| \ll \frac{e^{-\pi|t|/2}}{x^{k-1/2}(1+|t|)^k}.$$

The lemma then follows by letting $k \to \infty$. ∎

Now we give Siegel's proof of Hamburger's theorem.

PROOF   For real $x > 0$ we consider the function

$$\phi(x) \stackrel{\text{def}}{=} \frac{1}{2\pi i}\int_{2-i\infty}^{2+i\infty}\pi^{-s/2}\Gamma(s/2)f(s)x^{-s}\,ds.$$

Since $f(s)$ is absolutely convergent for $\Re(s) = 2$, we obtain from Mellin's integral that this is

$$\phi(x) = \sum_{n=1}^{\infty}a(n)\frac{1}{2\pi i}\int_{2-i\infty}^{2+i\infty}\Gamma(s/2)(\pi n^2 x)^{-s/2}\,ds$$

$$= 2\sum_{n=1}^{\infty}a(n)e^{-\pi n^2 x}.$$

The integral that defines $\phi(x)$ is now replaced by its equivalent form (4.33) to get

$$\phi(x) = \frac{1}{2\pi i}\int_{2-i\infty}^{2+i\infty}\pi^{-(1-s)/2}\Gamma((1-s)/2)g(1-s)x^{-s/2}\,ds.$$

We want to move the line of integration from $\sigma = 2$ to $\sigma = -1 - \alpha$. We observe first that $f(s)$ is bounded on the line $\sigma = 2$ and $g(1-s)$ is bounded on $\sigma = -1 - \alpha$. Now, by (4.33) and (4.34) we have

$$|g(1-s)| = \left|\pi^{-s+1/2}\frac{\Gamma(s/2)}{\Gamma((1-s)/2)}f(s)\right| \ll |t|^{3/2}, \qquad \text{for } \Re(s) = 2.$$

The Phragmén-Lindelöf principle and Cauchy's residue theorem give the new formula

$$\phi(x) = \frac{1}{2\pi i} \int_{-\alpha-1-i\infty}^{-\alpha-1+i\infty} \pi^{-(1-s)/2}\Gamma((1-s)/2)g(1-s)x^{-s/2}\,ds + \sum_{v=1}^{m} R_v,$$

where $R_1, \ldots, R_m$ are the residues at the poles, say $s_1, \ldots, s_m$. Using the expansion

$$x^{-s/2} = \sum_{m=0}^{\infty} x^{-s_v/2} \left(-\frac{1}{2}\log x\right)^k \frac{(s - s_v)^k}{k!}, \quad 1 \leq v \leq m,$$

we see that

$$\sum_{v=1}^{m} R_v = \sum_{v=1}^{m} x^{-s_v/2} Q_v(\log x) = Q(x),$$

where the $Q_v(\log x)$ are polynomials in $\log x$. Again by using Mellin's formula we have

$$\phi(x) = \frac{1}{\sqrt{x}} \sum_{n=1}^{\infty} b(n) \int_{-\alpha-1-i\infty}^{-\alpha-1+i\infty} \Gamma((1-s)/2)(\pi n^2 x^{-1})^{-(1-s)/2}\,ds + Q(x)$$

$$= \frac{2}{\sqrt{x}} \sum_{n=1}^{\infty} b(n) e^{-\pi n^2/x} + Q(x).$$

Finally we obtain the transformation formula

$$\sum_{n=1}^{\infty} a(n) e^{-\pi n^2 x} = \frac{1}{\sqrt{x}} \sum_{n=1}^{\infty} b(n) e^{-\pi n^2/x} + \frac{1}{2}Q(x). \tag{4.35}$$

We multiply both sides by $e^{-\pi t^2 x}$, $t > 0$, and integrate over $(0, \infty)$. We apply the integral formula

$$\left(\frac{\pi}{p}\right)^{1/2} e^{-2\sqrt{ap}} = \int_0^{\infty} e^{-pt} \frac{e^{-a/t}}{\sqrt{t}}\,dt$$

to the first integral on the right hand side and obtain

$$\sum_{n=1}^{\infty} \frac{a(n)}{\pi} \cdot \frac{1}{t^2 + n^2} = \sum_{n=1}^{\infty} b(n) \cdot \frac{e^{-2\pi nt}}{t} + \frac{1}{2}\int_0^{\infty} Q(x) e^{-\pi t^2 x}\,dx,$$

where the last integral is a sum of terms of the form

$$\int_0^{\infty} x^a (\log x)^b e^{-\pi t^2 x}\,dx.$$

Since the b's are integers and $\Re(a) > -1$, integration by parts shows that this integral is a sum of terms of the form $t^\alpha (\log x)^\beta$. Hence

$$\sum_{n=1}^{\infty} a(n) \left\{ \frac{1}{t+in} + \frac{1}{t-in} \right\} - \pi t H(t) = 2\pi \sum_{n=1}^{\infty} b(n) e^{-\pi nt},$$

where $H(t)$ is a sum of terms of the form $t^\alpha (\log x)^\beta$. Now, the series on the left is a meremorphic function with poles at $\pm in$. But the function on the right is periodic with period i and, by analytic continuation, so is the function on the left. Therefore the residues at ki and $(k+1)i$ are equal, that is, $a(k+1) = a(k) = \cdots = a(1)$ and

$$f(s) = a_1 \sum_{n=1}^{\infty} n^{-s} = a_1 \zeta(s).$$

From the functional equation (4.33) we get $g(1-s) = a_1 \zeta(1-s)$. This completes the proof of Hamburger's theorem. ∎

4.6 Exercises

1. Prove that $(1-x)^{24} \equiv (1-x^8)(1-x^{16}) \pmod 4$ and hence obtain that $\tau(n) \equiv a(n) \pmod 4$, where τ is the Ramanujan arithmetical function defined by

$$q \prod_{n=1}^{\infty} (1-q^n)^{24} = \sum_{n=1}^{\infty} \tau(n) q^n.$$

2. If p is a prime $\equiv 1 \pmod 8$, show that

$$\left(\frac{1+i}{p} \right) = \left(\frac{1+d}{p} \right),$$

where i is a solution of $x^2 + 1 \equiv 0 \pmod p$ and d is a solution of $x^2 - 2 \equiv 0 \pmod p$, and hence show that for these primes the polynomial $x^4 - 2x^2 - 1$ has the same type of splitting as the polynomial $x^4 - 2x^2 + 2$.

3. Use Gauss' criteria for the biquadratic character of 2 to find the complete splitting of $x^4 - 2$ over the finite field $\mathbb{F}_p$ for all primes p.

4. Use the definition

$$q \prod_{n=1}^{\infty} (1-q^n)^{24} = \sum_{n=1}^{\infty} \tau(n) q^n$$

to show that the coefficients satisfy $|\tau(n)| \ll n^c$ with some positive constant c.

5. Prove that if $a(n)$ is defined by

$$\phi(z) = q \prod_{n=1}^{\infty} \left(1 - q^{8n}\right) \left(1 - q^{16n}\right) = \sum_{n=1}^{\infty} a(n)q^n,$$

then

$$\sum_{v=0}^{\infty} a(p^v)T^v = \frac{1}{1 - a(p)T + \left(\frac{-8}{p}\right)T^2}.$$

6. Show that $\phi(z) = \eta(8z)\eta(16z)$ and use the functional equation for $\eta(z)$ given in Lemma 4.2 to show that

$$\phi\left(\frac{-1}{128z}\right) = \frac{\sqrt{128}}{i} z\phi(z).$$

7. Use the Mellin transform to show that the Euler product

$$L(s, \phi) = G_{\mathbb{R}}(s)G_{\mathbb{R}}(s+1) \prod_{p \, \text{prime}} \left(1 - a(p)p^{-s} + \left(\frac{-8}{p}\right)p^{-2s}\right)^{-1},$$

where $G_{\mathbb{R}}(s) = \pi^{-s/2}\Gamma(s/2)$, has an analytic continuation to the whole s-plane and satisfies the functional equation

$$L(s, \phi) = f(\phi)^{1/2-s}L(1-s, \phi),$$

where $f(\phi) = 128$.

8. Let

$$q \prod_{n=1}^{\infty} \left(1 - q^n\right) \left(1 - q^{23n}\right) - \sum_{n=1}^{\infty} b(n)q^n.$$

Show that $b(n)$ is multiplicative. (Hint: use $\tau(n)$ to show that

$$b(p^\lambda n) \equiv b(p)b(p^{\lambda-1}n) - \left(\frac{-23}{p}\right)b(p^{\lambda-2}n), \quad \lambda \geq 2, \quad \gcd(p, n) = 1.)$$

9. Show

$$q \prod_{n=1}^{\infty} \left(1 - q^n\right) \left(1 - q^{23n}\right) = \frac{1}{2}\left\{ \sum_{m,n\in\mathbb{Z}} q^{m^2+mn+6n^2} - \sum_{m,n\in\mathbb{Z}} q^{2m^2+mn+3n^2} \right\}.$$

10. Prove that if $f(x)$ is a monic polynomial of degree 3 with integer coefficients, and if α, β, and γ are its roots over the complex numbers, then its discriminant

$$D_f = (\alpha - \beta)^2(\beta - \gamma)^2(\gamma - \alpha)^2$$

is a rational number, which is an integer only when the three roots α, β, and γ are rational numbers.

11. Prove that if the polynomial $f(x)$ of degree 3 with integer coefficients splits into 3 linear factors in the finite field $\mathbb{F}_p$, then the discriminant D_f is a quadratic residue modulo p.

12. If $f(x)$ is a cubic polynomial with integer coefficients and is irreducible when considered as a polynomial in $\mathbb{F}_p[x]$, then the extension obtained by adjoining the roots of $f(x)$ to $\mathbb{F}_p$, that is

$$K = \mathbb{F}_p[x]/(f(x)),$$

where $(f(x))$ is the ideal generated in $\mathbb{F}_p[x]$ by all the multiples of $f(x)$, is a cyclic extension of degree 3 of the finite field $\mathbb{F}_p$. Deduce from this that the polynomial $x^2 - D_f$ in $\mathbb{F}_p[x]$ is reducible, that is, D_f is a quadratic residue modulo p.

13. Prove that if $x^2 - D_f$ is irreducible over $\mathbb{F}_p$, then necessarily $f(x)$ splits as a product of a linear factor and an irreducible quadratic factor.

Chapter 5

Hecke's Theory of Modular Forms

5.1 Introduction

In this chapter we present an outline of the elementary theory of modular forms. From the beginning we emphasize those aspects of the theory which are closely connected with combinatorial number theory: additive vs. multiplicative problems. This outline is intended to serve the student as a guide to the more advanced books on the subject, such as Shimura's text [101]. Most of the material presented is quite standard and can be derived with only a good knowledge of elementary number theory and the rudiments of the theory of functions of one complex variable. The exceptions are Sections 5.10–5.13, where we have included material that is not readily available and can only be found scattered in the literature. In Section 5.10 we compute the Fourier coefficients of Poincaré series in terms of Bessel functions and Kloosterman sums. In Section 5.11 we use the results of Section 5.10 and the well-known Weil estimate for Kloosterman sums to obtain an upper bound for the Ramanujan τ-function by a method that is applicable to the Fourier coefficients of modular forms of nonintegral weight, a case not covered by Deligne's bound. The Eichler-Selberg trace formula, an exact formula for the trace of Hecke operators, is described in Section 5.12. Finally, Section 5.13 is a brief description of the connection between ℓ-adic Galois representations attached to eigenforms of Hecke operators and the Ramanujan conjecture.

In the following, $\mathbb{C}$, $\mathbb{R}$, $\mathbb{Q}$ and $\mathbb{Z}$ denote respectively the field of complex numbers, the field of real numbers, the field of rational numbers and the ring of rational integers. Recall that $\Re(z)$ and $\Im(z)$ denote the real and imaginary parts of the complex number z, respectively. When R is a commutative ring with unity, we let $SL(2, R)$ denote the set of 2×2 matrices with coefficients in R and determinant 1. Two of these matrices are the identity matrix 1_2 (or simply 1) and its negative -1_2 (or simply -1). Thus $SL(2, \mathbb{R})/\{1, -1\}$ or $SL(2, \mathbb{R})/\{\pm 1\}$ denotes the set of 2×2 matrices with real coefficients and determinant 1 modulo the subgroup $\{1_2, -1_2\}$.

The classical theory of modular forms deals mainly with functions defined on the Poincaré upper half plane

$$\mathcal{H} = \{z \in \mathbb{C} : \Im(z) > 0\}.$$

5.2 *Modular Group* Γ *and Its Subgroup* $\Gamma_0(N)$

The group of all holomorphic automorphisms of $\mathcal{H}$ is the set of all linear fractional transformations

$$z \longmapsto \sigma(z) = \frac{az + b}{cz + d}, \qquad z \in \mathcal{H},$$

where a, b, c and d are real numbers and $ad - bc = 1$. (See Section 4.3.) This group is isomorphic to $SL(2, \mathbb{R})/\{1, -1\}$. Modular forms will be defined mainly with respect to the modular group, which consists of all linear fractional transformations of $\mathcal{H}$ of the form

$$z \longmapsto \sigma(z) = \frac{az + b}{cz + d},$$

where a, b, c and d are integers and $ad - bc = 1$. Henceforth, we shall denote this group by Γ and will identify it with the quotient group $SL(2, \mathbb{Z})/\{1, -1\}$. Often, when there is no danger of confusion, we shall identify an element α in Γ with one of its matrix representatives. Γ is a discrete subgroup of $SL(2, \mathbb{R})/\{1, -1\}$ and acts discontinuously on $\mathcal{H}$. As a group, Γ is generated by the linear transformations

$$T : z \longmapsto z + 1, \qquad S : z \longmapsto -\frac{1}{z},$$

whose corresponding matrices are

$$\begin{pmatrix} 1 & 1 \\ 0 & 1 \end{pmatrix}, \qquad \begin{pmatrix} 0 & -1 \\ 1 & 0 \end{pmatrix}.$$

Clearly, we have $S^2 = 1$ and $(ST)^3 = 1$. In fact, a simple topological argument will show that Γ has a presentation of the form $< S, T; S^2, (ST)^3 >$, that is, it is the free product of the cyclic group of order 2 generated by S and the cyclic group generated by ST.

The quotient of the upper half plane $\mathcal{H}$ by the modular group Γ, suitably compactified by adding the "cusp" at infinity, is a Riemann surface of genus zero which via the absolute modular invariant can be thought of as an unramified cover of the projective line $\mathbb{P}^1(\mathbb{C})$. Recall that the absolute invariant $j(z)$, which defines the mapping

$$j : \mathcal{H}/\Gamma \cup \mathbb{Q}/\Gamma \longrightarrow \mathbb{P}^1(\mathbb{C}),$$

is the quotient of $(E_4(z))^3$, the third power of the Eisenstein series $E_4(z)$ of weight 4 (see Section 5.4), by the Ramanujan modular form $\Delta(z)$, normalized so that the residue at ∞ is 1. From the point of view of algebraic geometry, more interesting objects arise when we consider arithmetically defined subgroups of Γ that are also of finite index in Γ. One such class of groups is the Hecke congruence subgroups, denoted by $\Gamma_0(N)$, for each positive integer N. The elements of $\Gamma_0(N)$ are defined by

$$z \longmapsto \sigma(z) = \frac{az + b}{cz + d}, \quad \sigma = \begin{pmatrix} a & b \\ c & d \end{pmatrix} \in \Gamma, \quad c \equiv 0 \pmod{N}.$$

The index of $\Gamma_0(N)$ in Γ is

$$\mu = N \prod_{p|N} \left(1 + \frac{1}{p}\right).$$

Unlike the principal congruence subgroups of Γ, $\Gamma_0(N)$ is not a normal subgroup. The group structure of $\Gamma_0(N)$ was studied by Rademacher in the early 1930s. He obtained a presentation of $\Gamma_0(N)$. His explicit determination of a complete set of generators for $\Gamma_0(N)$ is of interest in connection with Weil's extension of Hecke's theorem to modular forms on $\Gamma_0(N)$ (see Weil [118].)

The quotient of the upper half plane $\mathcal{H}$ by $\Gamma_0(N)$ can be compactified by suitably adding the cusps of $\Gamma_0(N)$ to $\mathcal{H}$. When $N = p$ is prime, the cusps are the points $\{\infty, 0\}$. For general N, the resulting Riemann surface $X_0(N)$ has genus

$$g = 1 + \frac{\mu}{12} - \frac{\nu_2}{4} - \frac{\nu_3}{3} - \frac{\nu_\infty}{2},$$

where

$$\nu_2 = \begin{cases} 0 & \text{if } N \equiv 0 \pmod 4 \\ \prod_{p|N} \left(1 + \left(\frac{-1}{p}\right)\right) & \text{otherwise,} \end{cases}$$

$$\nu_3 = \begin{cases} 0 & \text{if } N \equiv 0 \pmod 9 \\ \prod_{p|N} \left(1 + \left(\frac{-3}{p}\right)\right) & \text{otherwise, and} \end{cases}$$

$$\nu_\infty = \sum_{d|N} \phi(\gcd(d, N/d)),$$

where ϕ is Euler's function. In ν_2 and ν_3 the quadratic symbols take the values

$$\left(\frac{-1}{p}\right) = \begin{cases} 0 & \text{if } p = 2 \\ 1 & \text{if } p \equiv 1 \pmod 4 \\ -1 & \text{if } p \equiv 3 \pmod 4 \end{cases}$$

$$\left(\frac{-3}{p}\right) = \begin{cases} 0 & \text{if } p = 3 \\ 1 & \text{if } p \equiv 1 \pmod 3 \\ -1 & \text{if } p \equiv 2 \pmod 3 \end{cases}$$

The details of the above computation of g can be found on page 25 of Shimura [101].

It can be shown that the Riemann surface $X_0(N)$ admits a model defined by equations with coefficients in the rational number field. This result is rather deep and depends on a study of the arithmetical properties of the Fourier coefficients of the holomorphic Eisenstein series for the group $\Gamma_0(N)$.

5.3 Fundamental Domains for Γ and $\Gamma_0(N)$

Let us recall the description of the fundamental domain for Γ given in Section 4.3. Using the translation T in Γ we can bring any complex number z in $\mathcal{H}$ to lie in the strip $|\Re(z)| \leq \frac{1}{2}$. From the relation $\Im\sigma(z) = |cz + d|^{-2}\Im(z)$, which holds for any transformation σ in Γ, we can show that any complex $z \in \mathcal{H}$ is equivalent to another complex number z' with the properties that $|z'| \geq 1$ and $|\Re(z')| \leq \frac{1}{2}$. Thus we can take as fundamental domain for Γ the set

$$D = \{z \in \mathcal{H}\,;\, |\Re(z)| \leq \frac{1}{2},\ |z| \geq 1\}.$$

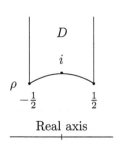

Fundamental Domain D for Γ.

The elliptic points of Γ are $i = (-1)^{1/2}$ and $\rho = (-1 + (-3)^{1/2})/2$. Strictly speaking, we should only take half of the boundary of the fundamental domain, but we avoid this here so as not to complicate the notation.

A fundamental domain $D(N)$ for $\Gamma_0(N)$ is easily obtained from that of Γ by taking a left coset decomposition of Γ modulo $\Gamma_0(N)$, say

$$\Gamma = \sigma_1\Gamma_0(N) \cup \cdots \cup \sigma_\mu\Gamma_0(N),$$

and putting

$$D(N) = \sigma_1 D \cup \cdots \cup \sigma_\mu D,$$

where care must be taken when identifying points on the boundary.

Example 5.1

The Riemann surface $X_0(11)$ has genus 1 and is the first example of an $X_0(N)$ with positive genus. The unique differential Ω of the first kind on $X_0(11)$ is

$$\Omega = \prod_{n=1}^{\infty} (1 - q^n)^2 \left(1 - q^{11n}\right)^2 dq = \sum_{n=1}^{\infty} a_n q^{n-1} dq$$

(see Shimura [101]). Affine models for $X_0(11)$ had been known for some time; the following simple one was found by Tate:

$$y^2 - y = x^3 - x^2.$$

A differential form defined on $\mathcal{H}$ and invariant under the action of Γ is given by

$$w = y^{-2} dx\, dy.$$

The w-volume of the fundamental domain D for Γ is $\pi/3$ (see Shimura [101], p. 44).

5.4 *Integral Modular Forms*

A function $f(z)$ defined in the upper half plane $\mathcal{H}$ is called an *integral modular form of weight $k \in \mathbb{Z}$ associated with the modular group* Γ if

(A) it is holomorphic in the upper half plane,

(B) it satisfies the functional identity

$$f\left(\frac{az + b}{cz + d}\right) = (cz + d)^k f(z), \quad \begin{pmatrix} a & b \\ c & d \end{pmatrix} \in \Gamma, \quad \text{and}$$

(C) it is bounded in the fundamental domain D.

The space of integral modular forms of weight k on Γ will henceforth be denoted by $M_k(\Gamma)$. Property (B) above implies that any $f(z) \in M_k(\Gamma)$ is invariant under the transformation $z \mapsto z + 1$ and therefore has a Fourier expansion of the form

$$f(z) = \sum_{n=0}^{\infty} a_n q^n, \quad q = \exp(2\pi i z).$$

If $a_0 = 0$, we call $f(z)$ a *cusp form*. In the following the subspace of cusp forms will be denoted by $M_k^0(\Gamma)$.

The Eisenstein series of weight $k > 2$ are defined by the infinite series

$$G_k(z) = \sum_{m,n}^{*} (mz + n)^{-k},$$

where the $*$ indicates that the summation is over all integers m and n not both zero. The Eisenstein series of even weight are the first nontrivial examples of integral modular forms for Γ. The functions of main interest from an arithmetical point of view are the seminormalized Eisenstein series with Fourier expansions

$$E_k(z) = \frac{1}{2}\zeta(1-k) + \sum_{n=1}^{\infty} \sigma_{k-1}(n)q^n, \quad q = \exp(2\pi iz),$$

where $\sigma_{k-1}(n)$ is the arithmetic function

$$\sigma_{k-1}(n) = \sum_{d|n} d^{k-1},$$

and $\zeta(1-k)$ is the value of the Riemann zeta function at the point $1-k$. Recall that $\zeta(1-k) \neq 0$ for k even and hence $E_k(z)$ is not a cusp form. For odd $k \geq 3$, $G_k(z)$ is trivially equal to zero. At present it is of considerable interest to understand the nature of the series

$$E_k^\dagger(z) = -\frac{1}{2}\zeta'(1-k) + \sum_{n=1}^{\infty} \left(n^{k-1} \log n\right) \frac{q^n}{1-q^n}$$

for k odd. The general problem of understanding the arithmetic nature of the coefficients in the Taylor expansion of the Riemann zeta function $\zeta(s)$ about the negative odd integers remains one of the most interesting and difficult questions concerning special values of zeta and L-functions, with important implications for the orders of the higher K-groups of the rings of integers of number fields.

If we consider the evaluation map δ_∞ whose value at a function $f(z) \in M_k(\Gamma)$ is $\delta_\infty(f) = f(\infty)$, we get the exact sequence

$$0 \longrightarrow M_k^0(\Gamma) \longrightarrow M_k(\Gamma) \xrightarrow{\delta_\infty} \mathbb{C} \longrightarrow 0,$$

from which there results the fundamental splitting of $M_k(\Gamma)$ as multiples of the Eisenstein series $E_k(z)$ plus the space of cusp forms:

$$M_k(\Gamma) = \mathbb{C}E_k \oplus M_k^0.$$

A weak form of the Riemann-Roch theorem, itself a simple generalization of the argument given in Lemma 4.3, for the Riemann sphere $\mathbb{P}^1(\mathbb{C})$ shows that $\dim_\mathbb{C} M_k(\Gamma)$ is finite. More precisely, if we take an $f(z) \in M_k(\Gamma)$ and put $n_t =$ the order of vanishing of $f(z)$ at t for $t = \infty$, i and ρ, and $n =$ the number of zeros of $f(z)$ properly counted with multiplicities in a fundamental domain D not including the points $\{i, \rho, \infty\}$, then by integrating $d\log f(z)$ on a suitable boundary of D we obtain the following exponential product formula:

$$\frac{k}{12} = n + n_\infty + \frac{n_i}{2} + \frac{n_\rho}{3}.$$

This diophantine equation has no solutions for $k < 0$ or k odd. It has a trivial solution for $k = 0$ and no solution for $k = 2$. For $k = 4, 6, 8$ and 10 it has only one solution, as shown in this table:

k	n	n_∞	n_i	n_ρ
4	0	0	0	1
6	0	0	1	0
8	0	0	0	2
10	0	0	1	1

From the existence of Eisenstein series for even $k \geq 4$ and the vanishing of $E_4(z)$ at $z = \rho$ and $E_6(z)$ at $z = i$ we get

$$\dim_{\mathbb{C}} M_k(\Gamma) = 1 \quad \text{for} \quad k = 4, 6, 8, 10.$$

For $k = 12$ the diophantine equation has four solutions corresponding to: the Eisenstein series $E_{12}(z)$, which is nonzero at $i\infty$, i and ρ;

$$\Delta(z) = q \prod_{n=1}^{\infty} (1 - q^n)^{24} = \sum_{n=1}^{\infty} \tau(n) q^n, \quad \text{where} \quad q = \exp(2\pi i z),$$

which vanishes at the cusp $i\infty$ and which is nonzero otherwise in the upper half plane; $(E_6(z))^2$, with a double zero at $z = i$; and $(E_4(z))^3$, with a triple zero at $z = \rho$. These four functions span a two-dimensional vector space, a fact which gives rise to several identities among divisor functions and coefficients of cusp forms. Examples are given below.

The fact that $\Delta(z)$ vanishes only at infinity implies that for $k > 12$, multiplication by $\Delta(z)$ gives an isomorphism

$$M_{k-12}(\Gamma) \xrightarrow{\sim} M_k^0(\Gamma)$$

between the space of integral modular forms of weight $k - 12$ and the space of cusp forms of weight k. A simple induction argument using the above isomorphism and the dimension of $M_k(\Gamma)$ for $k = 2, 4, 6, 8, 10$ and 12 gives

$$\dim_{\mathbb{C}} M_k(\Gamma) = \begin{cases} \lfloor \frac{k}{12} \rfloor + 1 & \text{if } k \not\equiv 2 \ (\mathrm{mod}\ 12) \\ \lfloor \frac{k}{12} \rfloor & \text{if } k \equiv 2 \ (\mathrm{mod}\ 12). \end{cases}$$

The dimension of the space of cusp forms is also given by

$$\dim_{\mathbb{C}} M_k^0(\Gamma) = \begin{cases} \lfloor \frac{k}{12} \rfloor & \text{if } k \not\equiv 2 \ (\mathrm{mod}\ 12) \\ \lfloor \frac{k}{12} \rfloor - 1 & \text{if } k \equiv 2 \ (\mathrm{mod}\ 12). \end{cases}$$

Here are some examples.

Example 5.2

$$M_{14}(\Gamma) = \mathbb{C}E_{14},$$
$$M_k(\Gamma) = \mathbb{C}E_k \oplus \mathbb{C}\Delta E_{k-12}, \quad k = 16, 18, 20, 22,$$
$$M_{24}(\Gamma) = \mathbb{C}E_{24} \oplus \mathbb{C}\Delta E_{12} \oplus \mathbb{C}\Delta^2.$$

The graded ring of integral modular forms for Γ is generated by the forms $E_4(z)$ and $E_6(z)$:

$$M_\Gamma = \mathbb{Q}\,[E_4(z), E_6(z)] = \bigoplus_{n=0}^{\infty} M_n(\Gamma),$$

where the grading is by weight and

$$E_k(z) = \frac{\zeta(1-k)}{2} + \sum_{n=1}^{\infty} \sigma_{k-1}(n)q^n, \qquad \text{where} \qquad q = \exp(2\pi i z),$$

is the Eisenstein series of weight k.

Exercise 1 at the end of this chapter shows that the Poincaré series of M_Γ is

$$P_\Gamma(t) = \sum_{n=0}^{\infty} (\dim_\mathbb{Q} M_n(\Gamma))\, t^n = \frac{1}{(1-t^4)(1-t^6)}.$$

A simple computation shows $\dim M_n(\Gamma) = 1$ for $n = 0, 4, 6, 8, 10,$ and $\dim M_{12}(\Gamma) = 2$. In fact, $M_{12}(\Gamma) = \{E_{12}(z), \Delta(z)\}$, where

$$\Delta(z) = q \prod_{n=1}^{\infty} (1 - q^n)^{24}$$

is Ramanujan's modular form. These facts may be used to obtain identities between the Fourier coefficients of modular forms. For example, the modular forms $E_4(z)^3$, $E_6(z)^2$ and $\Delta(z)$ are linearly dependent over the field of rational numbers. Making explicit the coefficients in such a linear relation gives the well-known identity

$$\tau(n) = \frac{65}{756}\sigma_{11}(n) + \frac{691}{756}\sigma_5(n) - \frac{691}{3}\sum_{m=1}^{n}\sigma_5(m)\sigma_5(n-m).$$

5.5   Modular Forms of Type $M_k(\Gamma_0(N), \chi)$ and Euler-Poincaré Series

Let χ be a Dirichlet character defined modulo N, that is, a homomorphism from the multiplicative group $(\mathbb{Z}/N\mathbb{Z})^*$ to the group of roots of 1, and extend its domain to the ring of integers in the usual way and then to $\Gamma_0(N)$ by putting

$$\chi(\gamma) = \chi(d) \quad \text{if} \quad \gamma = \begin{pmatrix} a & b \\ c & d \end{pmatrix}.$$

DEFINITION 5.1   *A function $f(z)$ defined in the upper half plane $\mathcal{H}$ is a modular form of nebentype (k, χ) for $k \in \mathbb{Z}$ if*

1. *$f(z)$ is holomorphic on $\mathcal{H}$,*

2. *$f\left(\dfrac{az+b}{cz+d}\right) = \chi(d)(cz+d)^k f(z)$, for $\gamma = \begin{pmatrix} a & b \\ c & d \end{pmatrix}$ in $\Gamma_0(N)$, and*

3. *$f(z)$ is holomorphic at all the cusps of $\Gamma_0(N)$.*

The third condition means that at every cusp of $\Gamma_0(N)$ there is a uniformizing parameter q in terms of which the expansion of $f(z)$ has the form

$$f = \sum_{n=0}^{\infty} a_n q^n.$$

The space of modular forms of nebentype (k, χ) will be denoted by $M_k(\Gamma_0(N), \chi)$. When χ is trivial we get a the space of functions called by Hecke of "principal type." The cusp forms in $M_k(\Gamma_0(N), \chi)$ are defined again as the kernel of the evaluation map

$$0 \longrightarrow M_k^0(\Gamma_0(N), \chi) \longrightarrow M_k(\Gamma_0(N), \chi) \longrightarrow \mathbb{C}^\sigma \longrightarrow 0,$$

where the third arrow consists of evaluating an $f \in M_k(\Gamma_0(N), \chi)$ at the σ cusps of $\Gamma_0(N)$. The dimension of the space $M_k(\Gamma_0(N), \chi)$ was worked out in special cases by Hecke. Cusp forms of weight $k = 2$ and trivial χ correspond to differential forms of the first kind on $X_0(N)$. In this case the dimension is equal to the genus of $X_0(N)$ which we stated in Section 5.2.

REMARK 5.1   We would like to mention briefly the difficult problem of computing explicitly the Poincaré polynomial for $\Gamma_0(N)$ with $\chi(-1) = -1$:

$$P_\chi(t, \Gamma_0(N)) = \sum_{k=0}^{\infty} (\dim_{\mathbb{C}} M_k(\Gamma_0(N), \chi))\, t^k.$$

The Riemann-Roch theorem for $X_0(N)$ implies a functional equation connecting $P_{\overline{\chi}}\left(\frac{1}{t}, \Gamma_0(N)\right)$ and $P_\chi(t, \Gamma_0(N))$, and it is hoped that a closer look at this relation may give some information about the troublesome term $M_1(\Gamma_0(N), \chi)$. This term is in fact related to the dimension of the space of cusp forms of weight 1 and nebentype χ which are the Mellin transforms of Artin L-functions of two-dimensional representations of Galois groups. The latter representations are classified into two categories, the first corresponding to induced representations and the second to noninduced ones. It is modular forms of the second category, the "noninduced" ones, that are difficult to count. ∎

We now give examples of Euler-Poincaré series in some simple cases. Let Γ be a group of linear fractional transformations of the upper half plane $\mathcal{H}$. Let g be the genus of $\mathcal{H}/\Gamma$, m be the number of inequivalent cusps of Γ, and $e_1, \ldots, e_r$ be the orders of the inequivalent elliptic elements of Γ. Let $\delta_{i,j}$ denote the Kronecker δ-function. We have the following result.

THEOREM 5.1 Formula for $P(t, \Gamma)$

If $-1_2 \in \Gamma$, then

$$P(t, \Gamma) = \sum_{k=0}^{\infty} \left(\dim_{\mathbb{C}} M_k^0(\Gamma) \right) t^k =$$

$$\delta_{0,m} + gt^2 + (2g - 2 + m)(2t^4 - t^6)(1 - t^2)^{-2} + (1 - g - m)t^4(1 - t^2)^{-1}$$

$$+ \sum_{i=1}^{r} \left(1 - t^{2e_i} \right)^{-1} (1 - t^2)^{-1} \sum_{k=1}^{e_i} \left(\left\lfloor \frac{k(e_i - 1)}{e_i} \right\rfloor - \left\lfloor \frac{(k - 1)(e_i - 1)}{e_i} \right\rfloor \right) t^{2k}.$$

The proof of this theorem is based on Shimura's formula for the dimension of $M_k^0(\Gamma)$, as in [101], p. 46, and Exercise 2 at the end of this chapter.

5.6   Hecke Operators

The quotient $\mathcal{H}/\Gamma$ uniformizes the analytic space of isomorphism classes of exact sequences

$$0 \longrightarrow L \longrightarrow \mathbb{C} \longrightarrow E \longrightarrow 0,$$

where L is a lattice and E is the associated elliptic curve. For a given positive integer n we define a modular correspondence by associating with the above exact sequence all nonisomorphic sequences of the form

$$0 \longrightarrow L^* \longrightarrow \mathbb{C} \longrightarrow E^* \longrightarrow 0,$$

where L^* runs over all sublattices of L of index n in L. The trace of this modular correspondence gives rise to a linear operator whose restriction to the space M_k of integral modular forms of weight k on $SL(2, \mathbb{Z})/\{\pm 1_2\}$ will be denoted by $T_k(n)$ and called the *Hecke operator*. An explicit construction of the above modular correspondences leads, up to a normalization constant, to the Hecke operator acting on modular forms $f(z)$ in M_k by the formula

$$T_k(n)f(z) = n^{k-1} \sum_{\substack{ad=n \\ 0 \le b < d}} d^{-k} f \left(\frac{az + b}{d} \right).$$

When $n = p$, a prime, we get the simpler formula

$$T_k(p)f(z) = p^{k-1}f(pz) + \frac{1}{p}\sum_{b=0}^{p-1} f\left(\frac{z+b}{p}\right).$$

The fundamental properties of the Hecke operators are:

H_1) The space M_k of integral modular forms of weight k is stable under the action of the Hecke operators $T_k(n)$ for all positive integers n,

H_2) $T_k(m)T_k(n) = \sum_{d|\gcd(m,n)} d^{k-1}T_k(mn/d^2)$,

H_2') $T_k(m)T_k(n) = T_k(mn)$ when m and n are relatively prime, and

H_3) For a prime p and a positive integer r we have

$$T_k(p)T_k(p^r) = T_k(p^{r+1}) + p^{k-1}T_k(p^{r-1}).$$

We proved these results in Section 4.3. The basic result in the classical theory of modular forms on $\Gamma = SL(2,\mathbb{Z})/\{\pm 1_2\}$ is the following theorem:

THEOREM 5.2 Hecke-Petersson Theorem

The operators $\{T_k(n) \, : \, n \in \mathbb{Z}^+\}$ generate a semisimple abelian subalgebra of $\mathrm{End}_{\mathbb{C}}(M_k)$.

Property H_1 can be stated in a sharper form by observing that the subspace of Eisenstein series in M_k is stable under the action of $T_k(n)$ for all n. We therefore reformulate H_1 as

H_1') The Hecke operators $\{T_k(n) \, : \, n \in \mathbb{Z}^+\}$ respect the splitting

$$M_k = \mathbb{C}E_k \oplus M_k^0.$$

The action of $T_k(n)$ on a modular form f whose expansion about the cusp at infinity is

$$f(z) = \sum_{m=0}^{\infty} a(m)q^m, \qquad q = \exp(2\pi i z),$$

is given by

$$T_k(n)f(z) = \sum_{m=0}^{\infty} q^n \sum_{d|\gcd(m,n)} d^{k-1}a\left(\frac{mn}{d^2}\right).$$

The Hecke polynomials are defined by

$$H_k^{(p)}(u) = \det\left(I - T_k(p)u + p^{k-1}Iu^2\right),$$

where I is the identity operator on M_k. The global Hecke operator is defined by the formal Dirichlet series

$$D(s) = \prod_p \left(I - T_k(p)p^{-s} + p^{k-1-2s}I\right)^{-1} = \sum_{n=1}^{\infty} n^{-s}T_k(n).$$

5.7 *Dirichlet Series and Their Functional Equation*

With every modular form $f(z)$ in M_k whose Fourier expansion is

$$f(z) = a_0 + \sum_{n=1}^{\infty} a_n q^n, \qquad q = \exp(2\pi i z),$$

we associate the Dirichlet series

$$\phi_f(s) = \sum_{n=1}^{\infty} a_n n^{-s}, \qquad \Re(s) > \frac{k+1}{2}.$$

Hamburger [46] proved in 1921 that the Riemann zeta function is characterized by its functional equation. One of Hecke's main contributions to the theory of modular forms was his generalization of Hamburger's theorem.

THEOREM 5.3 Hecke's Theorem
A function $f(z)$ holomorphic in $\mathcal{H}$ is in M_k if and only if the function

$$\Lambda_f(s) = (2\pi)^{-s}\Gamma(s)\phi_f(s) = \int_0^{\infty} \{f(iy) - f(i\infty)\} y^{s-1} dy$$

satisfies the functional equation

$$\Lambda_f(s) = i^k \Lambda_f(k - s).$$

Weil [118] obtained in 1967 a generalization of Hecke's theorem to the more general situation of modular forms on $\Gamma_0(N)$, which has lead to many new developments and has thrown light on the relationship between modular forms and infinite group representation. The class of such results in the theory of automorphic forms and modular forms is usually called "converse Hecke theorems."

5.8 *The Petersson Inner Product*

The subspace of cusp forms M_k^0 on Γ has the structure of a finite dimensional Hilbert space. More precisely, we have a bilinear pairing

$$<, > : M_k^0 \times M_k^0 \longrightarrow \mathbb{C}$$

defined by the positive and nondegenerate Petersson inner product

$$<f, g> = \int_D f(z)\overline{g(z)} y^{k-2} \, dx \, dy \qquad (z = x + iy),$$

where D is a fundamental domain for the quotient space $\mathcal{H}/\Gamma$. The theorem of Hecke and Petersson, Theorem 5.2, is an immediate consequence of the identity

$$< T_k(n)f, g > \ = \ < f, T_k(n)g >,$$

which implies that the Hecke operators $T_k(n)$ are Hermitian. The principal axis theorem applied to the Hilbert space consisting of cusp forms M_k^0 metrized by the Petersson inner product $< \, , \, >$ implies that we can select a basis for M_k^0 consisting of simultaneous eigenfunctions for the Hecke operators. Henceforth we assume M_k^0 is equipped with a basis made up of all normalized eigenfunctions $\{f_j(z) \, : \, j \, = \, 1, 2, \ldots, \mu\}$, where $\mu \, = \, \dim_{\mathbb{C}} M_k^0$, whose Fourier expansions have the form

$$f_j(z) = \sum_{n=1}^{\infty} a_n(f_j)q^n, \qquad a_1(f_j) = 1.$$

When we do not want to stress the dependence of the coefficients on $f_j(z)$ we simply write a_n instead of $a_n(f_j)$. The above normalization fixes the eigenvalue of the Hecke operator $T_k(n)$, which is then given by the n-th coefficient, namely

$$T_k(n)f_j(z) = a_n(f_j)f_j(z).$$

The seminormalized Eisenstein series introduced in Section 5.4 not only has the property of being an eigenfunction of all Hecke operators but also has associated with it a Dirichlet series

$$\phi_{E_k}(s) = \sum_{n=1}^{\infty} \sigma_{k-1}(n)n^{-s}, \qquad \Re(s) > k,$$

with the remarkable property that

$$\phi_{E_k}(s) = \prod_p \left(1 - \sigma_{k-1}(p)p^{-s} + p^{k-1-2s}\right)^{-1},$$

which is also equal to the product $\zeta(s)\zeta(s+1-k)$.

The following theorem due to Hecke is the basic result that makes it possible to apply techniques from the analytic theory of numbers to problems about the coefficients of modular forms:

THEOREM 5.4 Hecke's Theorem
The Dirichlet series

$$\phi_f(s) = \sum_{n=1}^{\infty} a_n(f)n^{-s}, \qquad \Re(s) > k,$$

associated with a modular form f has an Euler product

$$\phi_f(s) = \prod_p \left(1 - a_p(f)p^{-s} + p^{k-1-2s}\right)^{-1}$$

if and only if it is a simultaneous eigenfunction of all the Hecke operators $\{T_k(n) : n = 1, 2, \ldots\}$.

The details of the proof of this important result can be found in any of the standard expositions of Hecke's work (see Shimura [101]). Special cases of the above theorem had been given by Mordell, who proved that

$$\phi_\Delta(s) = \sum_{n=1}^{\infty} \tau(n) n^{-s} = \prod_p \left(1 - \tau(p) p^{-s} + p^{11-2s}\right)^{-1},$$

thus establishing that Ramanujan's arithmetical function $\tau(n)$ is a multiplicative arithmetic function. (See Section 4.3.)

The theory of modular forms on the Hecke congruence subgroups and of given nebentype is far more complex than the theory we have outlined above. We have only recently begun to understand the rather special case of modular forms of principal type. The analogue of Hecke's theorem characterizing modular forms on $\Gamma_0(N)$ by the kind of functional equations satisfied by their associated Dirichlet series was proved by Weil in his celebrated paper [118] which we highly recommend to the serious student of modular forms. The parallel theory of Hecke operators on the space of modular forms for $\Gamma_0(N)$ was begun by Hecke and completed in a satisfactory manner by Atkin and Lehner in their paper [4], where the reader can find a detailed exposition of this topic.

5.9   The Method of Poincaré Series

In this section we give a brief presentation of Petersson's theory of Poincaré series and give as an example the explicit form for the coefficients of modular forms. Most of these results are classical and can be found in the literature and are given here for the sake of completeness; most of the results will be used in later chapters, and particularly in Chapter 6.

By restricting our considerations to the modular group $\Gamma = SL(2, \mathbb{Z})/\{\pm 1_2\}$ we will hopefully achieve a simplicity absent in more classical treatments of this subject. Nevertheless, the fundamental problems of the theory, such as the vanishing of the Poincaré series do require for their understanding a more general treatment. For example, we can give a fairly satisfactory solution to the problem of the vanishing of Poincaré series for the Hecke congruence subgroups $\Gamma_0(N)$ and weight 2 in terms of the geometry of the compactified Riemann surface $X_0(N) = \mathcal{H}/\Gamma_0(N)$, for example, when the genus of the latter is 1, as is the case with $X_0(11)$. (See Exercise 5.)

The definition of the subgroup Γ_∞ is $\left\{ \begin{pmatrix} 1 & n \\ 0 & 1 \end{pmatrix} : n \in \mathbb{Z} \right\}$. For each element L in Γ we define the function

$$J_L(z) = (cz + d)^{-2}, \quad \text{where} \quad L = \begin{pmatrix} a & b \\ c & d \end{pmatrix}.$$

The matrix identity

$$\begin{pmatrix} 1 & n \\ 0 & 1 \end{pmatrix} \begin{pmatrix} a & b \\ c & d \end{pmatrix} = \begin{pmatrix} a+nc & b+nd \\ c & d \end{pmatrix},$$

which corresponds to the linear fractional transformation

$$\frac{(a+cn)z + (b+dn)}{cz+d} = \frac{az+b}{cz+d} + n,$$

shows that the functions $J_L(z) = (cz+d)^{-2}$ and $\exp(2\pi i L(z))$ are well defined on the quotient $\Gamma_\infty \backslash \Gamma$.

DEFINITION 5.2   *The Poincaré series of weight k and index $m \in \mathbb{Z}^+$ for Γ is*

$$\phi_m(z;k) = \sum_{L \in \Gamma_\infty \backslash \Gamma} \exp(2\pi i m L(z)) J_L(z)^{k/2}.$$

We first prove the existence of such series. For this we need the following result.

LEMMA 5.1
   The series

$$\sideset{}{'}\sum_{m,n \in \mathbb{Z}} |mz+n|^{-\ell},$$

where the prime means we have omitted the term with $m = n = 0$, converges uniformly on compact subsets of $\mathcal{H}$ whenever $\ell > 2$.

PROOF   For z in a compact region R of the upper half plane we define the lattice $\Lambda = \{mz+n : m, n \in \mathbb{Z}\}$. For each $r = 0, 1, 2, \ldots$ define a partition of the complex plane by annular regions

$$D_r = \{z \in \mathbb{C} : r \le |z| < r+1\}.$$

We clearly have

$$\text{Card}\,(\Lambda \cap D_r) \le Ar,$$

where the constant A depends only on the compact set R. Therefore we have

$$\sum_{\Omega \in \Lambda,\,\Omega \ne 0} |\Omega|^{-\ell} = \sum_{r=0}^{\infty} \sum_{\Omega \in D_r \cap \Lambda,\,\Omega \ne 0} |\Omega|^{-\ell}$$

$$\le \sum_{\Omega \in D_0 \cap \Lambda,\,\Omega \ne 0} |\Omega|^{-\ell} + \sum_{r=1}^{\infty} r^{-\ell} \text{Card}(D_r \cap \Lambda)$$

$$\le A_0(R) + A \sum_{r=1}^{\infty} r^{1-\ell} < \infty$$

if $\ell > 2$. Here $A_0(R)$ is a constant that depends only on the compact set R. This proves the lemma. ∎

We take as a set of coset representatives for $\Gamma_\infty \backslash \Gamma$ the linear fractional transformations corresponding to the set of matrices

$$\begin{pmatrix} * & * \\ c & d \end{pmatrix}, \ 0 \le c < \infty, \ -\infty < d < \infty, \ \gcd(c,d) = 1, \ d = 1 \text{ if } c = 0.$$

The starred entries are occupied by just one pair of integers (a,b) corresponding to a solution of the diophantine equation $ad - bc = 1$.

A typical term in the Poincaré series corresponding to the transformation $L(z) = \frac{az+b}{cz+d}$ can be bounded by

$$\left| \exp(2\pi i m L(z))(cz+d)^{-k} \right| = \exp\left(-\frac{2\pi m \Im(z)}{|cz+d|^2} \right) |cz+d|^{-k}$$

$$\le |cz+d|^{-k}$$

for any $m \ge 0$. Using the above coset representatives for $\Gamma_\infty \backslash \Gamma$ we see that the Poincaré series $\phi_m(z;k)$ is bounded by

$$|\phi_m(z;k)| \le {\sum_{m,n\in\mathbb{Z}}}' |mz+n|^{-k}.$$

Hence Lemma 5.1 shows that $\phi_m(z;k)$ exists and defines a holomorphic function in the upper half plane. The delicate case of $k = 2$ can be handled by introducing a convergence factor $|J_L(z)|^S$ and showing that the series

$$\sum_{L\in\Gamma_\infty\backslash\Gamma} \exp(2\pi i m L(z)) J_L(z)^{k/2} |J_L(z)|^s$$

converges for $\Re(s) > 0$ and has a limit as $s \to 0$. These considerations are not relevant here since $M_2^0(\Gamma) = 0$.

Using the uniform convergence of the Poincaré series on compact subsets of $\mathcal{H}$ we can check immediately the formal identity

$$\phi_m\left(\frac{az+b}{cz+d}; k \right) = (cz+d)^k \phi_m(z;k).$$

To establish that the Poincaré series $\phi_m(z;k)$ are integral modular forms of weight $k > 2$ for the modular group Γ, it remains to show that they are holomorphic at the cusp at $i\infty$. When $m = 0$ we get that $\phi_0(z;k)$ is essentially a multiple of the Eisenstein series. Thus we consider the case $m \ge 1$. Since $\Gamma_\infty \backslash \Gamma$ contains only one translation we can write

$$\phi_m(z;k) = e^{2\pi i m z} + \sum_{L\in\Gamma_\infty\backslash\Gamma, \, c>0} \exp(2\pi i m L(z))(cz+d)^{-k},$$

where the sum runs over representatives of $\Gamma_\infty \backslash \Gamma$ with $c > 0$. Using the annular regions defined above, we have

$$\left| \sum_{L \in \Gamma_\infty \backslash \Gamma, \, c > 0} e^{2\pi i m L(z)} (cz + d)^{-k} \right| \leq \sum_{\Omega \in D_0 \cap \Lambda, \, \Omega \neq 0} |\Omega|^{-k} + \sum_{r=1}^{\infty} \sum_{\Omega \in D_r \cap \Lambda} |\Omega|^{-k}.$$

For large $\Im(z)$ and $r \geq 1$, we can check easily that

$$\mathrm{Card}(D_r \cap \Lambda) \leq \frac{9r}{\Im(z)}.$$

Hence, we have that

$$|\phi_m(z; k)| \leq e^{-2\pi m \Im(z)} + \frac{9}{\Im(z)} \sum_{r=1}^{\infty} r^{1-k}.$$

From this inequality we finally obtain that for $m \geq 1$,

$$\lim_{\Im(z) \to \infty} \phi_m(z; k) = 0.$$

When $m = 0$, we clearly have

$$\phi_0(z; k) = 1 + \sum_{L \in \Gamma_\infty \backslash \Gamma, \, c > 0} (cz + d)^{-k},$$

and thus $\lim_{\Im(z) \to \infty} \phi_0(z; k) = 1$. Hence we have proved the following result.

THEOREM 5.5 Poincaré Series is a Modular Form
The Poincaré series

$$\phi_m(z; k) = \sum_{L \in \Gamma_\infty \backslash \Gamma} \exp(2\pi i m L(z))(cz + d)^{-k}, \quad m \in \mathbb{Z}^+,$$

is an integral modular form of weight k for Γ and a cusp form when $m \geq 1$.

Let $f(z)$ be a cusp form of weight k and let

$$f(z) = \sum_{n=1}^{\infty} a_n q^n, \quad q = \exp(2\pi i z),$$

be its Fourier expansion about the cusp at infinity. In the next chapter an important role will be played by the Petersson inner product formula which we state formally as

THEOREM 5.6 Petersson Fundamental Formula
We have

$$< f, \phi_m > = \int_{\mathcal{H}/\Gamma} f(z)\overline{\phi_m(z;k)} y^{k-2} dx\, dy$$

$$= \Gamma(k-1)(4\pi m)^{1-k} a_m,$$

where a_m is the m-th coefficient in the Fourier expansion of $f(z)$ and $\phi_m(z;k)$ is the corresponding Poincaré series of index m. (Here, $\Gamma(k-1) = (k-2)!$.)

PROOF A suitable set of coset representatives of $\Gamma_\infty \backslash \Gamma$ can be selected so that the strip

$$S = \{z = x + iy \, : \, -\frac{1}{2} \le x \le \frac{1}{2}, \, 0 < y < \infty\} = \bigcup_{L \in \Gamma_\infty \backslash \Gamma} LD,$$

where D is the canonical fundamental domain for Γ. We now consider the double integral

$$I = \iint_S f(z) e^{-2\pi i m \bar{z}} y^{k-2} dy\, dx$$

and observe that this is equal to

$$\sum_{L \in \Gamma_\infty \backslash \Gamma} \iint_{LD} f(z) e^{-2\pi i m \bar{z}} y^{k-2} dy\, dx.$$

The linear transformation

$$z = L(w) = \frac{aw+b}{cw+d}, \qquad w = u + iv,$$

and its corresponding Jacobian factors

$$\Im(z) = \Im(w)|cw+d|^{-2}, \qquad dx\, dy = |cw+d|^{-4} du\, dv$$

lead to the following change of variables in the integral above.

$$I = \sum_{L \in \Gamma_\infty \backslash \Gamma} \iint_D f(L(w)) \exp(-2\pi i m \overline{L(w)}) v^{k-2} |cw+d|^{-2k} du\, dv$$

$$= \sum_{L \in \Gamma_\infty \backslash \Gamma} \iint_D f(L(w))(cw+d)^{-k} \overline{(cw+d)}^{-k} \exp(\overline{2\pi i m L(w)}) v^{k-2} dv\, du$$

$$= \iint_D f(w) \sum_{L \in \Gamma_\infty \backslash \Gamma} \overline{(cw+d)}^{-k} \exp(\overline{2\pi i m L(w)}) v^{k-2} du\, dv$$

$$= \iint_D f(w)\overline{\phi_m(w;k)} v^{k-2} du\, dv = < f, \phi_m > .$$

The interchange in the order of summation and integration is justifiable by the uniform and absolute convergence of the Poincaré series. We have also used the identity $f(L(w))(cw + d)^{-k} = f(w)$.

On the other hand we observe that the integral

$$
\begin{aligned}
I &= \int_{-\frac{1}{2}}^{\frac{1}{2}} \int_0^\infty f(z) e^{-2\pi i m \bar{z}} y^{k-2} dx\, dy \\
&= \int_{-\frac{1}{2}}^{\frac{1}{2}} \int_0^\infty \sum_{n=1}^\infty a_n e^{-2\pi i n z} e^{-2\pi i m \bar{z}} y^{k-2} dx\, dy \\
&= \sum_{n=1}^\infty a_n \int_{-\frac{1}{2}}^{\frac{1}{2}} e^{2\pi i (n-m) x} dx \int_0^\infty e^{-2\pi(n+m)y} y^{k-2} dy \\
&= a_m \int_0^\infty e^{-4\pi m y} y^{k-1-1} dy = \Gamma(k-1)(4\pi m)^{1-k} a_m.
\end{aligned}
$$

This completes the proof of Theorem 5.6. ∎

Remark. The formula given in Theorem 5.6 for the Petersson inner product $< f, \phi_m >$ is known in the literature as "the fundamental formula."

From Theorem 5.6 the following important result, also due to Petersson, is obtained.

THEOREM 5.7 The Completeness Theorem
The space of cusp forms of weight k for the modular group is spanned by the Poincaré series $\phi_m(z; k)$, $m \geq 1$.

PROOF Since $\dim_{\mathbb{C}} M_k^0(\Gamma) < \infty$, the Poincaré series $\phi_m(z; k)$, $m \geq 1$, span a finite dimensional closed subspace M of $M_k^0(\Gamma)$. Using the Petersson inner product we get the following orthogonal decomposition

$$ M_k^0(\Gamma) = M \oplus M^\perp. $$

Now, the fundamental formula shows that a cusp form f is in the orthogonal complement $M^\perp$ of M if and only if all its Fourier coefficients vanish, that is to say, if f is identically zero. Hence the theorem follows. ∎

5.10   Fourier Coefficients of Poincaré Series

The theorems we proved in Section 5.9 will now be used to derive the Fourier expansion of the Poincaré series $\phi_m(z; k)$ for $m = 1, 2, \ldots$.

Put

$$ \phi_m(z; k) = \sum_{n=1}^\infty c_{m,n} q^n, \qquad q = \exp(2\pi i z). $$

Consider the double coset decomposition $R = \Gamma_\infty \backslash \Gamma / \Gamma_\infty$. A set of distinguished coset representatives for R can be taken to be

$$R^+ = \{1_2\} \cup \left\{ \begin{pmatrix} a & b \\ c & d \end{pmatrix} : c \geq 1,\ 0 \leq d < c,\ \gcd(d,c) = 1,\ 0 \leq a < c \right\}$$
$$= \{1_2\} \cup R.$$

This is a simple consequence of the matrix identity

$$\begin{pmatrix} a & b \\ c & d \end{pmatrix} \begin{pmatrix} 1 & n \\ 0 & 1 \end{pmatrix} = \begin{pmatrix} a & b + na \\ c & d + nc \end{pmatrix}.$$

Therefore we can write

$$\Gamma_\infty \backslash \Gamma = \{1_2\} + \bigcup_{1 \in R,\, c > 0} \bigcup_{q=-\infty}^{\infty} LU^q.$$

From the definition of the Poincaré series we have

$$\phi_m(z; k) = \sum_{T \in \Gamma_\infty \backslash \Gamma} (cz + d)^{-k} \exp(2\pi i m T(z))$$

$$= e^{2\pi i m z} + \sum_{T \in \Gamma_\infty \backslash \Gamma,\, T \neq 1_2} (cz + d)^{-k} \exp(2\pi i m T(z))$$

$$= e^{2\pi i m z} + \sum_{L \in R} \sum_{q=-\infty}^{\infty} \exp(2\pi i L U^q(z)) J_{LU^q}(z)^{k/2}$$

$$= e^{2\pi i m z} + \sum_{L \in R,\, c > 0} \sum_{q=-\infty}^{\infty} (cz + cq + d)^{-k} \exp\left(2\pi i m \frac{az + aq + b}{cz + cq + d}\right),$$

where $L = \begin{pmatrix} a & b \\ c & d \end{pmatrix}$. We now use the identity

$$\frac{az + aq + b}{cz + cq + d} = \frac{a}{c} - \frac{1}{c^2(z + q + d/c)}$$

to obtain

$$\phi_m(z; k) = e^{2\pi i m z} +$$

$$\sum_{L \in R,\, c > 0} e^{2\pi i m a/c} \sum_{q=-\infty}^{\infty} (cz + cq + d)^{-k} \exp\left(-\frac{2\pi i m}{c^2(z + q + d/c)}\right).$$

If we expand the last exponential in a power series, we get

$$\phi_m(z;k) = e^{2\pi imz} +$$

$$\sum_{L\in R,\, c>0} e^{2\pi ima/c} c^{-k} \sum_{q=-\infty}^{\infty} \sum_{w=0}^{\infty} \left(\frac{-2\pi im}{c^2}\right)^w \left(z+q+\frac{d}{c}\right)^{-w-k} \cdot \frac{1}{w!}$$

$$= e^{2\pi imz} +$$

$$\sum_{L\in R,\, c>0} e^{2\pi ima/c} c^{-k} \sum_{w=0}^{\infty} \left(\frac{-2\pi im}{c^2}\right)^w \frac{1}{w!} \sum_{q=-\infty}^{\infty} \left(z+q+\frac{d}{c}\right)^{-w-k}.$$

Using the well-known Lipschitz summation formula,

$$\sum_{n=-\infty}^{\infty} (-1)^{k-1}(k-1)!(u+n)^{-k} = \sum_{n=1}^{\infty} (2\pi i n)^{k-1} e^{2\pi i n u},$$

we can write the last sum as

$$\sum_{q=-\infty}^{\infty} \left(z+q+\frac{d}{c}\right)^{-w-k} = \frac{(-2\pi i)^{w+k-1}}{(w+k-1)!} \sum_{q=1}^{\infty} q^{w+k-1} e^{2\pi i q(z+d/c)}.$$

Thus we have

$$\phi_m(z;k) = e^{2\pi imz} + \sum_{L\in R,\, c>0} e^{2\pi ima/c} c^{-k} \sum_{w=0}^{\infty} \left(\frac{-2\pi im}{c^2}\right)^w \frac{1}{w!} \times$$

$$\frac{(-2\pi i)^{w+k-1}}{(w+k-1)!} \sum_{n=1}^{\infty} n^{w+k-1} e^{2\pi i n(z+d/c)},$$

which we can also write in the simpler form

$$e^{2\pi imz} + 2\pi i^k m^{(1-k)/2} \sum_{L\in R,\, c>0} e^{2\pi ima/c} c^{-1} \times$$

$$\sum_{n=1}^{\infty} n^{(1-k)/2} e^{2\pi i n(d/c)+2\pi inz} \sum_{w=0}^{\infty} \frac{(-1)^w \left(\frac{4\pi}{e}\sqrt{nm}\right)^{2w+k-1}}{w!(w+k-1)!}$$

$$= e^{2\pi imz} + 2\pi i^k m^{(1-k)/2} \sum_{n=1}^{\infty} \sum_{L\in R,\, c>0} e^{2\pi i(ma+nd)/c} c^{-1} n^{(1-k)/2} \times$$

$$J_{k-1}\left(\frac{4\pi\sqrt{nm}}{c}\right) e^{2\pi inz},$$

where $J_{k-1}(z)$ is the Bessel function. Using the representation of R given above we can now write

$$\phi_m(z;k) = e^{2\pi imz} + \sum_{n=1}^{\infty} \left(2\pi i^k m^{(1-k)/2} n^{(k-1)/2} \sum_{c=1}^{\infty} \frac{S_c(m,n)}{c} \times \right.$$

$$\left. J_{k-1}\left(\frac{4\pi\sqrt{nm}}{c}\right)\right) e^{2\pi inz},$$

where

$$S_c(m, n) = \sum_{\substack{d=1 \\ \gcd(d,c)=1,\, ad \equiv 1 \ (\mathrm{mod}\ c)}}^{c} e^{2\pi i(ma+nd)/c}$$

is the Kloosterman sum. We can also write the last expression for $\phi_m(z; k)$ in the closed form

$$\phi_m(z; k) = \sum_{n=1}^{\infty} \left(\delta_{m,n} + 2\pi i^k \sum_{c=1}^{\infty} \frac{S_c(m, n)}{c} J_{k-1}\left(\frac{4\pi\sqrt{nm}}{c} \right) \right) \left(\frac{n}{m} \right)^{(k-1)/2} e^{2\pi i n z},$$

where $\delta_{m,n} = 1$ if $m = n$ and 0 if $m \neq n$. This completes the derivation of the Fourier expansion of Poincaré series.

REMARK 5.2   Note that the condition $ad \equiv 1 \ (\mathrm{mod}\ c)$ in the definition of $S_c(m, n)$ results from the determinant condition $ad - bc = 1$.   ∎

We now give an interesting application of the Fourier expansion of $\phi_m(z; k)$.

THEOREM 5.8 Ramanujan τ-Function
We have for the Ramanujan arithmetical function τ,

$$\tau(m)\tau(n) = \frac{(4\pi)^{11}}{10!} <\Delta, \Delta> (mn)^{11/2} \left(\delta_{m,n} + 2\pi \sum_{c=1}^{\infty} \frac{S_c(m, n)}{c} J_{11}\left(\frac{4\pi\sqrt{nm}}{c} \right) \right),$$

where $<\Delta, \Delta>$ is the Petersson inner product of $\Delta(z)$ with itself.

PROOF   Since $\dim_{\mathbb{C}} M_{12}^0(\Gamma) = 1$, we must have for any $m \geq 1$

$$\phi_m(z; k) = c(m)\Delta(z).$$

The fundamental formula now gives

$$\frac{11!}{(4\pi m)^{11}} \tau(m) = <\Delta, \phi_m> = \bar{c}(m)<\Delta, \Delta>.$$

Hence,

$$c(m) = \frac{11!\,\tau(m)}{(4\pi m)^{11}<\Delta, \Delta>},$$

or equivalently,

$$\frac{11!\,\tau(m)\Delta(z)}{(4\pi m)^{11}<\Delta, \Delta>} = \phi_m(z; k).$$

By comparing the Fourier expansions of both sides we then obtain the equality claimed in the theorem.   ∎

5.11 A Classical Bound for the Ramanujan τ-Function

Although in this book we shall use the full strength of Deligne's proof of the Ramanujan-Petersson conjecture, it may not be without interest to illustrate with the least amount of effort how the best estimates for $\tau(n)$ were gotten prior to Deligne's work. The three main ingredients in these estimates are

1. The Petersson explicit formula (see Theorem 5.8 with $m = 1$, $n = p$),

$$\tau(p) = \frac{(4\pi)^{11}}{10!} <\Delta, \Delta> p^{11/2} \left(2\pi \sum_{q=1}^{\infty} S_q(1, p) \frac{J_{11}(4\pi\sqrt{p}/q)}{q} \right).$$

2. Weil's estimate for Kloosterman sums (Weil [117], p. 207)

$$S_q(1, m) \ll q^{1/2+\varepsilon}.$$

3. Estimates for the Bessel function (Watson [116], p. 194 ff.)

$$J_{11}(x) \ll \min\left(x^{11}, \frac{1}{\sqrt{x}} \right).$$

If we now partition the infinite sum in the Petersson explicit formula in two parts, one containing the terms with $q \leq p^{1/2}$ and the other sum containing terms with $q > p^{1/2}$ and using the estimates in 2 and 3 above, we get

$$\sum_{q=1}^{\infty} S_q(1, p) J_{11}(4\pi\sqrt{p}/q)/q \ll p^{1/4+\varepsilon}.$$

We thus obtain

THEOREM 5.9 Size of Ramanujan τ Function
For all primes p,
$$\tau(p) \ll p^{11/2+1/4+\varepsilon}.$$

Clearly, the same sort of estimation can be carried out for the coefficients of other cusp forms.

5.12 The Eichler-Selberg Trace Formula

The purpose of this section is to describe the explicit computation of the trace of the Hecke operators obtained by Selberg [96], p. 80 and Eichler [28], p. 134.

We shall give the formula only for the case $\Gamma = SL(2, \mathbb{Z})/\{\pm 1_2\}$. It reads as follows

$$\operatorname{tr} T_k(n) = -\sum_{(\rho, \bar{\rho})} \sum_A \frac{h_A}{w_A} F^{(k-2)}(\rho, \bar{\rho}) - \sigma'_{k-1}(n)$$

$$+ \delta(\sqrt{n}) \left(\frac{k-1}{12} n^{k/2-1} - n^{(k-1)/2} \right),$$

where $F^{(k-2)}(\rho, \bar{\rho}) = \left(\rho^{k-1} - \bar{\rho}^{k-1} \right) / (\rho - \bar{\rho})$, $(\rho, \bar{\rho})$ runs over all pairs of mutually conjugate irrational quadratic integers with norm n, A runs over all orders of imaginary quadratic fields, such that $\rho \in A$, h_A is the class number of A, w_A is the number of roots of unity contained in A, $\sigma'_{k-1}(n) = \sum_{d|n, d<\sqrt{n}} d^{k-1}$, and $\delta(\sqrt{n}) = 1$ or 0 according as $\sqrt{n}$ is an integer or not. When $k = 2$ we must add $\sigma(d) = \sum_{d|n} d$ to the right hand side of the above formula. Since $\dim_{\mathbb{C}} M_k^0(\Gamma) = 0$ for $k = 2, 4, 6, 8, 10$, we have $\operatorname{tr} T_k(n) = 0$ in these cases. The first nontrivial value of the trace of the Hecke operator occurs for $k = 12$, where we obtain this simple expression for the value of $\tau(p)$ when p is prime,

$$\tau(p) = -\sum_{|m|<\sqrt{4p}} \frac{h(4p - m^2)}{w_\Delta} M_{p,m} - 1 + p,$$

where the class number $h(4p - m^2)$ of the imaginary quadratic field $\mathbb{Q}\left(\sqrt{m^2 - 4p}\right)$ is to be weighted by the appropriate number of roots of one, that is,

$$w_\Delta = \begin{cases} 6 & \text{if } \Delta = -3, \\ 4 & \text{if } \Delta = -4, \\ 2 & \text{if } \Delta < -4, \text{ and} \end{cases}$$

$$M_{p,m} = m^{10} - 18m^8 p + 35m^6 p^2 - 50m^4 p^3 + 25m^2 p^4 - 2p^5.$$

A simple calculation also gives the congruence

$$\tau(p) + 1 \equiv -\sum_\Delta \frac{h(\Delta)}{w_\Delta} \Delta^5 \pmod{p},$$

where the summation is over all negative discriminants of the form $\Delta = m^2 - 4p$.

5.13 ℓ-Adic Representations and the Ramanujan Conjecture

The results of this section are due mainly to Deligne and Serre (Deligne [22]). Prior to the work of Deligne, Ihara [55] had worked out a scheme which in principle reduced the Ramanujan conjecture about the correct bound for $\tau(n)$

to the Weil conjectures concerning the zeros of the zeta functions of varieties. In Ihara's approach there remained three main problems: the first had to do with the desingularization of the fiber variety obtained by Ihara, the second had to do with the proper contribution to the Hecke polynomial coming from the "middle" cohomology groups and the third, and perhaps most crucial, was the actual proof of the Weil conjectures. All these three problems were successfully dealt with by Deligne. We now state these results for the case of the Ramanujan arithmetical function.

For each prime ℓ, let K_ℓ denote the maximal extension of the rationals unramified outside ℓ and for a prime $p \neq \ell$, let F_p be the conjugacy class in $\mathrm{Gal}(K_\ell/\mathbb{Q})$ which is the inverse of the class containing the Frobenius element ϕ_p corresponding to the prime p. The results of Deligne are:

THEOREM 5.10 Deligne's Theorem on Ramanujan's τ-Function
A) For each ℓ there exists a continuous representation

$$\rho_\ell : \mathrm{Gal}(K_\ell/\mathbb{Q}) \longrightarrow GL_2(W_\ell),$$

such that for each prime $p \neq \ell$,

$$1 - \tau(p)x + p^{11}x^2 = \det\left(1_2 - \rho_\ell(F_p)x\right),$$

where W_ℓ is a two-dimensional vector space over the ℓ-adic numbers $\mathbb{Q}_\ell$.
B) The roots of the polynomial

$$H_{12}^{(p)}(x) = 1 - \tau(p)x + p^{11}x^2$$

have absolute value $p^{-11/2}$, that is, $|\tau(p)| < 2p^{11/2}$.

REMARK 5.3   Part B of Deligne's theorem implies $\tau(p) = \pi_p + \overline{\pi}_p$ with $\pi_p = p^{11/2}\exp(i\vartheta_p)$, ϑ_p real.   ∎

As an interesting application of part A of Deligne's theorem, Swinnerton-Dyer [107] has shown that the known congruences for the Ramanujan function $\tau(n)$ are the only possible congruences. (See Bourbaki's exposé 416, June 1972 [98]).

5.14   Exercises

1. Derive the following expression for the Euler-Poincaré polynomial for Γ:

$$P_\Gamma(t) = \sum_{k=0}^{\infty} (\dim_{\mathbb{C}} M_k(\Gamma))\, t^k = \frac{1}{(1-t^4)(1-t^6)}.$$

2. If a and b are relatively prime integers with $0 \le a < b$, then

$$\sum_{k=1}^{\infty} \left\lfloor \frac{ka}{b} \right\rfloor t^k = \frac{1}{(1-t^b)(1-t)} \left(\sum_{k=1}^{b} \left(\left\lfloor \frac{ka}{b} \right\rfloor - \left\lfloor \frac{(k-1)a}{b} \right\rfloor \right) t^k \right).$$

3. Prove that the Kloosterman sum $S_c(m,n)$ satisfies a multiplicative relation of the type

$$S_a(m,n)S_b(m',n') = S_{ab}(m'',n'')$$

where a and b are relatively prime and m'', n'', are constants that depend linearly on m, n, and m', n'.

4. For $c = p^\alpha$ and $\alpha \ge 2$, show that the Kloosterman sum $S_{p^\alpha}(m,n)$ can be evaluated in simple terms using the quadratic Gauss sums

$$G_n(a) = \sum_{s=1}^{n} \exp\left\{ \frac{2\pi i}{n} \cdot as^2 \right\}.$$

5. (The vanishing of Poincaré series). In analogy with the argument given in Theorem 5.8, prove that the Poincaré series $\phi_m(z;2)$ on the group $\Gamma_0(11)$ and the modular form

$$f(z) = q \prod_{n=1}^{\infty} (1-q^n)^2 (1-q^{11n})^2 = \sum_{n=1}^{\infty} a(n)q^n$$

satisfy the relation

$$\phi_m(z;2) = C \cdot a(m)f(z),$$

with C a nonzero constant. Here $\phi_m(z;2)$ is to be understood as the limiting value of the Poincaré series modified by the Hecke convergence factor $|J_L(z)|^s$ as $s \mapsto 0$ (See remarks following the proof of Lemma 5.1).

6. (Continuation of Exercise 5.) Develop an efficient algorithm for computing the coefficients $a(m)$ of the modular form $f(z)$ using ideas similar to those that appear in the proof of Lemma 4.4. Calculate the first 100 coefficients $a(1), a(2), \ldots, a(100)$ and verify that $p = 29$ is the first prime for which $a(p) = 0$. Such primes are called supersingular primes for the elliptic curve

$$y^2 - y = x^3 - x^2,$$

and their number is infinite.

7. This exercise considers the complexity of computing the Ramanujan tau function, defined by

$$q \prod_{n=1}^{\infty} (1-q^n)^{24} = \sum_{n=1}^{\infty} \tau(n)q^n.$$

Let $w(-3) = 3$, $w(-4) = 2$ and $w(D) = 1$ for $D < -4$. Set $h'(D) = h(D)/w(D)$, where $h(D)$ is the number of classes of binary quadratic forms $Q = ax^2 + bxy + cy^2$ with $b^2 - 4ac = D$, if $D \equiv 0$, $1 \pmod 4$. Otherwise, put $h(D) = 0$. Define the Hurwitz class number of a positive integer N by

$$H(N) = \sum_{d^2 \mid N} h' \left(-\frac{N}{d^2} \right).$$

Selberg [96] proved that for p prime,

$$\tau(p) = - \sum_{0 < t \le \sqrt{4p}} P(t, p) H(4p - t^2) + \frac{1}{2} p^5 H(4p) - 1,$$

where

$$P(t, p) = t^{10} - 9t^8 p + 28t^6 p^2 - 35t^4 p^3 + 15t^2 p^4 - p^5.$$

Estimate the time complexity (number of arithmetic operations) in computing $\tau(n)$. (Note: D.X. Charles [16] has proved that $\tau(n)$ can be computed by a randomized algorithm in time $O(n^{1/2+\varepsilon})$ for every $\varepsilon > 0$.)

Chapter 6

Representation of Numbers as Sums of Squares

6.1 Introduction

The problem of representing a positive integer as a sum of s squares has a long and interesting history, partly recounted in Dickson's monumental history of number theory [24], volume II, pages 225–339. The two best-known results of elementary number theory which have played a significant role in the development of various approaches to the problem are Fermat's theorem that any prime of the form $p = 4m + 1$ is representable in a unique way as a sum of of two squares and Lagrange's theorem that every positive integer is the sum of at most four squares of integers. As we have seen in Chapter 2, each of these results comes equipped with an explicit formula for the number of possible representations. It has been the goal of many number theorists to obtain similar exact formulas for the number of representations of an integer as the sum of s squares for $s > 4$. We shall see in this chapter, using the Hecke theory of modular forms, to what extent such a goal is possible.

In this chapter we present an outline of those aspects of the problem connected with the theory of modular forms. The central result is the derivation of the so-called singular series as an Euler product consisting of the quotient of two classical Dirichlet L-functions, whose values are expressible in terms of generalized Bernoulli numbers (see Theorems 6.3 and 6.4). We precede our discussion by recalling some of the well-known formulas related to the problem for $s = 2$, 3, 4, 5, 6, 7 and 8. The chapter ends with a brief description (in Section 6.7) of the relation between Liouville's methods and the theory of elliptic modular forms.

In this development, the theory of modular forms of half integral weight, including the representation theoretic interpretation of the Jacobi theta form

$\vartheta(z)^s$, plays a major role, particularly in the case of odd s. This part of the theory is lucidly presented in Gelbart's monograph [36].

For more about modular forms and the arithmetical properties of the coefficients appearing in their q-series expansions, the reader can consult the monograph of Ono [80].

6.2   The Circle Method and Poincaré Series

We will consider the function $r_s(n)$ which counts the number of solutions of

$$n = x_1^2 + x_2^2 + \cdots + x_s^2$$

in integers x_i. In the following we fix the notation $q = \exp(2\pi i z)$ and introduce the generating function

$$\vartheta(z) = \sum_{n=-\infty}^{\infty} q^{n^2}$$

and observe that

$$\vartheta(z)^s = 1 + \sum_{n=1}^{\infty} r_s(n)q^n.$$

Arguments of a combinatorial nature involving the function $\vartheta(z)^s$ lead to expressions of the following type:

$$r_2(n) = 4 \sum_{2\ell+1|n} (-1)^\ell, \tag{6.1}$$

$$r_4(n) = 2^3 \cdot 3^\delta \sum_{2\ell+1|n} (2\ell + 1), \qquad \delta = \begin{cases} 1 & \text{if } n \text{ is even} \\ 0 & \text{otherwise.} \end{cases} \tag{6.2}$$

$$r_6(n) = 4 \sum_{2\ell+1|n} (-1)^\ell \left\{ \left(\frac{2n}{2\ell + 1} \right)^2 - (2\ell + 1)^2 \right\}, \tag{6.3}$$

$$r_8(n) = 2^4 \sum_{d|n} (-1)^{n+d} d^3, \tag{6.4}$$

$$r_{10}(n) = \frac{4}{5} \sum_{2\ell+1|n} (-1)^\ell \left\{ \left(\frac{2n}{2\ell + 1} \right)^4 - (2\ell + 1)^4 \right\} + \frac{8}{5} \sum_{a^2+b^2=n} (a + ib)^4, \tag{6.5}$$

$$r_{12}(n) = 8 \sum_{d|n} (-1)^{n-1+d+(n/d)} d^5 + 2 \sum (a + ib + jc + kd)^4, \tag{6.6}$$

where the second sum in the last formula runs over all integral quaternions $a + ib + jc + kd$ whose norm is n.

$$r_{24}(n) = \frac{16}{691} \sum_{d|n} (-1)^{n+d} d^{11} + \frac{128}{691} \left\{ (-1)^{n-1} 259\tau(n) - 512\tau\left(\frac{n}{2}\right) \right\}, \tag{6.7}$$

where

$$\Delta(z) = q \prod_{n=1}^{\infty} (1 - q^n)^{24} = \sum_{n=1}^{\infty} \tau(n) q^n$$

is Ramanujan's modular function and $\tau(x) = 0$ if x is not an integer.

The formulas (6.1), (6.2) and (6.4) are derived by equating the coefficients in the following well-known ([50], page 133) identities of Jacobi:

$$\vartheta(z)^2 = 1 + 4 \left(\frac{q}{1-q} - \frac{q^3}{1-q^3} + \frac{q^5}{1-q^5} - \frac{q^7}{1-q^7} + \cdots \right), \qquad (6.8)$$

$$\vartheta(z)^4 = 1 + 8 \left(\frac{q}{1-q} + \frac{2q^2}{1+q^2} + \frac{3q^3}{1-q^3} + \frac{4q^4}{1+q^4} + \cdots \right), \qquad (6.9)$$

$$\vartheta(z)^8 = 1 + 16 \left(\frac{1^3 q}{1+q} - \frac{2^3 q^2}{1-q^2} + \frac{3^3 q^3}{1+q^3} + \cdots \right). \qquad (6.10)$$

(Formula (6.10) is Jacobi's entry for April 24, 1828.)

We will give below an identity from which the other expressions also follow.

Early in the development of the theory it was realized that the problem of finding a closed expression for $r_s(n)$ when s is odd is much more subtle than the corresponding case for s even. We hope to convince the reader that the difficulty lies deeper than had been supposed till now, and that conceptually the case of odd s requires a deeper analysis than is possible for even s. We also hope to explain why Hardy's attempt ([47], page 190) to treat the two cases on an equal footing was doomed to fail in establishing exact representations for $r_s(n)$.

The earliest known result in the case of odd s is that of Gauss, who proved that (in a different but equivalent notation)

$$r_3(n) = \frac{A\sqrt{n}}{\pi} \sum_{m=1}^{\infty} \left(\frac{-n}{m} \right) \frac{1}{m}. \qquad (6.11)$$

Much later Minkowski and Smith gave arithmetical proofs of the cases $s = 5$ and $s = 7$:

$$r_5(n) = \frac{Bn\sqrt{n}}{\pi^2} \sum_{m=1}^{\infty} \left(\frac{n}{m} \right) \frac{1}{m^2}, \qquad (6.12)$$

$$r_7(n) = \frac{Cn^2\sqrt{n}}{\pi^3} \sum_{m=1}^{\infty} \left(\frac{-n}{m} \right) \frac{1}{m^3}. \qquad (6.13)$$

In the above formulas we have put (for n odd)

$$A = \begin{cases} 24 \\ 16 \\ 24 \\ 0 \end{cases}, \quad B = \begin{cases} 80 \\ 160 \\ 112 \\ 160 \end{cases}, \quad C = \begin{cases} 448 & \text{if } n \equiv 1 \pmod 8 \\ 560 & \text{if } n \equiv 3 \pmod 8 \\ 448 & \text{if } n \equiv 5 \pmod 8 \\ 592 & \text{if } n \equiv 7 \pmod 8 \end{cases}.$$

There are corresponding formulas for even n which depend on the highest power of 2 dividing n. Examples of these formulas will be made explicit when we develop the theory of the singular series. The impatient reader can skip ahead to Theorem 6.3.

In his brief note ([47], page 192) Hardy gave the following three power series identities from which the formulas (6.11), (6.12) and (6.13) could be derived.

$$\vartheta(z)^3 = 1 + 8\left\{\sum_{\substack{n=1 \\ n \text{ odd}}}^{\infty} \frac{(-1)^{(n-1)/2}}{n} \sum_j \sum_{m=0}^{\infty} (mn+j)^{1/2} q^{mn+j}\right. \tag{6.14}$$

$$\left. + \sum_{\substack{n=2 \\ n \text{ even}}}^{\infty} \frac{1}{n} \sum_j \sum_{m=0}^{\infty} (-1)^{m+v}(mn+j)^{1/2} q^{mn+j}\right\},$$

$$\vartheta(z)^5 = 1 + \frac{32}{5}\left\{\sum_{\substack{n=1 \\ n \text{ odd}}}^{\infty} \frac{1}{n^2} \sum_j \sum_{m=0}^{\infty} (mn+j)^{3/2} q^{mn+j}\right. \tag{6.15}$$

$$\left. - \sum_{\substack{n=2 \\ n \text{ even}}}^{\infty} \frac{1}{n^2} \sum_j \sum_{m=0}^{\infty} (-1)^{m+v}(mn+j)^{3/2} q^{mn+j}\right\},$$

$$\vartheta(z)^7 = 1 + \frac{256}{15}\left\{\sum_{\substack{n=1 \\ n \text{ odd}}}^{\infty} \frac{(-1)^{(n-1)/2}}{n^3} \sum_j \sum_{m=0}^{\infty} (mn+j)^{5/2} q^{mn+j}\right. \tag{6.16}$$

$$\left. - \sum_{\substack{n=2 \\ n \text{ even}}}^{\infty} \frac{1}{n^3} \sum_j \sum_{m=0}^{\infty} (-1)^{m+v}(mn+j)^{5/2} q^{mn+j}\right\}.$$

In Formula (6.15), j runs through a complete set of least positive residues of $0, 1^2, 2^2, \ldots, (n-1)^2$ modulo n, each taken with the appropriate multiplicity, and vn is the multiple of n subtracted to get each such residue. In Formulas (6.14) and (6.16), j and v are as for (6.15) except that when n is even, j is the residue of $n/2, n/2+1^2, \ldots, n/2+(n-1)^2$. Hardy's subsequent work ([48]) made use of the circle method in an attempt to give a unified treatment for the cases $s = 5, 6, 7$ and 8. The paper also contains an asymptotic estimate for the case $s > 8$. To state Hardy's main results we first introduce the concept of a *singular series*. This is formally defined for a positive integer s as

$$\rho_s(n) = \frac{n^{s/2-1}\pi^{s/2}}{\Gamma(s/2)} \sum_{m=1}^{\infty} \sum_{\substack{h=1 \\ \gcd(h,m)=1}}^{m} \left(\frac{S_{h,m}}{m}\right)^s e^{-2\pi i n h/m}, \tag{6.17}$$

where

$$S_{h,m} = \sum_{j=1}^{m} e^{2\pi i h j^2 / m}.$$

We record here for future reference some for the basic properties of the Gaussian sum $S_{h,m}$:

$$S_{h,mm'} = S_{hm,m'} S_{hm',m} \quad \text{when} \quad \gcd(m, m') = 1.$$

This multiplicativity reduces the evaluation of $S_{h,m}$ to the case of m a prime power. In fact it is well known that for

$$
\begin{array}{llll}
m = 2, & & S_{h,m} = 0, & \\
m = 2^{2\nu+1}, & \nu > 0 & S_{h,m} = 2^{\nu+1} e^{\pi i h/4}, & \\
m = 2^{2\nu}, & \nu > 0 & S_{h,m} = 2^{\nu}(1 + i^h), & i = \sqrt{-1}, \\
m = p, & & S_{h,m} = \left(\frac{h}{p}\right) i^{(p-1)^2/4} \sqrt{p}, & \\
m = p^{2\nu+1}, & & S_{h,m} = p^{\nu} S_{h,p}, & \\
m = p^{2\nu}, & & S_{h,m} = p^{\nu}, &
\end{array}
$$

where (h/p) is the Legendre-Jacobi symbol. It is fairly easy to deduce from the multiplicativity of the Gaussian sum $S_{h,m}$ that the series

$$S(n) = \sum_{m=1}^{\infty} A_m, \quad \text{where} \quad A_1 = 1, \quad A_m = \sum_{\substack{h=1 \\ \gcd(h,m)=1}}^{m} \left(\frac{S_{h,m}}{m}\right)^s e^{-2\pi i n h/m},$$

has an Euler product, that is, $A_{mm'} = A_m A_{m'}$ whenever $\gcd(m, m') = 1$. We can now state Hardy's main result from [48].

THEOREM 6.1 Hardy's Theorem on $r_s(n)$
We have $r_s(n) = \rho_s(n)$ for $s = 5$, 6, 7 and 8, and for $s > 8$ we have

$$r_s(n) = \rho_s(n) + O(n^{s/4}).$$

The basic idea of the circle method as applied by Hardy in the proof of this theorem consists in observing that the essential part of the generating function $\vartheta(z)^s$ is to some degree determined by its behavior at all the rational points on the real axis. This latter behavior is in turn derived from the theta transformation formulas. In one respect the solution given by Hardy is somewhat unsatisfactory in that it leaves open the all too important question of the arithmetical nature of $r_s(n)$. For example, it does not explain why $r_s(n)$ fails to be multiplicative in general. As we shall see later, and this is the main point we want to make in this chapter, this difficulty arises because the generating function $\vartheta(z)^s$ belongs to three quite different spaces depending on the residue class to which s belongs modulo 4. We also note that from an arithmetic point of view it is more natural to consider $\vartheta(z)^s$ as a modular function on the Hecke group $\Gamma_0(4)$, or a double covering of it, than on the traditional theta group Γ_ϑ.

The circle method of Hardy was later improved by Petersson [84]. To gain a better perspective of the new approach, let us describe briefly the nature of Hardy's method from the point of view of modular forms. Recall that the theta group Γ_ϑ consists of all linear fractional transformations of the upper half plane into itself and is generated by

$$T : z \longmapsto z + 2, \qquad S : z \longmapsto -\frac{1}{z}.$$

As is well known ([63], page 46), the theta function $\vartheta(z)$ satisfies the functional equation

$$\vartheta(w(z)) = \chi_\vartheta(w)(cz + d)^{1/2}\vartheta(z), \quad \text{with} \quad w(z) = \frac{az + b}{cz + d} \in \Gamma_\vartheta,$$

where $-\frac{\pi}{2} < \arg z^{1/2} \le \frac{\pi}{2}$ and for any integer a, $z^{a/2} = (z^{1/2})^a$; $\chi_\vartheta(w)$ is independent of z and $|\chi_\vartheta| = 1$.

In order to state the value of the multiplier system χ_ϑ we must extend the meaning of the Jacobi-Legendre symbol. When n is an odd integer and j is an integer we put

$$\left(\frac{j}{n}\right)^* \overset{\text{def}}{=} \left(\frac{j}{|n|}\right) \quad \text{and} \quad \left(\frac{j}{n}\right)_* \overset{\text{def}}{=} \left(\frac{j}{n}\right)^* \sigma_{j,n},$$

where

$$\sigma_{j,n} = \begin{cases} -1 & \text{if } j < 0 \text{ and } n < 0, \\ +1 & \text{otherwise.} \end{cases}$$

Recall that if $1_2 = \begin{pmatrix} 1 & 0 \\ 0 & 1 \end{pmatrix}$ and $S = \begin{pmatrix} 0 & -1 \\ 1 & 0 \end{pmatrix}$, then the linear transformation

$$w(z) = \frac{az + b}{cz + d} \in \Gamma_\vartheta \tag{6.18}$$

belongs to the theta group Γ_ϑ if and only if

$$w \equiv 1_2 \pmod{2} \text{ or } w \equiv S \pmod{2}.$$

In particular, one of c, d is even and the other is odd.

Our proof of the following theorem, which gives the value of the multiplier system χ_ϑ, follows the argument given by Petersson in [85], Appendix E, pages 269–270. A different proof is given by Knopp on page 51 of [63].

THEOREM 6.2 Value of the Theta Multiplier
Let $\zeta_4 = \sqrt{i}$. For $w \in \Gamma_\vartheta$ given by (6.18), we have

$$\chi_\vartheta(w) = \begin{cases} \left(\frac{c}{d}\right)_* \zeta_4^{d-1} & \text{if } w \equiv 1_2 \pmod{2}, \\ \left(\frac{d}{c}\right)^* \zeta_4^{-c} & \text{if } w \equiv S \pmod{2}. \end{cases}$$

PROOF   We note first that

$$w(z) = \frac{(-a)z + (-b)}{(-c)z + (-d)},$$

and hence we may assume without loss of generality that $c > 0$.

Here is the motivation behind the proof. If the linear transformation

$$z \mapsto w(z) = \frac{az + b}{cz + d}$$

has a fixed point z_f, which lies in the upper half plane so that

$$z_f = \frac{(a-d) + \sqrt{(a+d)^2 - 4}}{2c}, \quad \text{where } |a+d| \le 1,$$

then, using that

$$cz_f + d = \left(\frac{a+d}{2}\right) + \sqrt{\left(\frac{a+d}{2}\right)^2 - 1},$$

and that

$$\vartheta(z_f) = \vartheta(w(z_f)) = \chi_\vartheta(w)\sqrt{cz_f + d}\,\vartheta(z_f), \tag{6.19}$$

we get, on canceling $\vartheta(z_f)$ from both sides,

$$1 = \chi_\vartheta(w)\sqrt{\left(\frac{a+d}{2}\right) + \sqrt{\left(\frac{a+d}{2}\right)^2 - 1}},$$

an expression that gives the value of $\chi_\vartheta(w)$ readily. Unfortunately, the fixed point z_f often lies on the real axis and in the last step we have to cancel the indefinite quantity $\vartheta(z_f)$ from both sides. To remedy this situation we investigate instead the asymptotic behavior of the theta function as we approach a rational number on the real axis from above. Knowledge of this asymptotic behavior, together with relation (6.19), will prove the theorem.

More precisely, given $w(z)$ as in (6.18), with $c > 0$, we put $z = -\frac{d}{c} + iy$, so that $cz + d = icy$ and $w(z) = \frac{a}{c} + \frac{i}{c^2 y}$. From (6.19), with z instead of z_f, we get, as $y \to 0+$,

$$\vartheta(w(z)) = 1 + o(1) = \chi_\vartheta(w)\sqrt{icy} \cdot \left\{1 + 2\sum_{m=1}^{\infty} e^{\pi i m^2 z}\right\} \tag{6.20}$$

$$= \zeta_4 \cdot \chi_\vartheta(w)\sqrt{cy} \left\{1 + 2\sum_{m=1}^{\infty} e^{\pi i m^2 \rho - \pi m^2 y}\right\},$$

where $\rho = -d/c$. We now split the summation into c arithmetic progressions

$$m = j + c\nu, \quad j = 1, 2, \dots, c, \quad \nu = 0, 1, 2, \dots,$$

and use that $cd \equiv 0 \pmod 2$ for any $w \in \Gamma_\vartheta$ to obtain

$$\sum_{m=1}^{\infty} e^{\pi i m^2 \rho - \pi m^2 y} = \sum_{j=1}^{c} e^{\pi i j^2 \rho} \sum_{\nu=0}^{\infty} e^{-\pi (j+c\nu)^2 y}.$$

It is a pleasant exercise on the use of the Euler-MacLaurin sum formula (Theorem 3.5) to show that for $y \to 0+$, $c > 0$ and any j, we have

$$\sum_{\nu=0}^{\infty} e^{-\pi(j+c\nu)^2 y} = \int_0^{\infty} e^{-\pi(j+ct)^2} dt + O(1) \qquad (6.21)$$

$$= \frac{1}{2c\sqrt{y}} + O(1).$$

We note that the O-term on the right side comes from the fact that the higher-order derivatives of the (monotonic) function

$$f(t) = \exp(-\pi(j + ct)^2 y)$$

are all polynomials in y. The value of the integral uses the well-known formula $\int_{-\infty}^{\infty} e^{-t^2} dt = \sqrt{\pi}$. (A direct argument using Riemann sums would also suffice to prove (6.21).)

Substituting the expression (6.21) in Equation (6.20) yields the relation

$$1 = \zeta_4 \cdot \chi_\vartheta(w) \cdot \frac{1}{\sqrt{c}} \cdot G(\rho),$$

where

$$G\left(\frac{n}{m}\right) \overset{\text{def}}{=} \sum_{j=1}^{m} \exp\left(\pi i \frac{n}{m} j^2\right)$$

is a quadratic Gauss sum. We thus obtain

$$\chi_\vartheta(w) = \zeta_4^{-1} \cdot \frac{\sqrt{c}}{G(\rho)}, \quad \text{where } \rho = -\frac{d}{c}.$$

It remains to show that the value of the quadratic Gauss sum is given by

$$\frac{G\left(\frac{n}{m}\right)}{\sqrt{m}} = \begin{cases} \left(\frac{n}{m}\right) \zeta_4^{1-m} & \text{if } m \equiv 1 \pmod 2, \\ \left(\frac{m}{n}\right) \zeta_4^n & \text{if } n \equiv 1 \pmod 2. \end{cases}$$

This fact is established using arguments like those we will give in Section 6.4.1. It is an interesting exercise to verify that the resulting expression for $\chi_\vartheta(w)$ in terms of the extended Jacobi-Legendre symbol agrees with the value found above. ∎

A fundamental domain for Γ_ϑ can be taken to be

$$D(\Gamma_\vartheta) = \{z = x + iy : y > 0, |z| \geq 1, -1 \leq \Re(z) \leq 1\}.$$

It contains two inequivalent cusps: one at infinity and the other at $z = -1$. More generally, for s a positive integer we can consider $\vartheta(z)^s$ as a modular form of type $\{\Gamma_\vartheta, s/2, \chi_\vartheta^s\}$, that is, $f(z) = \vartheta(z)^s$ is regular in $D(\Gamma_\vartheta)$ and satisfies the functional equation

$$f(w(z)) = \chi_\vartheta(w)^s(cz+d)^{s/2}f(z), \quad \text{with} \quad w(z) = \frac{az+b}{cz+d} \in \Gamma_\vartheta.$$

The Petersson-Riemann-Roch formula can be used to show that this latter space of functions, which is finite dimensional, contains no functions that vanish at both cusps $z = -1$ and $z = \infty$ when $1 \le s \le 8$. Of the two possible Eisenstein series of type $\{\Gamma_\vartheta, s/2, \chi_\vartheta^s\}$, for $s \ge 5$, one vanishes at $z = i\infty$ and the other is given by

$$E_s(z) = \frac{1}{2} \sum_{T = \begin{pmatrix} * & * \\ c & d \end{pmatrix}} \chi_\vartheta(T)^{-s}(cz+d)^{-s/2},$$

where the summation is over any maximal system of matrices T of Γ_ϑ having different second rows (see [63], page 73). The q-expansion of $E_s(z)$ is given ([63], page 74) by

$$E_s(z) = 1 + \sum_{m=1}^{\infty} \rho_s(m)q^m,$$

where $\rho_s(m)$ is the singular series introduced before.

The main idea in the proof of Hardy's theorem is to realize that $E_s(z)$ represents a good approximation to $\vartheta(z)^s$ in the sense that

$$g(z) = \vartheta(z)^s - E_s(z)$$

is a cusp form, that is, it vanishes at the cusps of Γ_ϑ. In fact, for $s = 5, 6, 7$ and 8, we have $g(z) \equiv 0$ since in these cases there are no cusp forms. If one then compares the q-expansions of $\vartheta(z)^s$ and $E_s(z)$, one is led to the equality $r_s(n) = \rho_s(n)$ for $s = 5, 6, 7$ and 8. The equality also holds for $s = 3$ and 4 but requires a more delicate analysis than that given by Hardy. To consider the cases with $s > 8$, Hardy observes that

$$c_s(n) \stackrel{\text{def}}{=} r_s(n) - \rho_s(n)$$

is the n-th coefficient of the q-expansion of the cusp form $f(z)$. One then uses the fact that $f(z)$ has weight $s/2$ to deduce that

$$c_s(n) = O(n^{s/4}).$$

In outline this is Hardy's procedure.

Petersson's contribution to the problem consisted in the construction of a basis for the space of modular forms of any Fuchsian type and in particular of type $\{\Gamma_\vartheta, s/2, \chi_\vartheta^s\}$, with s even. For each basis element Petersson could then calculate explicitly the q-expansion. This leads then in principle to exact formulas for $r_s(n)$. Rather than stating the most general result we give an example of the type of formula that results from Petersson's analysis.

Example 6.1

For odd n we have

$$r_{24}(n) = \rho_{24}(n) + \frac{2^{22} \cdot 37 \cdot \pi^{12} <\Delta, \Delta> n^{11/2}}{3^4 \cdot 5^2 \cdot 691} \sum_{c=1}^{\infty} \frac{S_c(n)}{c} J_{11}\left(\frac{4\pi\sqrt{n}}{c}\right),$$

where $<\Delta, \Delta>$ is the Petersson inner product of $\Delta(z)$ with itself, $J_{11}(z)$ is the Bessel function and

$$S_c(n) = \sum_{\substack{h=1 \\ \gcd(h,c)=1}}^{c} \exp\left(\frac{2\pi i}{c}(h + h'n)\right), \quad \text{where } hh' \equiv 1 \pmod{c},$$

is a Kloosterman sum.

Expansions similar to that given above are now possible for any real s (see [77], page 135). The advantage in using Petersson's method of Poincaré series over that of Hardy consists in applying estimates for $S_c(n)$ to get corresponding estimates for $r_s(n) - \rho_s(n)$. For example, one can deduce from Weil's estimate of Kloosterman sums the result

$$r_s(n) = \rho_s(n) + O(n^{(s-1)/4+\varepsilon}),$$

for even s and $\varepsilon > 0$. A slightly weaker result also follows for odd s.

The rest of the chapter is divided into five sections. In Sections 6.3 and 6.4 we consider the problem of computing explicitly in finite terms the singular series. In Section 6.5 we develop, using the modern theory of modular forms, explicit formulas for $r_s(n)$ when s is even and n is a power of a prime, and for s odd when n is a fixed multiple of the square of a prime power (see Theorems 6.8 and 6.9, respectively). In Section 6.6 we discuss some open problems and give several examples that seem to suggest where to search for a more satisfactory answer to the problem than that presented here. In Section 6.7 we outline the connection between Liouville's elementary method and the theory of elliptic modular forms.

6.3   Explicit Formulas for the Singular Series

The appearance of the singular series as the n-th coefficient in the q-expansion of a normalized Eisenstein series suggests that as an arithmetic function it should have a form close to the ordinary divisor function; in particular it should be a rational number or at worst an algebraic number. A close look at $\rho_s(n)$ in the simplest cases $s = 3, 4, 5, 6, 7$ and 8 does indeed give weight to our supposition. In the two propositions below we prove the fact, already implicit in the literature, that the singular series is, apart from a relatively simple divisor function, the value of a generalized Bernoulli number. The key point in the computation is the realization that once the singular series has been suitably normalized it has an analytic continuation in the weight $s/2$ and so the explicit value of zetas at the negative integers can be used to give $\rho_s(n)$ an explicit value.

We treat two cases depending on the parity of s.

THEOREM 6.3 $\rho_s(n)$ For Even s
For even s and a positive integer n we have

$$\rho_s(n) = -s(2n)^{(s/2)-1}(-1)^{((s/2)-\delta)/2}\frac{L_2(0,\rho_s)}{B_{s/2,\psi_s}}\left\{\sum_{\substack{d|n \\ n \text{ odd}}} \psi_s(d)d^{1-(s/2)}\right\},$$

where $\delta = (s/2) \bmod 2$ and $\psi_s(a) = ((-1)^{s/2}/a)$. Here $B_{s/2,\psi_s}$ is the generalized Bernoulli number defined by

$$\sum_{a=1}^{\mathfrak{f}} \frac{\psi_s(a)ze^{az}}{e^{2\mathfrak{f}}-1} = \sum_{k=0}^{\infty} B_{k,\psi_s} \cdot \frac{z^k}{k!},$$

where $\mathfrak{f}$ is the conductor of ψ_s. For $n = 2^e n'$ with n' odd, $L_2(0,\rho_s)$ is given by, with $T = 2^{1-(s/2)}$,

$$1 + \Theta_0 \sum_{2 \leq k \leq e} T^{k-1} + \Theta_1 T^e + \Theta_2 T^{e+1},$$

and

$$\Theta_i = \begin{cases} 0 & \text{if } s - 2^{3-i}n' \equiv 2 \text{ or } 6 \pmod 8, \\ (-1)^{(s-2^{3-i}n')/4} & \text{if } s - 2^{3-i}n' \equiv 0 \text{ or } 4 \pmod 8, \end{cases}$$

If $e = 1$, we put $L_2(0,\rho_s) = 1$.

THEOREM 6.4 $\rho_s(n)$ For Odd s and Square Free n
For odd s and square-free n with $(-1)^{(s-1)/2}n \equiv 1 \pmod 4$, we have

$$\rho_s(n) = \frac{L_2(0,\rho_s)}{1-2^{1-s}}\left\{2^{(s-1)/2} - \left(\frac{(-1)^{(s-1)/2}n}{2}\right)\right\}(-1)^{(s^2-1)/8} \cdot \frac{B_{(s-1)/2,\chi}}{B_{s-1}},$$

where $B_{(s-1)/2,\chi}$ is the generalized Bernoulli number defined by

$$\sum_{a=1}^{\mathfrak{f}} \frac{\chi(a)ze^{az}}{e^{\mathfrak{f}z}-1} = \sum_{k=0}^{\infty} B_{k,\chi} \cdot \frac{z^k}{k!}$$

and B_{s-1} is the Bernoulli number defined by

$$\frac{ze^z}{e^z - 1} = \sum_{k=0}^{\infty} B_k \cdot \frac{z^k}{k!}.$$

The expression $L_2(0, \rho_s) = H_2$ is the polynomial in 2^s given above.

The character χ in Theorem 6.4 depends on the parity of $(s-1)/2$ and will be defined later (as χ_{n^*} in Equation (6.42)).

REMARK 6.1   These two theorems will be proved in Section 6.4 where we will make a complete evaluation of the singular series for arbitrary n. ∎

REMARK 6.2   The Bernoulli numbers used in Theorem 6.4 are the same as those studied in Chapter 3, except that $B_1 = +\frac{1}{2}$ here. ∎

REMARK 6.3   Some numerical examples will be given in Section 6.4.
∎

REMARK 6.4   If $(-1)^{(s-1)/2} n \not\equiv 1 \pmod 4$, then the formula for $\rho_s(n)$ has to be modified by multiplying it by the factor $(1/4)^{(s/2)-1}$. ∎

REMARK 6.5   The explicit formulas of Theorem 6.3 can be written more simply as follows. For $s = 2k$, with odd k, we have

$$\rho_s(n) = \rho_{2k}(n) = \frac{4}{|E_{k-1}|} \sum_{d|n} \chi(d) \left\{ \left(\frac{2n}{d}\right)^{k-1} + \chi(k) d^{k-1} \right\}, \quad (6.22)$$

where $\chi(a) = (-1/a)$ (the Jacobi symbol) and E_{k-1} is the Euler number defined at the end of Chapter 3. For $s = 2k$ with even k we have

$$\rho_s(n) = \rho_{2k}(n) = \frac{2k}{(2^k - 1)|B_k|} \sum_{d|n} d^{k-1} (-1)^{(d+k/2)(n/d-1)}, \quad (6.23)$$

where B_k is the Bernoulli number defined in Chapter 3.

The explicit formulas (6.22) and (6.23) above can be derived from the q-Lambert expansions of of the Eisenstein series $E_s(z)$ (not the Euler polynomial) given by Mordell ([74], page 96). Suppose first that $s = 2k$ and k is odd. Then we have

$$\frac{1}{2}|E_{k-1}|E_{2k}(z) = \frac{1}{2}|E_{k-1}| + \sum_{n=1}^{\infty} \frac{2^k n^{k-1} q^n}{1 + q^{2n}} + 2\chi(k) \sum_{n=1}^{\infty} \frac{\chi(n) n^{k-1} q^n}{1 - q^n}.$$

Now if $|q| < 1$, then we have

$$\sum_{n=1}^{\infty} \frac{n^{k-1}q^n}{1+q^{2n}} = \sum_{n=1}^{\infty} n^{k-1}q^n \sum_{v=0}^{\infty}(-1)^v q^{2nv}$$

$$= \sum_{n=1}^{\infty}\sum_{v=0}^{\infty}(-1)^v n^{k-1}q^{n(2v+1)}$$

$$= \sum_{n=1}^{\infty}\sum_{m=1}^{\infty} n^{k-1}\chi(m)q^{nm}$$

$$= \sum_{n=1}^{\infty}\left(\sum_{d|n}\chi(d)\left(\frac{n}{d}\right)^{k-1}\right)q^n.$$

Similarly we have

$$\sum_{n=1}^{\infty}\frac{\chi(n)n^{k-1}q^n}{1-q^n} = \sum_{n=1}^{\infty}\left(\sum_{d|n}\chi(d)d^{k-1}\right)q^n.$$

When these two formulas are substituted into the Mordell expansion we get

$$\frac{1}{2}|E_{k-1}|E_{2k}(z) = \frac{1}{2}|E_{k-1}| + 2\sum_{n=1}^{\infty}\left(\sum_{d|n}\chi(d)\left(\frac{2n}{d}\right)^{k-1}\right)q^n$$

$$+ 2\chi(k)\sum_{n=1}^{\infty}\left(\sum_{d|n}\chi(d)d^{k-1}\right)q^n.$$

We now divide through by $\frac{1}{2}|E_{k-1}|$ and compare the n-th coefficients in the q-expansion above with those in $E_{2k}(z)$ to obtain

$$\rho_{2k}(n) = \frac{4}{|E_{k-1}|}\sum_{d|n}\chi(d)\left(\left(\frac{2n}{d}\right)^{k-1} + \chi(k)d^{k-1}\right).$$

To deal with the case of even k we use Mordell's q-Lambert expansion ([74], page 96):

$$(2^k-1)|B_k|E_{2k}(z) = (2^k-1)|B_k| + 2k\sum_{n=1}^{\infty}\frac{n^{k-1}q^n}{1-(-1)^{k/2+n}q^n}.$$

By a transformation similar to the one performed above we have

$$\sum_{n=1}^{\infty}\frac{n^{k-1}q^n}{1-(-1)^{k/2+n}q^n} = \sum_{n=1}^{\infty}\left(\sum_{d|n}d^{k-1}(-1)^{(d+k/2)(n/d-1)}\right)q^n.$$

We then divide through by $(2^k - 1)|B_k|$ and equate the n-th coefficients of the corresponding q-expansion to obtain

$$\rho_{2k}(n) = \frac{2k}{(2^k - 1)|B_k|} \sum_{d|n} d^{k-1}(-1)^{(d+k/2)(n/d-1)}.$$

This ends the remark. ∎

Theorem 6.3 clearly accounts for the simple form taken by $r_s(n)$ in the examples given in the previous section. They are obtained by using the well-known values for the Euler and Bernoulli numbers. (Note that $E_2 = -1$ and $E_4 = 5$. The Bernoulli numbers may be found in a table following Theorem 3.2.)

REMARK 6.6   Theorems 6.3 and 6.4 open the way for a closer arithmetic study of the singular series $\rho_s(n)$. In particular we can now study the p-adic analytic continuation of $\rho_s(n)$ in the variable s in the sense of Iwasawa and Serre. ∎

REMARK 6.7   We are hopeful that the methods of this section apply to the more general singular series studied by Hardy and Littlewood. In particular one may try to evaluate the singular series that arises in the cubic Waring problem (see Hardy and Littlewood [51])

$$n = x_1^3 + x_2^3 + \cdots + x_s^3$$

by means of Hurwitz numbers. ∎

REMARK 6.8   The results of Theorems 6.3 and 6.4 clearly show to what extent the singular series is made up of multiplicative functions. ∎

6.4   The Singular Series

The purpose of this section is to make a complete determination of the singular series $\rho_s(n)$, to prove Theorems 6.3 and 6.4, and to give examples of the explicit formulas obtained.

6.4.1   Quadratic Gaussian Sums

Let $f(\xi)$ be a complex-valued function of the real variable ξ defined in $0 \leq \xi \leq 1$ and continuously differentiable there. If we set

$$\hat{f}(n) = \int_0^1 f(\xi)e^{2\pi i n\xi}\,d\xi,$$

then we have ("the Fourier series theorem")

$$\sum_{n\in\mathbb{Z}} \hat{f}(n)e^{-2\pi in\xi} = \begin{cases} f(\xi) & \text{for } 0 < \xi < 1 \\ \frac{f(0)+f(1)}{2} & \text{for } \xi = 0. \end{cases}$$

In particular we have

$$\frac{f(0)+f(1)}{2} = \lim_{N\to\infty} \sum_{n=-N}^{N} \int_0^1 f(\eta)e^{2\pi in\eta}d\eta.$$

We apply this result to prove the following theorem.

THEOREM 6.5 Evaluation of the Gauss Sum
For n a positive integer we have

$$\sum_{j=1}^{n} e^{-2\pi ij^2/n} = \frac{1+i^n}{1+i}\sqrt{n}.$$

PROOF   It will be convenient to work with the sum

$$S \overset{\text{def}}{=} \sum_{j=1}^{n} e^{2\pi ij^2/n},$$

which is the sum we wish to evaluate, but with i replaced by $-i = i^{-1}$.
Let s be an integer and ξ a real variable. Consider the function

$$f_s(\xi) = e^{2\pi i(\xi+s)^2/n}.$$

Note first that applying the Fourier series theorem to $f_s(\xi)$ gives

$$\begin{aligned} S &= \sum_{s=0}^{n-1} \left\{ \frac{f_s(0)+f_s(1)}{2} \right\} \\ &= \sum_{s=0}^{n-1} \lim_{N\to\infty} \sum_{h=-N}^{N} \int_0^1 e^{2\pi i\{(s+\eta)^2/n+h\eta\}}d\eta \\ &= \lim_{N\to\infty} \sum_{h=-N}^{N} \sum_{s=0}^{n-1} \int_s^{s+1} e^{2\pi i\{\eta^2/n+h\eta\}}d\eta \\ &= \lim_{N\to\infty} \sum_{h=-N}^{N} \int_0^n e^{2\pi i\{\eta^2/n+h\eta\}}d\eta, \quad \eta = nx \\ &= n \lim_{N\to\infty} \left\{ \sum_{h=-N}^{N} \int_0^1 e^{2\pi in\{x^2+hx\}}dx \right\}. \end{aligned}$$

On the other hand, consider the integral

$$\gamma = \int_{-\infty}^{\infty} e^{2\pi i y^2}\, dy.$$

After making the change of variable of integration $y = \sqrt{n}x$ and partitioning the real axis according to

$$(i) \quad \mathbb{R} = \bigcup_{h=-\infty}^{\infty} [h, h+1] \quad \text{and} \quad (ii) \quad \mathbb{R} = \bigcup_{h=-\infty}^{\infty} [h - (1/2), h + (1/2)],$$

we obtain

$$(i) \quad \frac{\gamma}{\sqrt{n}} = \sum_{h=-\infty}^{\infty} \int_{h}^{h+1} e^{2\pi i n y^2}\, dy$$

$$= \sum_{h=-\infty}^{\infty} \int_{0}^{1} e^{2\pi i n (y+h)^2}\, dy$$

$$= \sum_{h=-\infty}^{\infty} \int_{0}^{1} e^{2\pi i n (y^2 + 2hy)}\, dy.$$

Similarly,

$$(ii) \quad \frac{\gamma}{\sqrt{n}} = \sum_{h=-\infty}^{\infty} \int_{h-1/2}^{h+1/2} e^{2\pi i n y^2}\, dy$$

$$= \sum_{h=-\infty}^{\infty} \int_{0}^{1} e^{2\pi i n (y+h-1/2)^2}\, dy$$

$$= \sum_{h=-\infty}^{\infty} \int_{0}^{1} e^{2\pi i n (y^2 + (2h-1)y)} i^n\, dy.$$

Adding (i) to i^{-n} times (ii) and comparing with the previously obtained formula for S we get

$$S = n \left\{ \frac{\gamma}{\sqrt{n}} + \frac{\gamma i^{-n}}{\sqrt{n}} \right\} = \sqrt{n}(1 + i^{-n})\gamma.$$

Substituting $n = 1$ we get $(1 + i^{-1})\gamma = 1$, and we have

$$S = \frac{1 + i^{-n}}{1 + i^{-1}} \sqrt{n},$$

which is equivalent to what we wanted to prove. ∎

We next study the sum

$$S(h, k) = \sum_{j=1}^{k} e^{2\pi i h j^2 / k}$$

for integers $k > 0$ and h with $\gcd(h, k) = 1$. We will prove that $S(h, k)$ enjoys a useful multiplicativity property.

THEOREM 6.6

If k_1 and k_2 are relatively prime positive integers and h is an integer relatively prime to both k_1 and k_2, then

$$S(hk_2, k_1)S(hk_1, k_2) = S(h, k_1 k_2). \tag{6.24}$$

The proof of Theorem 6.6 uses this elementary fact.

LEMMA 6.1

Let k and k' be relatively prime positive integers. Let j run through a complete set of residues modulo k and j' run through a complete set of residues modulo k'. Then $kj' + k'j$ runs through a complete set of residues modulo kk'.

PROOF   The lemma is equivalent to the statement that the kk' integers $kj' + k'j$ are all incongruent modulo kk'. If this were not so, then we would have

$$kj' + k'j \equiv ki' + k'i \pmod{kk'}.$$

This implies $k'j \equiv k'i \pmod{k}$, which gives $j \equiv i \pmod{k}$ because $\gcd(k', k) = 1$. Similarly, $j' \equiv i' \pmod{k'}$. Hence the kk' numbers are all incongruent and must form a complete set of residues modulo kk'.   ∎

We now prove Theorem 6.6.

PROOF   The product $S(hk_2, k_1)S(hk_1, k_2)$ on the left side of (6.24) is

$$\sum_{j, j'} e^{2\pi i \{hk_2 j^2 / k_1 + hk_1 j'^2 / k_2\}},$$

where j and j' run over complete sets of residues modulo k_1 and k_2, respectively. The expression in braces in the exponent satisfies

$$\frac{hk_2 j^2}{k_1} + \frac{hk_1 j'^2}{k_2} \equiv \frac{h(k_2 j + k_1 j')^2}{k_1 k_2} \pmod 1.$$

Letting $J = k_2 j + k_1 j'$ and using Lemma 6.1, we see that the above sum is

$$\sum_{J \bmod k_1 k_2} e^{2\pi i h J^2 / k_1 k_2},$$

which is precisely the right side of (6.24). ∎

Clearly, if $k = 1$, we have $S(h, 1) = 1$.

If $k > 1$ and $k = \prod_{p \mid k} p^{\ell_p}$ is the factorization of k into a product of powers of distinct primes, then repeated application of (6.24) yields

$$S(1, k) = \prod_{p \mid k} S\left(\frac{k}{p^{\ell_p}}, p^{\ell_p}\right).$$

We will now show that if $\ell \geq 2$ and p does not divide $2h$, then

$$S(h, p^\ell) = p S(h, p^{\ell-2}). \tag{6.25}$$

The integers J in the set $\{0, 1, \ldots, p^\ell - 1\}$ can be parametrized by $J = p^{\ell-1} j + j'$, with j in the set $\{0, 1, \ldots, p-1\}$ and j' in the set $\{0, 1, \ldots, p^{\ell-1} - 1\}$. Since

$$\frac{h J^2}{p^\ell} = \frac{h(p^{\ell-1} j + j')^2}{p^\ell} \equiv \frac{2h j j'}{p} + \frac{h j'^2}{p^\ell} \quad (\bmod 1),$$

the sum $S(h, p^\ell)$ can be represented as a double sum,

$$\sum_{j'=0}^{p^{\ell-1}-1} e^{2\pi i h j'^2 / p^\ell} \sum_{j=0}^{p-1} e^{2\pi i j (2 j' h / p)}.$$

Here the inner sum is p if $2 j' h \equiv 0 \ (\bmod\ p)$, that is, if $p \mid j'$; otherwise it is 0. Hence the double sum is

$$\sum_{\nu=0}^{p^{\ell-2}-1} e^{2\pi i h \nu^2 / p^{\ell-2}} \cdot p = p S(h, p^{\ell-2}).$$

This proves (6.25).

We thus obtain by induction on ℓ, for p odd and $\ell = 2\ell' + \varepsilon$, $\varepsilon \in \{0, 1\}$, that

$$\frac{S(h, p^\ell)}{\sqrt{p^\ell}} = \left(\frac{S(h, p)}{\sqrt{p}}\right)^\varepsilon. \tag{6.26}$$

If we let $\zeta = e^{2\pi i / p}$, and let j and j' run through the quadratic residues (QR) and quadratic nonresidues (QNR) modulo p, respectively, in the set $\{1, 2, \ldots, p-1\}$, then with $\gcd(h, p) = 1$, we have

$$\sigma \overset{\text{def}}{=} \sum_{J=1}^{p-1} \left(\frac{J}{p}\right) e^{2\pi i h J / p} = \sum_{j \ \mathrm{QR}} \zeta^{jh} - \sum_{j' \ \mathrm{QNR}} \zeta^{j'h}.$$

On the other hand we have

$$1 + \sum_{j \text{ QR}} \zeta^{jh} + \sum_{j' \text{ QNR}} \zeta^{j'h} = \frac{1 - \zeta^{ph}}{1 - \zeta} = 0.$$

Thus

$$\sum_{j=1}^{p-1} \zeta^{-hj^2} = 2 \sum_{j \text{ QR}} \zeta^{-hj} = \sum_{j \text{ QR}} \zeta^{-hj} - \sum_{j' \text{ QNR}} \zeta^{-hj'} - 1,$$

or

$$S(h,p) = 1 + \sum_{j=1}^{p-1} \zeta^{-hj^2} = \sum_{j=1}^{p-1} \left(\frac{j}{p} \right) \zeta^{-hj}$$

$$= \left(\frac{h}{p} \right) \sum_{j=1}^{p-1} \left(\frac{j}{p} \right) \zeta^{-j} = \left(\frac{h}{p} \right) \frac{1 + i^p}{1 + i} \cdot \sqrt{p}$$

by Theorem 6.5. Hence,

$$\sum_{j=1}^{p-1} \zeta^{-hj^2} = \left(\frac{h}{p} \right) \frac{1 + i^p}{1 + i} \sqrt{p}$$

and

$$\frac{S(h,p)}{\sqrt{p}} = \left(\frac{h}{p} \right) \frac{1 + i^p}{1 + i}.$$

Note that for odd p,

$$\frac{1 + i^p}{1 + i} = i^{-(p-1)^2/4}.$$

In the case $p = 2$ we clearly have $S(h,2) = 0$. The values of $S(h,k)$ for $k = 2^u$ are easy to calculate. For example, we can verify directly that

$$S(h,k) = \begin{cases} 0 & \text{if } k = 2, \\ 2(1 + i^h) & \text{if } k = 2^2, \\ 4e^{\pi ih/4} & \text{if } k = 2^3. \end{cases}$$

If $u \geq 3$, $k = 2^u$ and $\gcd(h,k) = 1$, then we can write

$$h \equiv a \cdot b^2 \pmod{2^u} \quad \text{with } a \in \{1,3,5,7\}. \tag{6.27}$$

This is a well-known consequence of a result in dyadic analysis. If $\alpha \in \mathbb{Z}_2$ and $\text{ord}_2(\alpha - 1) \geq 3$, then $\alpha \in (\mathbb{Z}_2)^2$. What we need here can actually be proved in an elementary way by observing that, since $\gcd(h, 2^u) = 1$, we can write

$$h \equiv a_0 + a_1 2 + \cdots + a_{u-1} 2^{u-1} \pmod{2^u}, \qquad a_i \in \{0,1\}$$
$$\equiv (a_0 + a_1 2 + a_2 2^2) C \pmod{2^u},$$

where $C \equiv 1 \pmod{2^u}$. We now note that since $C \equiv 1 \pmod{2^u}$, C is a quadratic residue modulo 8, that is, there is an x so that $x^2 \equiv C \pmod 8$. By successive Newton approximations we can construct an integer b, such that $C \equiv b^2 \pmod{2^u}$. With $a = a_0 + a_1 2 + a_2 2^2$ (in which $a_0 = 1$ because h is odd), we have proved (6.27).

In (6.27), it is obvious that $\gcd(b, 2^u) = 1$.

From the definition

$$S(h, 2^u) = \sum_{j \bmod 2^u} e^{2\pi i h j^2 / 2^u},$$

we deduce easily that

$$S(ab^2, 2^u) = \sum_{j \bmod 2^u} e^{2\pi i a (bj)^2 / 2^u} = \sum_{j \bmod 2^u} e^{2\pi i a j^2 / 2^u} = S(a, 2^u),$$

whenever $\gcd(b, 2^u) = 1$. This says that $S(h, 2^u)$ is periodic of period 8 in its first argument:

$$S(h, 2^u) = S(h', 2^u) \text{ whenever } h \equiv h' \pmod{2^3}, \quad u \geq 3. \tag{6.28}$$

If $u \geq 3$, then $h \equiv ab^2 \pmod{2^u}$, where $a \in \{1, 3, 5, 7\}$ and b is a square modulo 2^u. Hence

$$S(h, 2^u) = \sum_{j \bmod 2^u} e^{2\pi i a (bj)^2 / 2^u} = S(a, 2^u).$$

For n an odd prime and $\gcd(h, n) = 1$, we have

$$\sum_{j=1}^{n} e^{2\pi i h j^2 / n} = \left(\frac{h}{n}\right) \sum_{j=1}^{n} e^{2\pi i j^2 / n}.$$

If $h \equiv x^2 \pmod n$, then the equality is obvious. But if $h \equiv x^2 \pmod n$ has no solution x, then the Legendre symbol $(h/n) = -1$ and it is required to prove that

$$\sum_{j=1}^{n} e^{2\pi i h j^2 / n} + \sum_{j=1}^{n} e^{2\pi i j^2 / n} = 0.$$

This is clear from $\sum_{k=1}^{n} e^{2\pi i k / n} = 0$, since j^2 runs through all the quadratic residues (twice) and hj^2 runs through all the quadratic nonresidues (twice) modulo n.

The multiplicative property already established in (6.24),

$$S(a, 2^u) S(2^u, a) = S(1, 2^u a),$$

yields for $h = ab^2$ and $u \geq 3$,

$$S(h, 2^u) = S(a, 2^u) = \frac{S(1, 2^u a)}{S(2^u, a)}.$$

Since

$$\sum_{j=1}^{a} e^{2\pi i 2^u j^2/a} = \left(\frac{2^u}{a}\right)\sqrt{a}\frac{(1+i)(1+i^{-a})}{2}$$

and

$$\sum_{j=1}^{2^u a} e^{2\pi i j^2/2^u a} = \sqrt{2^u a}\frac{(1+i)(1+i^{-a2^u})}{2},$$

we have

$$\frac{S(h, 2^u)}{\sqrt{2^u}} = \left(\frac{2}{h}\right)^u \cdot (1 + i^h). \tag{6.29}$$

A direct calculation for $u = 2$ shows that (6.29) also holds in that case.

6.4.2 Ramanujan Sums

Here we prove some identities related to Ramanujan's sum $c_q(n)$. If we define

$$c_q(n) = \sum_{\substack{1 \le h \le q \\ \gcd(h,q)=1}} e^{(2\pi i/q)hn},$$

then we have

$$\frac{\sigma_{1-s}(n)}{\zeta(s)} = \sum_{q=1}^{\infty} \frac{c_q(n)}{q^s}.$$

It can be proved directly from the definition of $c_q(n)$ (see Section 16.6 of [52]) that $c_q(n)$ is multiplicative in q and

$$c_q(n) = \frac{\phi(q)}{\phi(q')}\mu(q'), \quad q' = \frac{q}{\gcd(n, q)}.$$

From the Dirichlet series definition of $c_q(n)$ we see readily that

$$\sum_{d|n} \frac{d}{d^s} \frac{1}{\zeta(s)} = \sum_{d|n} \frac{d}{d^s} \prod_{p}\left(1 - \frac{1}{p^s}\right) = \prod_{p}\left(1 + \sum_{u=1}^{\infty} \frac{c_{p^u}(n)}{(p^u)^s}\right).$$

This in turn implies that

$$1 + \sum_{u=1}^{\infty} \frac{c_{p^u}(n)}{(p^u)^s} = \left(1 - \frac{1}{p^s}\right) \sum_{0 \le u \le \text{ord}_p(n)} p^u \left(\frac{1}{p^u}\right)^s.$$

If we put $T = 1/p^s$, then the equation above can be written more simply as

$$1 + \sum_{u=1}^{\infty} c_{p^u}(n)T^u = (1 - T) \sum_{0 \le u \le \text{ord}_p(n)} p^u T^u.$$

This is equivalent to

$$1 + \sum_{u=1}^{\infty} c_{p^u}(n) a^u T^u = (1 - aT) \left\{ \frac{1 - (apT)^{\mathrm{ord}_p(n)+1}}{1 - (apT)} \right\}.$$

These identities will play an important role in the evaluation of the Fourier transform of the quadratic Gaussian sum in the next section.

6.4.3 Fourier Transforms of Gaussian Sums

Let n be a fixed positive integer. We define the Fourier transform of the quadratic Gaussian sum by

$$A_k = \sum_{\substack{h=1 \\ \gcd(h,k)=1}}^{k} \left(\frac{S(h,k)}{k} \right)^s \cdot e^{-2\pi i n h/k}.$$

In this section we will evaluate A_{p^u} when p is prime and u is a positive integer. There are two cases, according as p is odd or $p = 2$.

6.4.3.1 The Case p odd

LEMMA 6.2
 Let p be an odd prime and u be a positive integer. Then

$$(p^u)^{s/2} A_{p^u} = \begin{cases} c_{p^u}(n) \left(\frac{(-1)^{s/2}}{p} \right) & \text{if } p^u \text{ is not square and } s \text{ is even} \\ -c_{p^u}(n)\sqrt{p} \left(\frac{(-1)^{(s-1)/2}v}{p} \right) & \text{if } p^u \text{ is not square and } s \text{ is odd} \\ c_{p^u}(n) & \text{if } p^u \text{ is square,} \end{cases}$$

where $n = p^{u-1} \cdot v$.

PROOF   We treat first the case $u = 1$, whose statement is equivalent to the following.

$$A_p = \begin{cases} p^{-s} \left(\sqrt{(-1)^{(p-1)/2} \cdot p} \right)^s \cdot c_p(n) & \text{if } s \text{ is even} \\ p^{-s+1} \left(\sqrt{(-1)^{(p-1)/2} \cdot p} \right)^{s-1} \cdot \left(\frac{n}{p} \right) & \text{if } s \text{ is odd.} \end{cases}$$

To prove this we note that since $S(h,p) = \left(\frac{h}{p} \right) \sqrt{(-1)^{(p-1)/2} \cdot p}$ and

$$\sum_{h=1}^{p-1} \left(\frac{h}{p} \right) e^{-2\pi i n h/p} = \left(\frac{n}{p} \right) \sqrt{p} \cdot \frac{1 + i^p}{1 + i}, \tag{6.30}$$

we have

$$A_p = p^{-s} \sum_{h=1}^{p-1} (S(h,p))^s \, e^{-2\pi i n h / p}$$

$$= p^{-s} \left(\sqrt{(-1)^{(p-1)/2} p} \right)^s \sum_{h=1}^{p-1} \left(\frac{h}{p} \right)^s e^{-2\pi i n h / p}.$$

The latter sum is $c_p(n)$ when s is even. Otherwise, its value is given by (6.30). Using that $\sqrt{(-1)^{(p-1)/2}} \cdot \frac{1+i^p}{1+i} = 1$, we obtain the desired result.

Suppose now that $u > 1$ and odd, say $u = 2v+1$. We then have $S(h, p^{2v+1}) = p^v S(h,p)$. In the sum defining A_{p^u} we put $h = pz + m$ with $z \in \{0, 1, \ldots, p^u - 1\}$ and $m \in \{1, 2, \ldots, p-1\}$. We thus have

$$A_{p^u} = (p^u)^{-s} \sum_{\substack{h=1 \\ \gcd(h, p^u)=1}}^{p^u} \left(S(h, p^{2v+1}) \right)^s \cdot e^{-2\pi i n h / p^u}$$

$$= p^{-(v+1)s} \sum_{\substack{h=1 \\ \gcd(h, p^u)=1}}^{p^u} (S(h,p))^s \cdot e^{-2\pi i n h / p^u}$$

$$= p^{-(v+1)s} \sum_{m} (S(m,p))^s \cdot e^{-2\pi i m n / p^u} \sum_{z} e^{-2\pi i n z / p^{u-1}}.$$

The inner sum vanishes unless $p^{u-1} | n$. But if $n = p^{u-1} \cdot \nu = p^{2v} \cdot \nu$, then

$$A_{p^u} = p^{-(v+1)s} p^{u-1} \sum_{m} (S(m,p))^s \, e^{-2\pi i \nu m / p}$$

$$= p^{-vs+u-1} \left\{ p^{-s} \sum_{m=1}^{p-1} (S(m,p))^s \, e^{-2\pi i \nu m / p} \right\}.$$

We now use (6.30) with $n = \nu$. Also when $\gcd(m,p) = 1$ we have

$$S(m,p) = \sum_{j=1}^{p} e^{-2\pi i m j^2 / p} = \left(\frac{m}{p} \right) \sqrt{(-1)^{(p-1)/2} p}.$$

Hence,

$$A_{p^u} = p^{-vs+u-1} p^{-s} \sum_{m=1}^{p-1} \left(\sqrt{(-1)^{(p-1)/2} p} \left(\frac{m}{p} \right) \right)^s e^{-2\pi i \nu m / p}$$

$$= p^{-vs+u-1-(s/2)} \left(\sqrt{(-1)^{(p-1)/2}} \right)^s \sum_{m=1}^{p-1} \left(\frac{m}{p} \right)^s e^{-2\pi i \nu m / p}.$$

If s is even, the last sum is simply $c_p(\nu)$. Thus we have

$$A_{p^u} = p^{-vs+u-1-(s/2)} \left(\frac{(-1)^{s/2}}{p} \right) c_p(\nu)$$

$$= p^{-vs-(s/2)} \left(\frac{(-1)^{s/2}}{p} \right) c_{p^u}(\nu).$$

If s is odd, using (6.30) we get

$$A_{p^u} = p^{-vs+u-1-(s/2)} \left(\frac{(-1)^{(s-1)/2}\,\nu}{p} \right) \sqrt{p},$$

which can also be written in the form

$$A p^u = -(p^u)^{-(s/2)} \left(\frac{(-1)^{(s-1)/2}\,\nu}{p} \right) \sqrt{p} \cdot c_{p^u}(n).$$

Now consider the case of u even, $u = 2v > 1$. In this case we claim

$$p^{vs} A_{p^{2v}} = c_{p^{2v}}(n).$$

In fact we have

$$A_{p^{2v}} = \left(p^{2v} \right)^{-s} \sum_{\substack{h=1 \\ \gcd(h,p^{2v})=1}}^{p^{2v}} \left(S(h, p^{2v}) \right)^s \cdot e^{-2\pi inh/p^{2v}}.$$

Since $S(h, 2^{2v}) = p^v$, we get

$$A_{p^{2v}} = \left(p^{2v} \right)^{-s/2} \sum_{\substack{h=1 \\ \gcd(h,p^{2v})=1}}^{p^{2v}} e^{-2\pi inh/p^{2v}} = p^{-vs} c_{p^{2v}}(n).$$

This completes the proof of Lemma 6.2.   ∎

We note that if we write $s = 2s' + \varepsilon$ and $u = 2u' + \delta$ with integer s' and u' and ε, $\delta \in \{0,1\}$, then the statement of the lemma can also be given as

$$(p^u)^{s/2} A_{p^u} = c_{p^u}(n) \left(\frac{(-1)^{s'\delta}}{p} \right) \left\{ - \left(\frac{\nu}{p} \right) \sqrt{p} \right\}^{\varepsilon\delta}.$$

6.4.3.2 The Case $p = 2$

LEMMA 6.3

We have $A_2 = 0$. Let $u > 1$ be an integer. Let $\zeta_8 = e^{2\pi i/8}$. Then

$$
\left(\sqrt{2^u}\right)^s A_n = \begin{cases}
0 & \text{if } u \text{ is even and } 2^{u+1} \nmid 8n \\
2^{(s/2)+u-2}\left\{\zeta_8^{s-\nu} + \zeta_8^{-(s-\nu)}\right\} & \text{if } u \text{ is even and } 2^{u+1}|8n \\
0 & \text{if } u \text{ is odd and } 2^u \nmid 8n \\
0 & \text{if } u \text{ is odd and } s \not\equiv \nu \pmod 4 \\
2^{(s/2)+u-2}\left\{\zeta_8^{s-\nu} + \zeta_8^{+(s-\nu)}\right\} & \text{if } u \text{ is odd and } s \equiv \nu \pmod 4,
\end{cases}
$$

where $8n = 2^u \cdot \nu$.

PROOF   First of all we have

$$
A_2 = \frac{1}{2^s} \sum_{\substack{h=1 \\ \gcd(h,2)=1}}^{2} (S(h,2))^s e^{-2\pi i n h/2} = \frac{1}{2^s}(S(1,2))^s e^{-\pi i n}
$$

$$
= \frac{1}{2^s}\left(\sum_{j=1}^{2} e^{2\pi i j^2/2}\right)^s e^{-\pi i n} = \frac{1}{2^s}\left\{e^{\pi i/1} + 1\right\}^s e^{-\pi i n} = 0.
$$

Now suppose u is even and ≥ 2. Here we have

$$
\frac{S(h, 2^u)}{\sqrt{2^u}} = \frac{S(h, 2^2)}{\sqrt{2^2}}.
$$

We put $h = 4z + m$ with $z \in \{0, 1, \ldots, 2^{u-2} - 1\}$ and $m \in \{1, 3\}$. Thus we have

$$
\left(\sqrt{2^u}\right)^s A_{2^u} = \sum_{\substack{h=1 \\ \gcd(h,2^u)=1}}^{2^u} \left(\frac{S(h, 2^u)}{\sqrt{2^u}}\right)^s e^{-2\pi i n h/2^u}
$$

$$
= \sum_{m} \left(\frac{S(m, 2^2)}{2}\right)^s e^{-2\pi i n m/2^u} \sum_{z=0}^{2^{u-2}-1} e^{-2\pi i n z/2^{u-2}}.
$$

The inner sum vanishes if $2^{u-2} \nmid n$, and it equals 2^{u-2} if $2^{u-2}|n$. Recall that $8n = 2^u \cdot \nu$. Then

$$
\left(\sqrt{2^u}\right)^s A_{2^u} = 2^{u-2}\left\{\left(\frac{S(1,4)}{2}\right)^s e^{-2\pi i n/2^u} + \left(\frac{S(3,4)}{2}\right)^s e^{-2\pi i n \cdot 3/2^u}\right\}
$$

$$
= 2^{u-2} 2^{-s}\left\{(S(1,4))^s e^{-2\pi i n/2^u} + (S(3,4))^s e^{-2\pi i n \cdot 3/2^u}\right\}.
$$

Since

$$\frac{S(1,4)}{\sqrt{2^2}} = 2^{1/2}\zeta_8 \quad \text{and} \quad \frac{S(3,4)}{\sqrt{2^2}} = 2^{1/2}\zeta_8^{-1},$$

we have

$$\left(\sqrt{2^u}\right)^s A_{2^u} = 2^{-s+(3s/2)+u-2} \left\{ \zeta_8^{(\nu-s)} + \zeta_8^{-(\nu-s)} \right\}$$

$$= 2^{u-2+(s/2)} \left\{ \zeta_8^{(\nu-s)} + \zeta_8^{-(\nu-s)} \right\}.$$

Now let u be odd and > 1. We have

$$\left(\sqrt{2^u}\right)^s A_{2^u} = \sum_{\substack{h=1 \\ \gcd(h,2^u)=1}}^{2^u} \left(\frac{S(h,2^u)}{\sqrt{2^u}}\right)^s e^{-2\pi i n h/2^u}.$$

We let $h = 8z + a$ with $z \in \{0,1,\ldots,2^{u-3}-1\}$ and $a \in \{1,3,5,7\}$. Then we have

$$\left(\sqrt{2^u}\right)^s A_{2^u} = \sum_a \left(\sqrt{2} \cdot e^{\pi i h/4}\right)^s e^{-2\pi i n a/2^4} \sum_z e^{-2\pi i n z/2^{u-3}}.$$

The inner sum vanishes unless $2^{u-3}|n$, and in that case it equals 2^{u-3} and we have

$$\left(\sqrt{2^u}\right)^s A_{2^u} = 2^{u-3} \sum_a \left(\sqrt{2}\right)^s e^{(\pi i a/4)-(2\pi i n a/2^u)}$$

$$= 2^{u-3+(s/2)} \sum_a e^{(2\pi i a/8)\{s-\nu\}}$$

$$= 2^{u-3+(s/2)} \sum_a \zeta_8^{\{s-\nu\}a}.$$

Putting $a = 2b+1$ with $b \in \{0,1,2,3\}$ we get

$$\left(\sqrt{2^u}\right)^s A_{2^u} = 2^{u-3+(s/2)} \sum_{b=0}^{3} \zeta_8^{\{s-\nu\}(1+2b)}$$

$$= 2^{u-3+(s/2)} \zeta_8^{\{s-\nu\}} \sum_{b=0}^{3} \zeta_8^{\{s-\nu\}2b}.$$

The inner sum vanishes unless $s - \nu \equiv 0 \pmod 4$, in which case it equals 4. Hence, when $s \equiv \nu \pmod 4$ we get

$$\left(\sqrt{2^u}\right)^s A_{2^u} = 2^{u-1+(s/2)} \zeta_8^{\{s-\nu\}}.$$

This completes the proof of Lemma 6.3. ∎

6.4.4 *Local Singular Series $L_p(w, \rho_s)$, s Odd and p Odd*

Let p be an odd prime and denote the local contribution to the singular series at p by

$$L_p(w, \rho_s) = 1 + \sum_{u=1}^{\infty} A_{p^u} \cdot \frac{1}{(p^u)^w}.$$

The goal of this section is to derive an explicit expression for this local factor.

Suppose p is an odd prime and s is a positive odd integer. From the defining relation for the Ramanujan function,

$$1 + \sum_{u=1}^{\infty} c_{p^u}(n)T^u = (1 - T) \sum_{u=0}^{\mathrm{ord}_p(n)} p^u \cdot T^u,$$

we get that, if p does not divide n, that is, $\mathrm{ord}_p(n) = 0$, then $c_{p^u} = 0$ for $u \geq 2$ and hence

$$A_{p^u} = 0 \quad \text{for} \quad u \geq 2.$$

Therefore,

$$L_p(w, \rho_s) = 1 + \left(\frac{(-1)^{(s-1)/2} \cdot n}{p} \right) p^{-(s-1)/2-w}. \tag{6.31}$$

Suppose now that n is square-free and $\mathrm{ord}_p(n) = 1$. Then we get

$$1 + \sum_{u=1}^{\infty} c_{p^u} T^u = (1 - T)(1 + pT),$$

and hence $c_{p^u} = 0$ for $u \geq 3$, and $c_{p^2} = -p$. Since $A_p = 0$ in this case, we have

$$L_p(w, \rho_s) = 1 - p^{1-s-2w} = \tag{6.32}$$
$$= \left(1 - \left(\frac{(-1)^{(s-1)/2} \cdot n}{p} \right) p^{-(s-1)/2-w} \right) \left(1 + \left(\frac{(-1)^{(s-1)/2} \cdot n}{p} \right) p^{-(s-1)/2-w} \right).$$

In the following we suppose that $n = 2^c p^e n'$, where $\gcd(n', 2p) = 1$. We consider two cases according as e is even or odd.

6.4.4.1 $e = 2b + 1$ is odd and $b > 0$

Since

$$1 + \sum_{u=1}^{\infty} c_{p^u} T^u = (1 - T) \sum_{u=0}^{\infty} p^u \cdot T^u,$$

we have $c_{p^u} = 0$ for $u > e + 1$. Therefore,

$$A_{p^u} = 0 \quad \text{for} \quad u > e + 1. \tag{6.33}$$

Since $e = 2b + 1$ is odd, p^{e+1} is a square and hence

$$\left(p^{e+1}\right)^{s/2} A_{p^{e+1}} = c_{p^{e+1}}(n) = -p^e.$$

If $u \le e$, then $n = p^{u-1}\left(p^{e+1-u} \cdot N\right) = p^{u-1} \cdot \nu$ with ν an integer divisible by p. Thus, if $u \le e$ and u is odd, then $p|\nu$ and

$$A_{p^u} = 0 \quad \text{for} \quad u \le e, \, u \text{ odd}. \tag{6.34}$$

Equations (6.33) and (6.34) imply that, with e odd, we have $A_{p^{2k+1}} = 0$ for all $k \ge 0$.

If u is even and $u \le e - 1$, then

$$(p^u)^{s/2} A_{p^u} = c_{p^u}(n) = p^{u-1}(p-1),$$

and

$$1 + \sum_{u=1}^{\infty} A_{p^u} \frac{1}{(p^u)^w} = 1 + \sum_{\substack{2 \le u \le e-1 \\ u \text{ even}}} (p^u)^{-(s/2)-w} p^u(p-1) - p^{-(e+1)((s/2)+w)} p^e$$

$$= 1 + \sum_{\substack{2 \le u \le e-1 \\ u \text{ even}}} \left(1 - \frac{1}{p}\right) \left\{ p^{u\{1-(s/2)-w\}} \right\} p^{e-(e+1)\{(s/2)+w\}}.$$

Let $T = p^{1-(s/2)}$. Then we have the geometric sum

$$\sum_{2 \le 2k \le E} T^{2k} = \sum_{1 \le k \le E/2} (T^2)^k = T^2 \left\{ \frac{T^E - 1}{T^2 - 1} \right\}. \tag{6.35}$$

Hence we have

$$L_p(w, \rho_s) = 1 + \left(1 - \frac{1}{p}\right) p^{2(1-(s/2)-w)} \frac{p^{(1-(s/2)-w)(e-1)} - 1}{p^{2(1-(s/2)-w)} - 1} - p^{e-(e+1)((s/2)+w)}$$

$$= 1 + \left(1 - \frac{1}{p}\right) p^{2-s-2w} \frac{p^{(2-s-2w)(e-1)/2} - 1}{p^{(2-s-2w)} - 1} - p^{-1-((e+1)/2)(s+w-2)}.$$

Letting $V = p^{2-s-2w}$, we get

$$L_p(w, \rho_s) = 1 + \left(1 - \frac{1}{p}\right) \cdot V \cdot \left\{ \frac{V^{(e-1)/2} - 1}{V - 1} \right\} - p^{-1} V^{(e+1)/2}$$

$$= \frac{1 - \frac{1}{p} V - V^{(e+1)/2} + \frac{1}{p} V^{1+(e+1)/2}}{1 - V}$$

$$= \frac{\left\{1 - \frac{1}{p} \cdot V\right\} \left\{1 - V^{(e+1)/2}\right\}}{1 - V}.$$

Substituting back the value $V = p^{2-s-2w}$ into the last expression, we get

$$L_p(w, \rho_s) = \frac{\left(1 - p^{1-s-2w}\right) \left(1 - p^{(2-s-2w)(e+1)/2}\right)}{1 - p^{2-s-2w}}.$$

This finishes the case of e odd.

6.4.4.2 $e = 2b$ is even and $b > 0$

In this case, we still have

$$c_{p^u}(n) = 0 \quad \text{if} \quad u > e + 1.$$

If $u = e + 1$, so that u is odd, then $n = p^{u-1} \cdot \nu = p^e \cdot \nu$, $p \nmid \nu$, and

$$(p^u)^{s/2} A_{p^u} = -c_{p^u}(n) \left(\frac{(-1)^{(s-1)/2} \cdot \nu}{p} \right) = p^e \cdot \left(\frac{(-1)^{(s-1)/2} \cdot \nu}{p} \right).$$

If $u \le e - 1$ and u is odd, then $n = p^{u-1}(p^{e+1-u} \cdot N) = p^{u-1} \cdot \nu$, and $p | \nu$, so that

$$A_{p^u} = 0 \quad \text{for } u \le e - 1, \quad u \text{ odd}.$$

Also, if $u = e + 1$, we obtain

$$A_{p^{e+1}} \cdot \frac{1}{(p^{e+1})^w} = p^{e-(e+1)s/2+(1/2)-(e+1)w} \left(\frac{(-1)^{(s-1)/2} \cdot \nu}{p} \right)$$

$$= p^{e-e(s/2)-(s-1)/2-(e+1)w} \left(\frac{(-1)^{(s-1)/2} \cdot \nu}{p} \right)$$

$$= p^{e-(e+1)s/2-(e+1)w+(1/2)} \left(\frac{(-1)^{(s-1)/2} \cdot \nu}{p} \right).$$

If $u > e + 1$, then $A_{p^u} = 0$. Thus $L_p(w, \rho_s) =$

$$1 + \sum_{\substack{u=1 \\ u \text{ even}}}^{e} A_{p^u} \cdot \frac{1}{(p^u)^w} + p^{e+(1/2)-(e+1)s/2-(e+1)w} \left(\frac{(-1)^{(s-1)/2} \cdot \nu}{p} \right)$$

$$= 1 + \sum_{\substack{u=2 \\ u \text{ even}}}^{e-2} A_{p^u} \cdot \frac{1}{(p^u)^w} + A_{p^e} \cdot \frac{1}{(p^e)^w} + \frac{p^{e+(1/2)-(e+1)s/2}}{(p^{e+1})^w} \left(\frac{(-1)^{(s-1)/2} \cdot \nu}{p} \right).$$

Since e is even and

$$A_{p^e} \cdot \frac{1}{(p^e)^w} = c_{p^e}(n)(p^e)^{-(s/2)-w}$$

$$= p^e \left(1 - \frac{1}{p} \right)(p^e)^{-(s/2)} p^{-ew}$$

$$= p^{e-es/2-ew} - p^{e-1-es/2-ew},$$

we get

$$L_p(w, \rho_s) = 1 + \sum_{\substack{u=2 \\ u \text{ even}}}^{e-2} c_{p^u}(n)(p^u)^{-(s/2)-uw} - p^{e-1-es/2-ew}$$

$$+ p^{e-es/2-ew} \left(1 + \left(\frac{\nu'}{p} \right) p^{-(s-1)/2-w} \right),$$

where we have put $\nu' = (-1)^{(s-1)/2} \cdot \nu$. Using the geometric sum

$$\left(1 - \frac{1}{p}\right) \sum_{1 \le k \le (e-2)/2} p^{(2-s)k-2kw} = \left(1 - \frac{1}{p}\right) p^{2-s-2w} \cdot \frac{p^{(2-s-2w)(e-2)/2} - 1}{p^{2-s-2w} - 1},$$

we can write

$$L_p(w, \rho_s) = 1 + \left(1 - \frac{1}{p}\right) p^{2-s-2w} \left\{ \frac{p^{(2-s-2w)(e-2)/2} - 1}{p^{2-s-2w} - 1} \right\} - p^{e-1-es/2-ew}$$
$$+ p^{e-es/2-ew} \left(1 + \left(\frac{\nu'}{p}\right) p^{-(s-1)/2-w}\right).$$

To simplify the last expression we put $T = p^{2-s-2w}$ and note that

$$1 + \left(1 - \frac{1}{p}\right) p^{2-s-2w} \left\{ \frac{p^{(2-s-2w)(e-2)/2} - 1}{p^{2-s-2w} - 1} \right\} - p^{e-1-es/2-ew}$$

$$= \frac{T - 1 + \left(1 - \frac{1}{p}\right) T \left(T^{(e-2)/2} - 1\right) - (T-1)p^{-1} \cdot T^{e/2}}{T - 1}$$

$$= \frac{1 - T^{e/2} - \frac{1}{p} T + \frac{1}{p} T \cdot T^{e/2}}{T - 1}$$

$$= \frac{\left(1 - \frac{1}{p} T\right)\left(1 - T^{e/2}\right)}{1 - T}.$$

Thus we get

$$L_p(w, \rho_s) = \frac{\left(1 - \frac{1}{p} p^{2-s-2w}\right)\left(1 - p^{(2-s-2w)(e/2)}\right)}{1 - p^{2-s-2w}}$$
$$+ p^{e-es/2-ew} \left(1 + \left(\frac{\nu'}{p}\right) p^{-(s-1)/2-w}\right),$$

an expression that holds true for $n = p^e \cdot \nu$ and $\nu' = (-1)^{(s-1)/2}\nu$.

Let us combine the cases of even and odd e into a single formula. If we put $e = 2e' + \delta$ with $\delta \in \{0,1\}$, we can finally write

$$L_p(w, \rho_s) = \frac{\left(1 - \frac{1}{p} p^{2-s-2w}\right)\left(1 - p^{(2-s-2w)(e+\delta)/2}\right)}{1 - p^{2-s-2w}}$$
$$+ (1 - \delta)p^{(2-s-2w)e/2} \left(1 + \left(\frac{\nu'}{p}\right) p^{-(s-1)/2-w}\right).$$

This completes the evaluation of the local factor $L_p(w, \rho_s)$ when p is odd and s is odd.

For later use we note that the above result can also be written in the following form. Let $\chi_{n^*}(p) = ((-1)^{(s-1)/2} \cdot n/p)$ and $\mathrm{ord}_p(n) = e = 2e' + \delta$ with $\delta \in \{0,1\}$, we then have

$$L_p(0, \rho_s) = (1 - p^{1-s}) \times$$

$$\left\{ 1 + \left(p^{2-s}\right)^1 + \cdots + \left(p^{2-s}\right)^{(e+\delta)/2-1} + (1 - \delta)\frac{\left(p^{2-s}\right)^{(e+\delta)/2}}{1 - \chi_{n^*}(p)p^{-(s-1)/2}} \right\}.$$

This formula will facilitate the calculation of the singular series in the general case when n is not necessarily square-free.

6.4.5 Local Singular Series $L_2(w, \rho_s)$, s Odd

If w is a complex number, we put

$$L_2(w, \rho_s) = 1 + \sum_{k=1}^{\infty} \frac{A_{2^k}}{(2^k)^w}.$$

The Euler product and corresponding Dirichlet series

$$L(w, \rho_s) = \prod_p L_p(w, \rho_s) = \sum_{k=1}^{\infty} \frac{A_k}{k^w}$$

converges for $\Re(w)$ sufficiently large to a meromorphic function, and, as we shall see, it has an analytic continuation for other values of w, including $w = 0$, which even make sense when s is small, say, $s = 1$ or 2. When dealing with the singular series for $s > 4$, there is no need to introduce the auxiliary complex variable w and in the formal calculation we may assume $w = 0$.

As in the case of p odd, our principal tool is Lemma 6.3. We recall that by direct calculation we have $A_2 = 0$. We suppose first that n is not divisible by 4, that is, $n = n'$ or $n = 2n'$ with n' odd.

(A) If n is odd and

(i) $2^{u+1} \nmid 8n$, then $a_{2^u} = 0$ when $u \geq 4$ is even.

(ii) $2^u \nmid 8n$, then $a_{2^u} = 0$ when $u > 3$ is odd.

Let us introduce the abbreviations

$$\zeta_8 = e^{2\pi i/8}, \quad \Theta^- = \frac{1}{2}\left\{ \zeta_8^{(s-2n)} + \zeta_8^{-(s-2n)} \right\}, \quad \Theta^+ = \frac{1}{2}\left\{ \zeta_8^{(s-n)} + \zeta_8^{+(s-n)} \right\}.$$

One verifies readily that

$$u = 0 : A_1 = 1,$$
$$u = 1 : A_2 = 0,$$
$$u = 2 : A_{2^2} = 2^{-(s/2)+1}\Theta^-,$$
$$u = 3 : A_{2^3} = \begin{cases} 0 & \text{if } n \not\equiv s \pmod 4, \\ 2^{(-(s/2+1))2}\Theta^+ & \text{if } n \equiv s \pmod 4, \end{cases}$$
$$u \geq 4 : A_{2^u} = 0.$$

In this case we have

$$L_2(w, \rho_s) = 1 + 0 + \frac{T}{2^{2w}}\Theta^-,$$

unless $n \equiv s \pmod 4$, in which case we get

$$L_2(w, \rho_s) = 1 + 0 + \frac{T}{2^{2w}}\Theta^- + \frac{T^2}{2^{3w}}\Theta^+.$$

Here and in the following we put

$$T = 2^{-(s/2)+1}.$$

(B) We suppose $n = 2n'$ with n' odd. Then, if $2^{u+1} \nmid 8n$, we have $A_{2^u} = 0$ for even $u \geq 4$ and, if $2^u \nmid 8n$, we have $A_{2^u} = 0$ for odd $u \geq 5$ If $8n = 2^3 \cdot \nu$, then ν is even and $s \not\equiv \nu \pmod 4$, and hence $A_{2^3} = 0$. We get in this case

$$L_s(w, \rho_s) = 1 + \frac{T}{2^{2w}}\Theta^-,$$

unless $n \equiv s \pmod 4$, in which case

$$L_s(w, \rho_s) = 1 + \frac{T}{2^{2w}}\Theta^- + \frac{T^2}{2^{3w}}\Theta^+.$$

We consider now the general case

$$n = 2^a \cdot n', \quad n' \text{ odd}, \quad a \geq 2.$$

(C) If u is odd and $2^u \nmid 2^{3+a}n'$, that is, $u > 3 + a$, then $A_{2^u} = 0$. If u is even and $2^{u+1} \nmid 2^{3+a}n'$, that is, $u > 2 + a$, then $A_{2^u} = 0$. If u is odd and $2^u \| 2^{3+a}n'$, that is, $u = 2 + a$, then $A_{2^u} = 0$, unless $s \equiv \nu \pmod 4$, in which case we have

$$A_{2^{3+a}} = T^{2+a}\Theta^+, \quad \nu = \frac{n}{2^a} = n'.$$

(D) If u is odd and $1 < u < 3 + a$, then $s - \nu \not\equiv 0 \pmod 4$ and $A_{2^u} = 0$. If u is even and $2 \leq u \leq 2 + a$, then

$$A_{2^u} = T^{u-1}\Theta^-.$$

We consider separately the cases $a = 2\ell$ and $a = 2\ell + 1$.

(E) From Lemma 6.3 we have

$$A_{2a+2} = \begin{cases} 0 & \text{if } a \text{ is odd,} \\ T^{a+1}\Theta^- & \text{if } a \text{ is even.} \end{cases}$$

(F) In a similar manner we have

$$A_{2a+1} = \begin{cases} T^a\Theta^- & \text{if } a \text{ is odd,} \\ 0 & \text{if } a \text{ is even.} \end{cases}$$

(G) Finally if u is even, we have

$$A_{2u} = T^{u-1}\Theta^- \quad \text{if } u < a + 1.$$

Note that in case (G) we have $\nu = 8n'$ and $\zeta_8^\nu = 1$. Hence,

$$A_{2u} = T^{u-1}\Theta^- \quad \text{if } u \text{ is even and } u \le a,$$

with $\Theta^- = (1/2)(\zeta_8^s + \zeta_8^{-s})$, a term independent of u or n.

We are now in a position to sum the series $L_2(w, \rho_s)$ in closed form:

$$L_2(w, \rho_s) = \sum_{u=0}^{a+3} \frac{A_{2u}}{(2^u)^w} = 1 + \sum_{\substack{2 \le u \le a \\ u \text{ even}}} \frac{A_{2u}}{(2^u)^w} + S,$$

where

$$S = \frac{A_{2u+1}}{(2^{u+1})^w} + \frac{A_{2u+2}}{(2^{u+2})^w} + \frac{A_{2u+3}}{(2^{u+3})^w}.$$

For the middle sum we have

$$\sum_{\substack{2 \le u \le a \\ u \text{ even}}} \frac{A_{2u}}{(2^u)^w} = \sum_{j=1}^{\ell} \Theta^- T^{2j-1} \cdot 2^{-2jw}$$

$$= \Theta^- \frac{1}{T} \sum_{j=1}^{\ell} (T^2 \cdot 2^{-2w})^j$$

$$= \Theta^- \frac{1}{T}(T^2 2^{-2w}) \left\{ \frac{1 - (T^2 2^{-2w})^\ell}{1 - (T^2 2^{-2w})} \right\}.$$

As for the sum S, we note that by (A), (B), (C), (D) and (E) we have

$$S = \Theta^- T^a 2^{-(a+1)w}$$

if a is odd or $s \not\equiv \nu \pmod 4$, where $\nu = 8n/2^{a+1} = 4n'$, with n' odd, and

$$\Theta^- = \frac{1}{2}\left\{ \zeta_8^{s-\nu} + \zeta_8^{-(s-\nu)} \right\}.$$

Otherwise we have

$$S = \Theta^- T^{a+1} 2^{-(a+2)w} + \Theta^+ T^{a+2} 2^{-(a+3)w}$$

if $a = 2\ell$ and $s \equiv \nu \pmod 4$, where $\nu = 8n/2^{3+a} = n'$ and

$$\Theta^+ = \frac{1}{2}\left\{\zeta_8^{n'-s} + \zeta_8^{(n'-s)}\right\}.$$

We wish to transform the expression obtained above for $L_2(w, \rho_s)$. Here we shall assume that $w = 0$. We prove the following result.

LEMMA 6.4

Let s be an odd positive integer and set $T = 2/2^{s/2}$. Let n be a positive integer and set $n = 2^a \cdot n'$ with n' odd. Put $a = 2\ell + \delta$ with $\delta \in \{0, 1\}$. Then $L_2(0, \rho_s)$ is given by this polynomial in T:

$$1 + 0 + \frac{1}{\sqrt 2}\left(\frac{s}{2}\right)\left\{T \cdot \frac{1 - T^{a-\delta}}{1 - T^2} + (-1)^{a+(s-n')(1-\delta)/2} T^{a+1-\delta}\right\}$$
$$+ (1 - \delta)(-1)^{(s-n')/4} T^{a+2-\delta},$$

with the understanding that $L_2(0, \rho_s) = 1$ if $a = 1$, and the last term is included only if $s - n' \equiv 0 \pmod 4$.

PROOF   Since s is odd, we must have $s \equiv 1, 3, 5,$ or $7 \pmod 8$, and one verifies directly that

$$\Theta^- = \frac{1}{2}\left(\zeta_8^s + \zeta_8^{-s}\right) = \frac{1}{\sqrt 2}\left(\frac{s}{2}\right),$$

where $(s/2)$ is the Gauss character $(s/2) = (-1)^{(s^2-1)/8}$. Using $a = 2\ell + \delta$, or equivalently, $a - \delta = 2\ell$, we get

$$1 + \Theta^- T\left\{\frac{1 - T^{2\ell}}{1 - T^2}\right\} = 1 + \frac{1}{\sqrt 2}\left(\frac{s}{2}\right) T\left\{\frac{1 - T^{a-\delta}}{1 - T^2}\right\}.$$

The Theta constants appearing in the expression $S = \Theta^- T^{a+1} + \Theta^+ T^{a+2}$ have the following values.

With $\nu = 8n/2^{a+1} = 4n'$, where n' is odd, we have $s - \nu \equiv 1, 3, 5,$ or $7 \pmod 8$ and

$$\Theta^- = \frac{1}{2}\left\{\zeta_8^{s-\nu} + \zeta_8^{-(s-\nu)}\right\} = \frac{1}{\sqrt 2}\left(\frac{s - 4n'}{2}\right) = \frac{1}{\sqrt 2}(-1)^{((s-4n')^2-1)/8}.$$

Now

$$\frac{(s - 4n')^2 - 1}{8} \equiv \frac{s^2 - 1}{8} + 1 \pmod 2,$$

and hence

$$\Theta^- = -\frac{1}{\sqrt{2}}\left(\frac{s}{2}\right) = -\frac{1}{\sqrt{2}}(-1)^{(s^2-1)/8}.$$

Similarly for Θ^+ we have, if $s \equiv \nu \pmod 4$, with $\nu = 8n/2^{a+3} = n'$ odd, $a+3$ odd,

$$\Theta^+ = \zeta_8^{s-n'} = \begin{cases} (-1)^{(s-n')/4} & \text{if } s - n' \equiv 0 \pmod 4, \\ 0 & \text{otherwise.} \end{cases}$$

The last two terms in the statement of the lemma come from the two possible terms that can appear in the sum S evaluated earlier, with the understanding that when $s \not\equiv n' \pmod 4$, then $(-1)^{(s-n')/2} = -1$. We note also that

$$\left(\frac{s-2n'}{2}\right) = (-1)^{(n'-1)/2}\left(\frac{s}{2}\right).$$

This completes the proof of Lemma 6.4. ∎

6.4.6 Local Singular Series $L_p(w, \rho_s)$, s *Even*

The case of p an odd prime is simple. By Lemma 6.2 we have for s even and p odd,

$$(p^u)^{s/2} A_{p^u} = \begin{cases} c_{p^u}(n) & \text{if } u \text{ is even} \\ c_{p^u}(n)\left(\frac{(-1)^{s/2}}{p}\right) & \text{if } u \text{ is odd.} \end{cases}$$

If we put

$$\chi_s(p) = \left(\frac{(-1)^{s/2}}{p}\right) \quad \text{and} \quad \chi_s(p^u) = (\chi_s(p))^u,$$

then we have

$$(p^u)^{s/2} A_{p^u} = c_{p^u}(n)\chi_s(p^u).$$

If we write $n = p^e \cdot n'$, where $p \nmid n'$, and $T = \chi_s(p)/p^{s/2}$, then we have

$$1 + \sum A_{p^u}(n) = \sum c_{p^u}(n)\left(\frac{\chi_s(p)}{p^{s/2}}\right)^u$$

$$= \left(1 - \frac{\chi_s(p)}{p^{s/2}}\right) \sum_{0 \le k \le \mathrm{ord}_p(n)} p^k \left(\frac{\chi_s(p)}{p^{s/2}}\right)^k$$

$$= (1 - T)\left\{\frac{1 - (pT)^{\mathrm{ord}_p(n)+1}}{1 - pT}\right\}.$$

Now we consider the case of $p = 2$.

LEMMA 6.5

Suppose s is an even integer, $n = 2^e \cdot n'$, with n' odd, and $T = 2/2^{s/2}$. We then have

$$L_2(0, \rho_s) = 1 + 0 + \Theta_0 \sum_{2 \le k \le e} T^{k-1} + \Theta_1 T^e + \Theta_2 T^{e+1},$$

where, for $i \in \{0, 1, 2\}$ we have put

$$\Theta_i = \begin{cases} 0 & \text{if } s - 2^{3-i} n' \equiv 2 \text{ or } 6 \pmod 8, \\ (-1)^{(s - 2^{3-i} n')/4} & \text{if } s - 2^{3-i} n' \equiv 0 \text{ or } 4 \pmod 8. \end{cases}$$

If $e = 1$, then $L_2(0, \rho_s) = 1$.

PROOF   Recall that Lemma 6.3 gives the following values for A_{2^u}: $A_1 = 1$ and $A_2 = 0$. Let $n = 2^e \cdot n'$ with n' odd. Let $\nu = 8n/2^u$. Let

$$\Theta^{\pm} = \frac{1}{2} \left\{ \zeta_8^{(s-\nu)} + \zeta_8^{\pm(s-\nu)} \right\}.$$

For u even we have

$$\frac{A_{2^u}}{\left(\frac{2}{2^{s/2}}\right)^{u-1}} = \begin{cases} 0 & \text{if } 2^{u+1} \nmid 8n, \\ \Theta^- & \text{if } 2^{u+1} | 8n, \end{cases} \tag{6.36}$$

while for u odd we have

$$\frac{A_{2^u}}{\left(\frac{2}{2^{s/2}}\right)^{u-1}} = \begin{cases} 0 & \text{if } 2^u \nmid 8n \text{ or if } s \not\equiv \nu \pmod 4, \\ \Theta^+ & \text{if } 2^u | 8n \text{ and } s \equiv \nu \pmod 4. \end{cases} \tag{6.37}$$

Throughout the discussion we assume s is even. We note first that $A_{2^u} = 0$ (a) if u is even and $u > 2 + e$, and (b) if u is odd and $u > 3 + e$. Hence, $A_{2^u} = 0$ for all $u > 3 + e$. Now note that if s is even and $u = 3 + e$ is odd, then

$$s \equiv \nu \equiv n' \pmod 4$$

cannot hold because n' is odd. Hence, $A_{2^u} = 0$ if $u \ge 3 + e$. Thus we have

$$L_2(0, \rho_s) = 1 + 0 + \sum_{2 \le k \le e} A_{2^k} + A_{2^{e+1}} + A_{2^{e+2}}.$$

For even $k \le e$ we have by (6.36) with $\nu = 8n/2^k$,

$$\frac{A_{2^k}}{\left(\frac{2}{2^{s/2}}\right)^{k-1}} = \frac{e^{2\pi i(s-\nu)/8} + e^{-2\pi i(s-\nu)/8}}{2}$$

$$= \begin{cases} 0 & \text{if } (s-\nu)/2 \equiv 1 \pmod 2, \\ (-1)^{(s-\nu)/4} & \text{if } (s-\nu)/2 \equiv 0 \pmod 2, \end{cases}$$

$$= \begin{cases} 0 & \text{if } s - \nu \equiv 2 \pmod 4, \\ (-1)^{(s-\nu)/4} & \text{if } s - \nu \equiv 0 \pmod 4, \end{cases}$$

$$= \begin{cases} 0 & \text{if } s - \nu \equiv 2 \text{ or } 6 \pmod 8, \\ (-1)^{(s-\nu)/4} & \text{if } s - \nu \equiv 0 \text{ or } 4 \pmod 8. \end{cases}$$

For odd $k \le e$ we have by (6.37),

$$\frac{A_{2^k}}{\left(\frac{2}{2^{s/2}}\right)^{k-1}} = e^{2\pi i(s-\nu)/8} = \begin{cases} (-1)^{(s-\nu)/4} & \text{if } s - \nu \equiv 0 \pmod 4, \\ 0 & \text{if } s - \nu \equiv 2 \pmod 4. \end{cases}$$

This is equivalent to

$$\frac{A_{2^k}}{\left(\frac{2}{2^{s/2}}\right)^{k-1}} = \begin{cases} (-1)^{(s-\nu)/4} & \text{if } s - \nu \equiv 0 \text{ or } 4 \pmod 8, \\ 0 & \text{if } s - \nu \equiv 2 \text{ or } 6 \pmod 8. \end{cases}$$

Since $\nu \equiv 0 \pmod 8$ we have that

$$\frac{A_{2^k}}{\left(\frac{2}{2^{s/2}}\right)^{k-1}} = \begin{cases} (-1)^{s/4} & \text{if } s - \nu \equiv 0 \text{ or } 4 \pmod 8, \\ 0 & \text{if } s - \nu \equiv 2 \text{ or } 6 \pmod 8. \end{cases}$$

This gives the value of Θ_0.

Now suppose $u = e + 1$. In this case $\nu = 4n'$. If $u = e + 1$ is even, then $2^{u+1} | 8n$ and

$$\frac{A_{2^u}}{\left(\frac{2}{2^{s/2}}\right)^{u-1}} = \Theta^- = \frac{e^{2\pi i(s-\nu)/8} + e^{-2\pi i(s-\nu)/8}}{2}$$

$$= \frac{i^{(s-\nu)/2} + i^{-(s-\nu)/2}}{2}$$

$$= \begin{cases} 0 & \text{if } (s-\nu)/2 \equiv 1 \pmod 2, \\ (-1)^{(s-\nu)/4} & \text{if } (s-\nu)/2 \equiv 0 \pmod 2, \end{cases}$$

$$= \begin{cases} 0 & \text{if } s - \nu \equiv 2 \text{ or } 6 \pmod 8, \\ (-1)^{(s-\nu)/4} & \text{if } s - \nu \equiv 0 \text{ or } 4 \pmod 8. \end{cases}$$

If $u = e + 1$ is odd, then $2^{u+1} | 8n$ and

$$\frac{A_{2^u}}{\left(\frac{2}{2^{s/2}}\right)^{u-1}} = \begin{cases} 0 & \text{if } s \not\equiv \nu \pmod 4, \\ \Theta^+ & \text{if } s \equiv \nu \pmod 4. \end{cases}$$

Since

$$\Theta^+ = \frac{1}{2}\left\{ e^{2\pi i(s-\nu)/8} + e^{+2\pi i(s-\nu)/8} \right\} = (-1)^{(s-\nu)/4},$$

$$\frac{A_{2^u}}{\left(\frac{2}{2^{s/2}}\right)^{u-1}} = \begin{cases} 0 & \text{if } s - \nu \equiv 2 \text{ or } 6 \pmod 8, \\ (-1)^{(s-\nu)/4} & \text{if } s - \nu \equiv 0 \text{ or } 4 \pmod 8. \end{cases}$$

This gives the value of Θ_1.

Now suppose $u = e + 2$. In this case $\nu = 2n'$. If $u = e + 2$ is even, then $2^{u+1}|8n$ and

$$\frac{A_{2^u}}{\left(\frac{2}{2^{s/2}}\right)^{u-1}} = \Theta^- = \frac{e^{2\pi i(s-\nu)/8} + e^{-2\pi i(s-\nu)/8}}{2}$$

$$= \frac{i^{(s-\nu)/2} + i^{-(s-\nu)/2}}{2}$$

$$= \begin{cases} 0 & \text{if } (s-\nu)/2 \equiv 1 \pmod 2, \\ (-1)^{(s-\nu)/4} & \text{if } (s-\nu)/2 \equiv 0 \pmod 2, \end{cases}$$

$$= \begin{cases} 0 & \text{if } s - \nu \equiv 2 \text{ or } 6 \pmod 8, \\ (-1)^{(s-\nu)/4} & \text{if } s - \nu \equiv 0 \text{ or } 4 \pmod 8. \end{cases}$$

If $u = e + 2$ is odd, then $2^u|8n$ and

$$\frac{A_{2^u}}{\left(\frac{2}{2^{s/2}}\right)^{u-1}} = \begin{cases} 0 & \text{if } s \not\equiv \nu \pmod 4, \\ \Theta^+ & \text{if } s \equiv \nu \pmod 4. \end{cases}$$

$$= \begin{cases} 0 & \text{if } s - \nu \equiv 2 \text{ or } 6 \pmod 8, \\ (-1)^{(s-\nu)/4} & \text{if } s - \nu \equiv 0 \text{ or } 4 \pmod 8. \end{cases}$$

This gives the value of Θ_2.

This completes the proof of 6.5. ∎

Example 6.2

The following table gives the values of the Thetas for $s \in \{2, 4, 6, 8\}$. Recall that $n = 2^e \cdot n'$ with n' odd. (For $s = 6$, we have $\Theta_2 = (-1)^{(3-n')/2} = -(-1)^{(1-n')/2}$.)

s	Θ_0	Θ_1	Θ_2
2	0	0	$+(-1)^{(1-n')/2}$
4	-1	1	0
6	0	0	$-(-1)^{(1-n')/2}$
8	1	-1	0

6.4.7   *Examples*

This section contains four extended examples to illustrate the use of the theory we have just developed.

6.4.7.1   *Sum of Two Squares*

As we have seen in earlier sections, the formula for the number $r_2(n)$ of representations of an integer n as a sum of two squares has a relatively simple structure and involves the divisors of n. Jacobi's combinatorial formula (6.8), which we write as

$$\left(\sum_{n=-\infty}^{\infty} q^{n^2}\right)^2 = 1 + 4\sum_{m=1}^{\infty} (-1)^{m-1} \frac{q^{2m-1}}{1-q^{2m-1}},$$

has on the left side the generating function for $r_2(n)$ and on the right side, after expanding the Lambert series as a power series in q, the arithmetic function expressing $r_2(n)$ in terms of the divisors of n. More precisely, we have

$$1 + \sum_{n=1}^{\infty} r_2(n)q^n = 1 + 4\sum_{m=1}^{\infty} \left\{ \sum_{\substack{d|m \\ d \text{ odd}}} (-1)^{(d-1)/2} \right\} q^m.$$

The discussion in this section is aimed at a proof of the identity

$$2r_2(n) = \rho_2(n),$$

first noted by Ramanujan.

We recall that the singular series is defined by

$$\rho_s(n) = \frac{n^{(s/2)-1}}{\pi^{-s/2}\Gamma(s/2)} \cdot S,$$

where $S = \sum_{k=1}^{\infty} A_k$, and

$$A_1 = 1, \qquad A_k = \sum_{\substack{h=1 \\ \gcd(h,k)=1}}^{k} \left(\frac{S(h,k)}{k}\right)^s e^{-2\pi i n h/k},$$

$$S(h,k) = \sum_{j=1}^{k} e^{2\pi i j^2 h/k}.$$

Since the A_k have been shown to be multiplicative in k, we have

$$S = \prod_p L_p(0, \rho_s),$$

where

$$L_p(0, \rho_s) = 1 + \sum_{k=1}^{\infty} A_{p^k}.$$

Using the results obtained earlier for $L_p(0, \rho_s)$, when s is even, we get

$$\rho_s(n) = \frac{n^{(s/2)-1}}{\pi^{-s/2}\Gamma(s/2)} \cdot L_2(0, \rho_s) \prod_{\substack{p \\ p \text{ odd}}} L_p(0, \rho_s)$$

$$= \frac{n^{(s/2)-1}}{\pi^{-s/2}\Gamma(s/2)} L_2(0, \rho_s) \prod_{\substack{p \\ p \text{ odd}}} \left(1 - \frac{\psi_s(p)}{p^{s/2}}\right) \prod_{\substack{p^e \| n \\ p \text{ odd}}} \left\{ \frac{1 - \left(\frac{\psi_s(p)}{p^{(s/2)+1}}\right)^{e+1}}{1 - \frac{\psi_s(p)}{p^{(s/2)+1}}} \right\},$$

Here the symbol $\psi_s(p) = ((-1)^{s/2}/p)$ is the Kronecker character arising from the first supplement to the law of quadratic reciprocity. Its value is given by

$$\psi_s(p) = (-1)^{\frac{p-1}{2} \cdot \frac{s}{2}}.$$

Its conductor is $\mathfrak{f} = \mathfrak{f}_{\psi_s} = 4$. We thus have

$$\rho_s(n) = \frac{1}{\pi^{-s/2}\Gamma(s/2)L(s/2,\psi_s)} \cdot L_2(0,\rho_s) \cdot n^{(s/2)-1} \left\{ \sum_{\substack{d|n \\ d \text{ odd}}} \psi_s(d)\frac{1}{d^{(s/2)-1}} \right\},$$

$$(6.38)$$

where $L(s/2, \psi_s)$ is the Dirichlet L-function associated to ψ_s.

We recall the well-known functional equation for Dirichlet L-functions in a format suitable for calculation. If ψ is a Dirichlet character of conductor $\mathfrak{f} = \mathfrak{f}_\psi$ and if we define $\delta \in \{0,1\}$ so that $\psi(-1) = (-1)^\delta$, then

$$\pi^{-s}\Gamma(s)L(s,\chi) = W_\chi \cdot \mathfrak{f}^{(1/2)-s} \cdot \frac{2^{s-1}L(1-s,\overline{\chi})}{\cos(\pi(s-\delta)/2)}.$$

The root number

$$W_\chi = \frac{\tau(\chi)}{\sqrt{\mathfrak{f}}\,i^\delta}, \quad \text{where } \tau(\chi) = \sum_{a=1}^{\mathfrak{f}} \chi(a)e^{2\pi i a/\mathfrak{f}},$$

has absolute value 1. If χ is real, its value is actually equal to 1, a result which is equivalent to the main statement of the law of quadratic reciprocity. (See Theorem 3.13 or page 104 of the Appendix of Iwasawa [56].)

In our case, since ψ_s is real-valued and of conductor $\mathfrak{f} = 4$, we obtain

$$\pi^{-s/2}\Gamma(s/2)L(s/2,\psi_s) = 2^{-s/2}\frac{L(1-s/2,\psi_s)}{\cos(\pi(s/2-\delta)/2)}.$$

Combining this formula with (6.38) gives

$$\rho_s(n) = \frac{2^{-s/2}L_2(0,\rho_s)\cos(\pi(s/2-\delta)/2)n^{(s/2)-1}}{L(1-s/2,\psi_s)} \left\{ \sum_{\substack{d|n \\ d \text{ odd}}} \psi_s(d)d^{-(s/2)+1} \right\}.$$

$$(6.39)$$

From the theory of generalized Bernoulli numbers we also need the following formula (see Theorem 3.14 or Iwasawa [56], Theorem 1, page 11):

$$L(1-(s/2),\psi_s) = -\frac{B_{s/2,\psi_s}}{s/2},$$

where

$$F_{\psi_s}(t) \overset{\text{def}}{=} \sum_{a=1}^{\mathfrak{f}} \psi_s(a) \cdot \frac{te^{at}}{e^{\mathfrak{f}t}-1} = \sum_{n=0}^{\infty} B_{n,\psi_s} \cdot \frac{t^n}{n!}.$$

The final formula is

$$\rho_s(n) = -s(2n)^{(s/2)-1} \frac{\cos(\pi(s/2-\delta)/2)}{B_{s/2,\psi_s}} L_2(0,\rho_s) \left\{ \sum_{\substack{d|n \\ d \text{ odd}}} \psi_s(d) d^{-(s/2)+1} \right\},$$

$$(6.40)$$

where $\psi_s(-1) = (-1)^\delta$, with $\delta \in \{0,1\}$ and $L_2(0,\rho_s)$ is given by Lemma 6.3.

We now specialize this formula to the case $s = 2$. Here $\psi_s(-1) = -1$ and hence $\delta = 1$. Thus $\cos(\pi(s/2-\delta)/2) = 1$. The expression for $L_2(0,\rho_s)$ from Lemma 6.5 is

$$L_2(0,\rho_s) = 1 + 0 + \Theta_0 \sum_{2 \le k \le \text{ord}_2(n)} T^{k-1} + \Theta_1 T^{\text{ord}_2(n)} + \Theta_2 T^{\text{ord}_2(n)+1},$$

where $T = 2/2^{s/2}$. For the case $s = 2$ we have from Example 6.2,

$$\Theta_0 = 0, \quad \Theta_1 = 0 \quad \text{and} \quad \Theta_2 = (-1)^{(n'-1)/2},$$

where $n = 2^{\text{ord}_2(n) \cdot n'}$, with n' odd. Hence

$$L_2(0,\rho_s) = 1 + (-1)^{(n'-1)/2}.$$

Since $\psi_s(d) = (-1)^{(d-1)/2}$ for d odd we get

$$\rho_2(n) = -2 \frac{1+(-1)^{(n'-1)/2}}{B_{1,\psi_2}} \left\{ \sum_{\substack{d|n \\ d \text{ odd}}} (-1)^{(d-1)/2} \right\}. \qquad (6.41)$$

It only remains to find the value of B_{1,ψ_2}. To do this we note that since the conductor of ψ_2 is 4 and $\psi_2(3) = -1$, we have

$$F_{\psi_2}(t) = \frac{te^t}{e^{4t}-1} + \frac{\psi_2(3)te^{3t}}{e^{4t}-1}$$

$$= \frac{te^t(e^{2t}+1)}{(e^{2t}+1)(e^{2t}-1)}$$

$$= \frac{te^t}{e^{2t}-1} = \sum_{n=0}^{\infty} B_{n,\psi_s} \cdot \frac{t^n}{n!}.$$

An easy calculation shows that for $s = 2$,

$$F_{\psi_2}(t) \equiv -\frac{t}{2} \pmod{t^2},$$

and hence $B_{1,\psi_2} = -1/2$. Substituting into (6.41) gives the final formula

$$\rho_2(n) = 4\left(1+(-1)^{(n'-1)/2}\right) \left\{ \sum_{\substack{d|n \\ d \text{ odd}}} (-1)^{(d-1)/2} \right\}.$$

Since $r_2(n) = 0$, unless $n' \equiv 1 \pmod 4$, we obtain that

$$\rho_2(n) = 2 \cdot r_2(n).$$

REMARK 6.9   This result is particularly striking because the evaluation of the singular series is tantamount to the identity

$$\left(\sum_{n=-\infty}^{\infty} e^{\pi i \tau n^2}\right)^s = 1 + \sum_{a/c} \gamma(a/c)^s (c\tau - a)^{-s/2}, \text{ for } s = 4, \ldots, 8,$$

where $\gamma(a/c)$ is an eighth root of unity which describes the behavior of the theta function $\vartheta(\tau) = \sum_{n=-\infty}^{\infty} e^{\pi i \tau n^2}$ as τ approaches the rational cusp a/c, that is to say,

$$(c\tau - a)^{1/2}\vartheta(\tau) \longrightarrow \gamma(a/c), \quad c > 0, \gcd(a,c) = 1.$$

To extend the validity of the identity above to values of $s < 4$, one normally introduces a convergence factor, an idea originally due to Hecke, and defines for z a complex variable

$$\Phi_s(\tau, z) = 1 + \sum_{a/c} \gamma(a/c)^s (c\tau - a)^{-s/2} |c\tau - a|^{-z},$$

where a/c runs over all rational numbers and $\Re(z) > 2 - (s/2)$. Determining $\rho_s(n)$ is equivalent to calculating the Fourier expansion of the function $\Phi_s(\tau, z)$. The analytic theory of automorphic forms can be used to prove that

$$\Phi_s(\tau, 0) = \vartheta(\tau)^s \quad \text{for } s = 1, 3, 4, 5, 6, 7, 8,$$

and $\Phi_2(\tau, 0) = 2\vartheta(\tau)^2$. (See Siegel [103].)   ■

6.4.7.2 *Sum of Eight Squares*

We consider first the simple case of n square-free. In this case we deduce easily from Lemma 6.4 that the singular series is

$$\mathfrak{S} = \frac{L_2(0, \rho_s)}{1 - \frac{1}{2^4}} \cdot \frac{1}{\zeta(4)} \prod_{p|n} \left(1 + \frac{1}{p^3}\right),$$

so that

$$\rho_8(n) = \frac{\pi^4}{\Gamma(4)} \cdot n^3 \cdot \frac{L_2(0, \rho_s) \cdot 16}{15} \cdot \frac{90}{\pi^4} \prod_{p|n} \left(1 + \frac{1}{p^3}\right).$$

We also have from Lemma 6.5 that

$$L_2(0, \rho_s) = \begin{cases} 1 & \text{if } n \text{ is odd,} \\ \frac{7}{8} & \text{if } n \text{ is even.} \end{cases}$$

Thus, $\rho_8(n) = 16 \sum_{d|n} d^3$ if n is odd. If n is even, $L_2(0, \rho_s) = 1 - 1/2^3$, and we have

$$\rho_8(n) = 16 \left(1 - \frac{1}{2^3}\right) \sum_{d|n} d^3,$$

or

$$\rho_8(n) = 16 \left(\sum_{\substack{d|n \\ d \text{ even}}} d^3 - \sum_{\substack{d'|n \\ d' \text{ odd}}} d'^3 \right).$$

Now suppose n is an arbitrary positive integer with factorization $n = 2^\alpha \prod_p p^{\text{ord}_p(n)}$. If s is even and p is an odd prime, then by Lemma 6.2,

$$(p^u)^{s/2} A_{p^u} = \begin{cases} c_{p^u}(n) \left(\frac{(-1)^{s/2}}{p} \right) & \text{if } p^u \text{ is not square,} \\ c_{p^u}(n) & \text{if } p^u \text{ is square,} \end{cases}$$

If we let $\varepsilon = (-1)^{s/2}$, then we can write this as

$$(p^u)^{s/2} A_{p^u} = \left(\frac{\varepsilon}{p} \right)^u c_{p^u}(n).$$

Then we have

$$1 + \sum_{u=1}^{\infty} A_{p^u} = 1 + \sum_{u=1}^{\infty} c_{p^u}(n) \left(\frac{\varepsilon}{p^{s/2}} \right)^u$$

$$= \left(1 - \frac{\varepsilon}{p^{s/2}}\right) \sum_{k=0}^{\text{ord}_p(n)} p^k \left(\frac{\varepsilon}{p^{s/2}} \right)^k,$$

or

$$L_p(0, \rho_s) = \left(1 - \frac{\varepsilon}{p^{s/2}}\right) \sum_{k=0}^{\text{ord}_p(n)} \left(\frac{\varepsilon}{p^{(s/2)-1}} \right)^k.$$

In particular, when $s = 8$, we have $\varepsilon = 1$ and

$$L_p(0, \rho_s) = \left(1 - \frac{\varepsilon}{p^4}\right) \sum_{k=0}^{\text{ord}_p(n)} \left(\frac{1}{p^3} \right)^k.$$

The value of $L_2(0, \rho_s)$ is obtained from Lemma 6.3. In fact, if $2^u > 2$, we have

$$\left(2^{(s/2)-1}\right)^{u-1} A_{2^u} = \begin{cases} 0 & \text{if } u \text{ is even and } \text{ord}_2(n) + 2 < u, \\ \frac{1}{2}\left(\zeta_8^{(s-\nu)} + \zeta_8^{-(s-\nu)}\right) & \text{if } u \text{ is even and } \text{ord}_2(n) + 2 \geq u, \\ 0 & \text{if } u \text{ is odd and } \text{ord}_2(n) + 3 < u, \\ 0 & \text{if } s \not\equiv 8n/2^u \pmod 4, \\ \frac{1}{2}\left(\zeta_8^{(s-\nu)} + \zeta_8^{(s-\nu)}\right) & \text{if } s \equiv 8n/2^u \pmod 4, \end{cases}$$

where $\nu = 8n/2^u$. From this we deduce easily that for $s = 8$,

$$L_2(0, \rho_s) = 1 + 0 + T + \cdots + T^{\mathrm{ord}_2(n)-1} - T^{\mathrm{ord}_2(n)},$$

where $T = 1/8$, with the understanding that $L_2(0, \rho_s) = 1$ when $\mathrm{ord}_2(n) = 0$. With the values for $L_p(0, \rho_s)$ obtained earlier, we have

$$\rho_8(n) = 16n^3 \left\{ 1 + 0 + \left(\frac{1}{8}\right) + \left(\frac{1}{8}\right)^2 + \cdots + \left(\frac{1}{8}\right)^{\mathrm{ord}_2(n)-1} - \left(\frac{1}{8}\right)^{\mathrm{ord}_2(n)} \right\} \times$$

$$\prod_{\substack{p \\ p \text{ odd}}} \left(1 + \frac{1}{p^3} + \cdots + \left(\frac{1}{p^3}\right)^{\mathrm{ord}_p(n)} \right),$$

with the understanding that the expression in curly braces is 1 when $\mathrm{ord}_2(n) = 0$. This expression agrees with the earlier expression obtained for n square-free. Thus in all cases

$$\rho_8(n) = 16 \left(\sum_{\substack{d|n \\ d \text{ even}}} d^3 - \sum_{\substack{d'|n \\ d' \text{ odd}}} d'^3 \right).$$

6.4.7.3   Sum of s Squares, s Odd

We treat first the case when n is square-free. In this case, using the values already obtained for the local factor $L_p(0, \rho_s)$ when p is odd and n is square-free, and using

$$L_\infty(0, \rho_s) = \frac{n^{(s/2)-1}}{\pi^{-s/2}\Gamma(s/2)},$$

we see that the singular series is given by

$$\rho_s(n) = L_\infty(0, \rho_s)L_2(0, \rho_s) \prod_{p\nmid n} \left\{ 1 + \chi_{n^*}(p)p^{-(s-1)/2} \right\} \times$$

$$\prod_{p|n} \left\{ 1 - \frac{1}{p^{s-1}} \right\}$$

$$= L_\infty(0, \rho_s)L_2(0, \rho_s) \prod_{\substack{p \\ p \text{ odd}}} \left\{ 1 - \frac{1}{p^{s-1}} \right\} \prod_{\substack{p\nmid n \\ p \text{ odd}}} \frac{1}{1 - \chi_{n^*}(p)p^{-(s-1)/2}}$$

$$= \frac{L_\infty(0, \rho_s)L_2(0, \rho_s)}{1 - \frac{1}{2^{s-1}}} \cdot \frac{L((s-1)/2, \chi_{n^*})}{\zeta(s-1)},$$

where

$$L\left(\frac{s-1}{2}, \chi_{n^*}\right) = \prod_{\substack{p \nmid n \\ p \text{ odd}}} \frac{1}{1 - \chi_{n^*}(p)p^{-(s-1)/2}}$$

$$= \sum_{\substack{m=1 \\ \gcd(m,2n)=1}}^{\infty} \left(\frac{(-1)^{(s-1)/2} \cdot n}{m}\right) \frac{1}{m^{(s-1)/2}}.$$

The determination of the primitive character associated to the Kronecker symbol

$$\chi_{n^*}(a) = \left(\frac{(-1)^{(s-1)/2} \cdot n}{a}\right) \tag{6.42}$$

is required in order to obtain explicit formulas for the singular series $\rho_s(n)$. We now review the basic facts about the Kronecker symbol needed to solve this problem.

Let $d \equiv 0$ or $1 \pmod 4$ and let d not be a perfect square. We denote by $\chi_d(n) = \left(\frac{d}{n}\right)$, the Kronecker symbol, defined as follows:

$$\left(\frac{d}{1}\right) = 1$$

$$\left(\frac{d}{p}\right) = 0 \text{ if } p \text{ is prime and } p|d,$$

$$\left(\frac{d}{2}\right) = \begin{cases} 1 & \text{if } d \equiv 1 \pmod 8, \\ -1 & \text{if } d \equiv 5 \pmod 8, \end{cases}$$

$$\left(\frac{d}{p}\right) = \text{Legendre's symbol } \left(\frac{d}{p}\right), \text{ if } p > 2, \ p \text{ prime, } p \nmid d,$$

$$\left(\frac{d}{m}\right) = \prod_{r=1}^{\nu} \left(\frac{d}{p_r}\right) \text{ for } m = \prod_{r=1}^{\nu} p_r \text{ with } p_r \text{ prime.}$$

Note that $\left(\frac{d}{2}\right)$ in the definition is Jacobi's symbol $\left(\frac{2}{|d|}\right)$ for odd d. See Landau [64] Definition 20, Chapter VI, page 70, for this definition.

The basic property of the Kronecker symbol is that it is a character modulo d, that is, $\chi_d(a) = 0$ if $\gcd(a, d) = 1$, $\chi_d(1) = 1$, $\chi_d(a)\chi_d(b) = \chi_d(ab)$ for all a and b, and $\chi_d(a) = \chi_d(b)$ whenever $a \equiv b \pmod d$.

We recall that a number $d \equiv 0$ or $1 \pmod 4$, and not a perfect square, is called a *fundamental discriminant* if it is not divisible by the square of any odd prime and is either odd or $\equiv 8$ or $12 \pmod{16}$.

The following basic result appears on page 219 of Landau [64].

THEOREM 6.7 Fundamental Discriminants
Every integer $d \equiv 0$ or $1 \pmod 4$ that is not a square can be written uniquely in the form $d = fm^2$, where m is a positive integer and f is a fundamental discriminant.

We shall make use of the following relation. If $d = fm^2$ as above, then

$$\prod_p \frac{1}{1 - \left(\frac{d}{p}\right)p^{-s}} = \prod_{q|m}\left(1 - \left(\frac{f}{q}\right)q^{-s}\right) \times \prod_p \frac{1}{1 - \left(\frac{f}{p}\right)p^{-s}}.$$

If f is a fundamental discriminant, the Kronecker symbol χ_f has conductor f.

Given an odd positive integer s and a nonsquare positive integer n, we define

$$d = \begin{cases} (-1)^{(s-1)/2}n & \text{if } (-1)^{(s-1)/2}n \equiv 1 \pmod 4 \\ 4(-1)^{(s-1)/2}n & \text{otherwise.} \end{cases}$$

By Theorem 6.7, we can write $d = m^2 f$ with f a fundamental discriminant. With this notation we then have

$$L\left(s, \left(\frac{(-1)^{(s-1)/2}n}{*}\right)\right) = \sum_{k=1}^\infty \left(\frac{(-1)^{(s-1)/2}n}{k}\right)\frac{1}{k^s}$$

$$= \prod_{p|m}\left(1 - \left(\frac{f}{p}\right)\frac{1}{p^s}\right)L(s, \chi_f),$$

with χ_f a primitive character of conductor f.

LEMMA 6.6
Let s be an odd integer greater than 1. Let χ be a real primitive character of conductor f. We then have

$$\frac{L\left(\frac{s-1}{2}, \chi\right)}{\zeta(s-1)} = \frac{\pi^{-s/2}\Gamma\left(\frac{s}{2}\right)}{f^{(s/2)-1}} \cdot 2^{(s-1)/2} \cdot (-1)^{(s^2-1)/8} \cdot \frac{B_{(s-1)/2,\chi}}{B_{s-1}},$$

where

$$\sum_{a=1}^f \frac{\chi(a)te^{at}}{e^{ft}-1} = \sum_{n=0}^\infty B_{n,\chi}\cdot\frac{t^n}{n!}$$

and

$$\frac{te^t}{e^t-1} = \sum_{n=0}^\infty B_n\cdot\frac{t^n}{n!}.$$

PROOF   We shall use the functional equations for the Riemann zeta function and for Dirichlet L-functions in the following well-known form (see Theorems 3.12 and 3.13 or Iwasawa [56], page 5),

$$\zeta(s) = \frac{\zeta(1-s)}{2(2\pi)^{-s}\Gamma(s)\cos(\pi s/2)},$$

$$L(s,\chi) = \frac{f^{(1/2)-s}2^{s-1}L(1-s,\chi)}{\pi^{-s}\Gamma(s)\cos(\pi(s-\delta)/2)},$$

where $\delta \in \{0,1\}$ satisfies $s \equiv \delta \pmod 2$. Note that in the functional equation for $L(s,\chi)$ we have used that the root number $W_\chi = 1$.

Substituting $s-1$ in $\zeta(s)$ and $(s-1)/2$ in $L(s,\chi)$, and using the duplication formula,

$$2^{s-2}\Gamma\left(\frac{s-1}{2}\right)\Gamma\left(\frac{s}{2}\right) = \pi^{1/2}\Gamma(s-1),$$

for the Gamma function, we obtain the expression

$$\frac{L\left(\frac{s-1}{2},\chi\right)}{\zeta(s-1)} = \frac{\pi^{-s/2}\Gamma\left(\frac{s}{2}\right)}{f^{(s/2)-1}} \cdot \frac{L\left(1-\left(\frac{s-1}{2}\right),\chi\right)}{\zeta(1-(s-1))} \cdot 2^{(s-1)/2-1} \cdot \frac{\cos\frac{\pi}{2}(s-1)}{\cos\frac{\pi}{2}\left(\frac{s-1}{2}-\delta\right)}.$$

The evaluation of the Riemann zeta function in terms of Bernoulli numbers and of the Dirichlet L-functions in terms of generalized Bernoulli numbers give (see Theorem 3.14 or Iwasawa [56], pages 11 and 13), for odd s,

$$L\left(1-\left(\frac{s-1}{2}\right),\chi\right) = -\frac{B_{(s-1)/2,\chi}}{\frac{s-1}{2}}, \qquad \zeta(1-(s-1)) = \frac{B_{s-1}}{s-1}.$$

The lemma follows readily from these observations by noting that

$$\frac{\cos\frac{\pi}{2}(s-1)}{\cos\frac{\pi}{2}\left(\frac{s-1}{2}-\delta\right)} = (-1)^{(s^2-1)/8}.$$

∎

If $d = (-1)^{(s-1)/2}n$ is a fundamental discriminant, that is,

$$d = \begin{cases} D & \text{if } D \equiv 1 \pmod 4, \\ 4D & \text{if } D \equiv 2, 3 \pmod 4, \end{cases}$$

and D is square-free, then the Kronecker symbol

$$\chi(a) = \left(\frac{d}{a}\right)$$

is a primitive character of conductor $|d| = f$. In this case we have

$$\sum_{\substack{a=1 \\ a \text{ odd}}}^{\infty} \left(\frac{(-1)^{(s-1)/2}n}{a}\right)\frac{1}{a^w} = \left(1 - \frac{(-1)^{(s-1)/2}n}{2}\right)\frac{1}{2^w}L(w,\chi).$$

Using the expression obtained earlier for $\rho_s(n)$, together with Lemma 6.6, we find that, when s is odd, we have $\rho_s(n) =$

$$\frac{L_2(0,\rho_s)}{1-\frac{1}{2^{s-1}}} \cdot \left\{ 1 - \left(\frac{(-1)^{(s-1)/2}n}{2} \right) \frac{1}{2^{(s-1)/2}} \right\} \cdot 2^{(s-1)/2} \cdot (-1)^{(s^2-1)/8} \cdot \frac{B_{(s-1)/2,\chi}}{B_{s-1}}.$$

REMARK 6.10   If $d = (-1)^{(s-1)/2}n$ is not square-free, then we use the representation $d = m^2 \cdot f$ and modify the above formula appropriately by the inclusion of a number of local factors corresponding to primes p that divide m. The resulting formula is in all its essential factors, such as $L_2(0,\rho_s)$ and $B_{(s-1)/2,\chi}/B_{s-1}$, similar to the one above.   ∎

6.4.7.4 *Sum of Five Squares*

To illustrate the explicit nature of the formula just obtained for the singular series $\rho_s(n)$ when s is odd, we now develop some particular cases useful in numerical calculations.

Calculation of the first three terms of the Taylor expansion gives

$$\frac{te^{at}}{e^{ft}-1} = \frac{1}{f}\left\{ 1 + \left(a - \frac{f}{2} \right)t + \frac{1}{2}\left(a^2 - af + \frac{1}{6}f^2 \right)t^2 + O(t^3) \right\}.$$

This formula gives us

$$B_{1,\chi} = \sum_{a=1}^{f} \chi(a)\frac{a}{f},$$

and

$$B_{2,\chi} = \frac{1}{f}\sum_{a=1}^{f} \chi(a)\left(a - \frac{f}{2} \right)^2.$$

From Lemma 6.4, we have with n square-free that

$$L_2(0,\rho_5) = \begin{cases} 1 - \frac{1}{4} + (-1)^{(n-5)/4} \cdot \frac{1}{8} & \text{if } n \equiv 5 \pmod 4, \\ 1 + \frac{1}{4} & \text{if } n \not\equiv 5 \pmod 4. \end{cases}$$

Thus,

$$L_2(0,\rho_5) = \begin{cases} \frac{7}{8} & \text{if } n \equiv 5 \pmod 8, \\ \frac{5}{8} & \text{if } n \equiv 1 \pmod 8, \\ \frac{5}{4} & \text{if } n \not\equiv 5 \pmod 4. \end{cases}$$

If $n^* = (-1)^{(s-1)/2}n > 0$ is a square-free fundamental discriminant, then $\rho_s(n) =$

$$\frac{L_2(0,\rho_s)}{1-\frac{1}{2^{s-1}}} \cdot \left\{ 1 - \left(\frac{n^*}{2} \right) \frac{1}{2^{(s-1)/2}} \right\} \cdot 2^{(s-1)/2} \cdot (-1)^{(s^2-1)/8} \cdot \frac{B_{(s-1)/2,\chi}}{B_{s-1}}.$$

When $n \not\equiv 5 \pmod 4$, the above expression has to be modified by the factor

$$\left(\frac{n}{4n}\right)^{(s/2)-1} = \frac{1}{2^{s-2}}$$

to account for the fact that the conductor of $\chi(a) = \left(\frac{n^*}{a}\right)$ is $4n$.

For $m \equiv 1 \pmod 4$ and square-free and $s = 5$, we have, using $B_4 = -1/30$, that

$$\rho_5(n) = \frac{L_2(0, \rho_5)}{\frac{15}{16}} \cdot \left\{1 - \left(\frac{n^*}{2}\right)\frac{1}{4}\right\} \cdot 4 \cdot 30 \cdot B_{2,\chi}.$$

If $n \equiv 1 \pmod 8$, then

$$\rho_5(n) = \frac{\frac{5}{8}}{\frac{15}{16}} \cdot \left\{1 - \frac{1}{4}\right\} \cdot 4 \cdot 30 \cdot B_{2,\chi}$$

$$= 60 \cdot \frac{1}{n} \sum_{a=1}^{n} \chi(a) \left(a - \frac{n}{2}\right)^2.$$

Thus, for example, $\rho_5(17) = 480$.

If $n \equiv 5 \pmod 8$, then

$$\rho_5(n) = \frac{\frac{7}{8}}{\frac{15}{16}} \cdot \left\{1 + \frac{1}{4}\right\} \cdot 4 \cdot 30 \cdot B_{2,\chi}$$

$$= 140 \cdot \frac{1}{n} \sum_{a=1}^{n} \chi(a) \left(a - \frac{n}{2}\right)^2.$$

Thus, for example, $\rho_5(13) = 560$.

If $n \not\equiv 5 \pmod 4$, then the conductor of $\left(\frac{n}{a}\right)$ is $4n$ and

$$\rho_5(n) = \frac{\frac{5}{4}}{\frac{15}{16}} \cdot 2^2 \cdot \left(\frac{1}{4}\right)^{3/2} \cdot 30 \cdot B_{2,\chi}$$

$$= 20 \cdot \frac{1}{4n} \sum_{a=1}^{4n} \chi(a) (a - 2n)^2.$$

Thus, for example, $\rho_5(3) = 80$.

6.5 Exact Formulas for the Number of Representations

When s is even, the function $r_s(n)$ is essentially a linear combination of multiplicative functions. This structure arises through the theory of Hecke operators. Here the prime 2 plays an exceptional role not because we are dealing with quadratic forms but rather because the level of the associated diagonal form is a multiple of 4. In the following the symbol p will always denote an odd prime. Now, due to the almost multiplicative nature of $r_s(n)$, it is sufficient, and this will become clear later on, to deal with the values of $r_s(p^\ell)$, and these values can be expressed as linear combinations of eigenvalues of the Hecke operators whose structure is well known. More precisely we prove the following result.

THEOREM 6.8 $r_s(p^{\ell-1})$ *For Even s*
If $s = 2k$ and p is an odd prime, then there exist μ algebraic numbers A_1, A_2, ..., A_μ depending only on k, such that for all $\ell \geq 1$,

$$r_s\left(p^{\ell-1}\right) = \rho_s\left(p^{\ell-1}\right) + \sum_{v=1}^{\mu} A_v \cdot \frac{\pi_v^\ell - {}^\sigma\pi_v^\ell}{\pi_v - {}^\sigma\pi_v}, \tag{6.43}$$

where the sum runs over μ algebraic numbers $\pi_1, \pi_2, ..., \pi_\mu$ that satisfy

$$\pi_v = p^{(k-1)/2+i\phi_p}, \qquad {}^\sigma\pi_v = p^{(k-1)/2+i\psi_p}, \tag{6.44}$$

and ϕ_p and ψ_p are real and $\phi_p + \psi_p \in \mathbb{Z}\dfrac{\pi}{\log p}$. \tag{6.45}

REMARK 6.11   In the above statement, and in the rest of this chapter, the notation ${}^\sigma\pi_v$ is taken to suggest that ${}^\sigma\pi_v$ is a Galois conjugate of the algebraic integer π_v. ∎

In the case when s is odd we no longer can make the assertion that, "$r_s(n)$ is essentially a linear combination of multiplicative functions." Nevertheless we can appeal to the theory of Shimura [102] concerning quadratic forms in an odd number of variables to get an approximation to Theorem 6.8 above. More precisely we prove this result.

THEOREM 6.9 $r_s(tp^{2\ell})$ *For Odd s*
If $s = 2k+1$, p is an odd prime and t is a square-free odd integer not divisible by p, then there exist μ algebraic numbers $C_1, C_2, ..., C_\mu$ depending only on k, and μ arithmetic functions $a_1(t), a_2(t), ..., a_\mu(t)$ depending only on t, such that for all $\ell \geq 1$,

$$r_s\left(tp^{2\ell}\right) = \tag{6.46}$$

$$\rho_s\left(tp^{2\ell}\right) + \sum_{v=1}^{\mu} C_v a_v(t)\left(\frac{\pi_v^{\ell+1} - {}^\sigma\pi_v^{\ell+1}}{\pi_v - {}^\sigma\pi_v} - \left(\frac{(-1)^k t}{p}\right) p^{k-1} \cdot \frac{\pi_v^\ell - {}^\sigma\pi_v^\ell}{\pi_v - {}^\sigma\pi_v}\right),$$

where the sum runs over μ algebraic numbers π_1, π_2, ..., π_μ that satisfy

$$\pi_v = p^{s/2-1+i\phi_p}, \qquad {}^\sigma\pi_v = p^{s/2-1-i\phi_p}, \qquad (6.47)$$

$$\text{and the } \phi_p \text{ are real.} \qquad (6.48)$$

As an easy consequence of Theorems 6.8 and 6.9 we get the following improvement to Hardy's result, Theorem 6.1.

COROLLARY 6.1

With the same notation as Theorems 6.8 and 6.9 we have for $s = 2k$,

$$r_s(p^\ell) = \rho_s(p^\ell) + \mathrm{O}\left(\ell p^{\ell(s/4-1/2)}\right), \qquad (6.49)$$

where the implied constant depends only on s; and for $s = 2k + 1$,

$$r_s(tp^{2\ell}) = \rho_s(tp^{2\ell}) + \mathrm{O}\left(\ell p^{2\ell(s/4-1/2)}\right), \qquad (6.50)$$

where the implied constant depends only on s and t but not on p or ℓ.

REMARK 6.12   Parts (6.43) and (6.44) of Theorem 6.8 were known before for a few values of s. In fact Hardy and Ramanujan had obtained (in a different notation)

$$r_{24}(p^{\ell-1}) = \frac{16}{691} \cdot \frac{p^{11\ell} - 1}{p^{11} - 1} + \frac{128 \cdot 259}{691} \cdot \frac{\pi_p^\ell - {}^\sigma\pi_p^\ell}{\pi_p - {}^\sigma\pi_p},$$

where $\pi_p + {}^\sigma\pi_p = \tau(p)$. (See Hardy [50], pages 155 and 165.) By a more delicate analysis than that of Hardy and Ramanujan, Rankin proved ([90], page 405) that

$$r_{20}(p^{\ell-1}) = \frac{8}{31} \cdot \frac{p^{9\ell} - 1}{p^9 - 1} + \frac{16}{93} \left\{ \frac{\pi_1^\ell - {}^\sigma\pi_1^\ell}{\pi_1 - {}^\sigma\pi_1} + 76 \cdot \frac{\pi_2^\ell - {}^\sigma\pi_2^\ell}{\pi_2 - {}^\sigma\pi_2} \right\},$$

where π_1 and π_2 and their σ-conjugates are complex numbers. Other examples will be given in Section 6.6. ∎

REMARK 6.13   In the introduction to his fundamental paper on p-adic spherical functions ([70], page 441), Mautner had hinted at the possibility of proving a result like that given by Theorem 6.8 parts (6.43) and (6.44). To the best of our knowledge this plan was never carried out. ∎

REMARK 6.14   The basic idea in the proof of Theorem 6.8 consists in showing, as we have already observed, that the cusp form $\vartheta(z)^s - E_s(z)$ is a linear combination of eigenfunctions of the Hecke operators associated to odd prime powers. A comparison of the coefficients in the q-expansions shows that $r_s(n)$ is a linear combination of functions that are multiplicative on the set of odd integers. The difficulty with the prime 2 arises because we are essentially dealing with the Hecke group $\Gamma_0(4)$. Finally, to get the formula of the theorem we identify the Fourier coefficients of eigenfunctions with p-adic spherical functions whose structure is well known and in this way we get parts (6.43) and (6.44). To get part (6.45) we must apply Deligne's proof of the Petersson-Ramanujan conjecture ([23], page 302).   ∎

REMARK 6.15   In [50], page 158, Section 9.3, Hardy remarked, "In order that $c_{2s}(n) = 0$ for all n, it is necessary and sufficient that $2s \leq 8$. In these cases only $r_{2s}(n)$ is a 'divisor function,' represented by the 'singular series' of Section 9.8. The 'reasons' for this have been shown in a new light recently by Siegel, *Annals of Math.* (2), 36 (1935), 527–606." Although not entirely incorrect, his remark is somewhat misleading, as Theorems 6.8 and 6.9 clearly show.   ∎

REMARK 6.16   Theorems 6.8 and 6.9 and a simple application of Kronecker's theorem on diophantine approximation make precise Ramanujan's empirical observation ([88], page 159, # 142),

$$r_s(n) - \rho_s(n) \neq o(n^{(s/4 - 1/2)}).$$

This shows in fact that Corollary 6.1 is best possible.   ∎

REMARK 6.17   The formulas of Theorems 6.8 and 6.9 contain crude versions of the interesting recursion relations for $r_s(n)$ obtained by Newman ([76], page 778).   ∎

We now prove Theorem 6.8.

PROOF   No new idea is needed in the proof beyond basic results that can be found in the literature, though not all in one place. For this reason and for the sake of the reader, we assemble here the necessary results. When $s = 2k$, the generating function is a modular function of type $M(\Gamma_0(4), k, \chi)$, where

$$\Gamma_0(4) = \left\{ \begin{pmatrix} a & b \\ c & d \end{pmatrix} \in SL(2, \mathbb{Z}) : c \equiv 0 \ (\text{mod } 4) \right\}$$

and the multiplier system χ is defined by one of the two quadratic characters

$$\chi : \Gamma_0(4) \longrightarrow \mathbb{T} = \{ z \in \mathbb{C} : |z| = 1 \},$$

$$\begin{pmatrix} a & b \\ c & d \end{pmatrix} \longmapsto \chi(d),$$

associated with the Gaussian field $\mathbb{Q}((-1)^{1/2})$. More precisely,

$$\chi(d) = \begin{cases} \left(\frac{-1}{d}\right) & \text{if } k \text{ is odd}, \\ \left(\frac{-1}{d}\right)^2 & \text{if } k \text{ is even}. \end{cases}$$

For a more precise definition of the quadratic residue symbol (c/d) see Shimura [102], page 442. The dependence of χ on k should be kept in mind in the following since it indicates that $\vartheta(z)^{2k}$ has two different multiplier systems depending on the parity of k. Recall that a modular form $f(z)$ is of type $M(\Gamma_0(4), k, \chi)$ if $f(z)$ is regular in the upper half plane, has a suitable q-expansion about each cusp of $\Gamma_0(4)$ and satisfies

$$f\left(\frac{az+b}{cz+d}\right) = \chi(d)(cz+d)^k f(z), \qquad \begin{pmatrix} a & b \\ c & d \end{pmatrix} \in \Gamma_0(4).$$

The dimension of the space of such functions is finite and can be computed by means of the Petersson-Riemann-Roch theorem. For example, when k is even we have

$$\dim M(\Gamma_0(4), k, \chi) = \frac{k}{2} + 1,$$

while when k is odd, the dimension of the space of cusp forms is

$$\mu = \frac{1}{4}(2k - 5 - \chi(P_1) - \chi(P_2) - \chi(P_3)),$$

where P_1, P_2 and P_3 are three nonequivalent parabolic elements of $\Gamma_0(4)$.

Hardy's estimate $r_s(n) - \rho_s(n) = O(n^{s/2})$ indicates that the function $f(z) = \vartheta(z)^s - E_s(z)$ is a cusp form on $\Gamma_0(4)$. The theory of Hecke operators acting on the space of cusp forms of type $M^0(\Gamma_0(N), k, \chi)$ is fairly well understood, having been worked out in great detail by Atkin-Lehner in [4] for the case of χ trivial and then by Winnie Li [67] for general χ. In the following we shall make essential use of several results from the paper of Winnie Li. Let us first recall some basic definitions. Let $M^-(\Gamma_0(4), k, \chi)$ denote the space generated by

1. cusp forms $f(z)$ of type $(\Gamma_0(1), k.\chi)$,

2. cusp forms $g(z)$ of type $(\Gamma_0(2), k.\chi)$, and

3. B_2-translates of those functions in 2.

Recall that if $g(z)$ is a cusp form of type $M^0(\Gamma_0(2), k.\chi)$ and has a q-expansion

$$g(z) = \sum_{n=1}^{\infty} a_n q^n,$$

then its B_2-translate is given by

$$g|B_2(z) = \sum_{n=1}^{\infty} a_n q^{2n}.$$

By $M^+(\Gamma_0(4), k, \chi)$ we denote the orthogonal complement in the Petersson inner product of $M^-(\Gamma_0(4), k, \chi)$ in $M^0(\Gamma_0(4), k, \chi)$. This gives a decomposition of the form

$$M^0(\Gamma_0(4), k, \chi) = M^+(\Gamma_0(4), k, \chi) \oplus M^-(\Gamma_0(4), k, \chi).$$

A *new form* in $M^0(\Gamma_0(4), k, \chi)$ is a nonzero form in $M^+(\Gamma_0(4), k, \chi)$ which is a common eigenfunction of all Hecke operators T_p, p and odd prime. The form $g(z)$ in $M^0(\Gamma_0(4), k, \chi)$ is *normalized* if $a_1 = 1$. Recall that the Hecke operator T_p for an odd prime p and

$$f(z) = \sum_{n=1}^{\infty} a(n) q^n$$

is defined by

$$(f|T_p)(z) = \sum_{n=1}^{\infty} \left(\chi(p) p^{k-1} a(n/p) + a(np) \right) q^n,$$

where $a(x) = 0$ if x is not an integer. The basic fact that we shall use from Winnie Li ([67], page 295) is that a new form is uniquely determined by its eigenvalues. In fact if $f(z)$ is normalized and $f|T_p = a(p)f$, $f|U_2 = a(2)f$, where U_2 is T_2 truncated so that only the terms with $a(2n)q^n$ in the series appear, then

$$f(z) = \sum_{n=1}^{\infty} a(n) q^n.$$

Now the space of old forms is left invariant under the action of the Hecke operators and so a basis can be selected consisting of eigenfunctions. Nevertheless, these are no longer uniquely determined by their eigenvalues. Let $f_1, \ldots, f_\alpha$ be a basis of $M^+(\Gamma_0(1), k, \chi)$ consisting of normalized new forms, $g_1, \ldots, g_\beta$ be a basis of $M^+(\Gamma_0(2), k, \chi)$ consisting of normalized new forms and $h_1, \ldots, h_\gamma$ be a basis of $M^+(\Gamma_0(4), k, \chi)$ consisting of normalized new forms. Now any cusp form in $M^0(\Gamma_0(4), k, \chi)$ is a linear combination of the above basis elements and the translates $g_1|B_2, \ldots, g_\beta|B_2$ ([67], page 295). In particular we have the linear combination

$$\vartheta(z)^{2k} - E_{2k}(z) = \sum_{v=1}^{\alpha} a_v f_z(z) + \sum_{v=1}^{\beta} \left(b_v g_v(z) + b'_v g_v|B_2(z) \right) + \sum_{v=1}^{\gamma} c_v h_v(z).$$

If we compare the n-th coefficients of the q-expansions of both sides of the above equation for n an odd integer we get that

$$r_{2k}(n) - \rho_{2k}(n) = \sum_{v=1}^{\mu} A_v a_v(n),$$

where each $a_v(n)$ is a function which is multiplicative on the set of odd integers, that is, $a_v(mn) = a_v(m)a_v(n)$ when m and n are odd and relatively prime. In particular we have

$$r_{2k}(p^{\ell}) = \rho_{2k}(p^{\ell}) + \sum_{v=1}^{\mu} A_v a_v(p^{\ell}).$$

At this point we could invoke the theory of p-adic spherical functions which gives the structure of the functions $a_v(p^{\ell})$. We prefer to derive our results by elementary considerations. By Theorem 3 of Winnie Li ([67], page 295) we have for each $a_v(p^{\ell})$ the "local factor" identity

$$\frac{T}{1 - a_v(p)T + \chi(p)p^{k-1}T^2} = \sum_{w=0}^{\infty} a_v(p^w)T^{w+1} = \sum_{\ell=1}^{\infty} \left(\frac{\pi_v^{\ell} - {}^{\sigma}\pi_v^{\ell}}{\pi_v - {}^{\sigma}\pi_v} \right) T^{\ell},$$

where $a_v(p) = \pi_v + {}^{\sigma}\pi_v$ and $\pi \cdot {}^{\sigma}\pi = \chi(p)p^{k-1}$. We then have

$$a_v(p^{\ell-1}) = \frac{\pi_v^{\ell} - {}^{\sigma}\pi_v^{\ell}}{\pi_v - {}^{\sigma}\pi_v}.$$

To complete the proof of the theorem we now use Deligne's proof of the Petersson-Ramanujan conjecture ([23], page 302, Theorem (8.2)) to remark that

$$\pi_v = p^{(k-1)/2+i\phi_p}, \qquad {}^{\sigma}\pi_v = p^{(k-1)/2+i\psi_p},$$

with ϕ_p and ψ_p real and $\phi_p + \psi_p \in \mathbb{Z}\frac{\pi}{\log p}$. This then completes the proof of Theorem 6.8. ∎

REMARK 6.18   The prime $p = 2$ can be treated by the same techniques, except that the corresponding expression for $a_v(2^{\ell})$ takes a different shape. (See [67], page 295, Theorem 3 (ii) and (iii).) ∎

REMARK 6.19   The coefficients $A_1, \ldots, A_{\mu}$ can be computed as ratios of Petersson inner products. ∎

Now we prove Theorem 6.9.

PROOF   The situation here is more complicated than what we have considered above, and the corresponding results are less complete. The proof of the theorem relies heavily on the theory of Shimura concerning modular forms of half integral weight [102]. The definitions and results we now describe are taken almost entirely from Shimura's paper.

Let $GL_2^+(\mathbb{R})$ be the group of all real 2×2 matrices with positive determinant, let $\mathbb{T} = \{z \in \mathbb{C} : |z| = 1\}$ denote the circle group, and let $\mathcal{H}$ be the upper half plane. As usual the action of an element $\alpha = \begin{pmatrix} a & b \\ c & d \end{pmatrix}$ on an element $z \in \mathcal{H}$ is denoted by $\alpha(z) = (az + b)/(cz + d)$. We let $\tilde{G}_0$ be the group consisting of all pairs $(\alpha, \phi(z))$ formed by an element $\alpha = \begin{pmatrix} a & b \\ c & d \end{pmatrix}$ of $GL_2^+(\mathbb{R})$ and a holomorphic function $\phi(z)$ on $\mathcal{H}$, such that

$$\phi(z)^2 = t(\det \alpha)^{-1/2}(cz + d)$$

with $t \in \mathbb{T}$. The group law is defined by the multiplication rule

$$(\alpha, \phi(z))(\beta, \psi(z)) = (\alpha\beta, \phi(\beta(z))\psi(z)).$$

We let

$$P : \tilde{G}_0 \longrightarrow GL_2^+(\mathbb{R})$$

be the natural projection map. For a complex-valued function $f(z)$ on $\mathcal{H}$, an integer κ and an element $\xi = (\alpha, \phi) \in \tilde{G}_0$, we define a function

$$(f\|[\xi]_\kappa)(z) = f(\alpha(z))\phi(z)^{-\kappa}.$$

We now consider the special group

$$\tilde{G} = \{(\alpha, \phi(z)) \in \tilde{G}_0 : \det \alpha = 1\}.$$

A subgroup Δ of $\tilde{G}$ is called *Fuchsian* if it satisfies the following three conditions:

1. $P(\Delta)$ is a discrete subgroup of $SL_2(\mathbb{R})$ and $P(\Delta)\backslash\mathcal{H}$ is of finite measure with respect to the invariant measure $y^{-2}\,dx\,dy$.

2. The only element in Δ of the form $(1, t)$ with $t \in \mathbb{T}$ is the identity.

3. If $-1 \in P(\Delta)$, then the inverse image of -1 by P in Δ is $\{-1, 1\}$.

Let κ be an odd integer. We call a meromorphic function $f(z)$ on $\mathcal{H}$ an *automorphic form of weight $\kappa/2$ with respect to Δ* if $f\|[\xi]_\kappa = f$ for all $\xi \in \Delta$ and f is meromorphic at each cusp of $P(\Delta)$. The precise meaning of this last condition is as follows: let s be a cusp of $P(\Delta)$ and let

$$\Delta_s = \{(\alpha, \phi(z)) \in \Delta : \alpha(s) = s\}.$$

Conditions 2 and 3 above imply that Δ_s is either free-cyclic, or the product of a free-cyclic group and a cyclic group of order 2 generated by $(-1, 1)$. Let

$\eta \in \Delta_s$ generate the free-cyclic part and $\rho = (\alpha, \phi(z))$ be an element of $\tilde{G}$, such that $\alpha(\infty) = s$. Then

$$\rho^{-1}\eta\rho = \left(\varepsilon \cdot \begin{pmatrix} 1 & h \\ 0 & 1 \end{pmatrix}, t\right),$$

with $\varepsilon = \pm 1$, $t \in \mathbb{T}$. By inverting the generator η if necessary, we may assume $h > 0$. Under these conditions t depends only on the $P(\Delta)$-equivalence class of s. Meromorphicity at the cusp s now means that in the Fourier expansion

$$f|[\rho]_\kappa = \sum_{n=-\infty}^{\infty} c_n e^{2\pi i(n+r)z/h},$$

$c_n = 0$ for all $n \le -N$ and some N, where r is a real number defined by $t^\kappa = e^{2\pi i r}$ and $0 \le r \le 1$.

The space $G_\kappa(\Delta)$ of integral forms of weight $\kappa/2$ with respect to Δ consists of those automorphic forms f which are holomorphic on $\mathcal{H}$ and for which $c_n = 0$ if $n < 0$. The space $S_\kappa(\Delta)$ of cusp forms of weight $\kappa/2$ with respect to Δ is the subspace of $G_\kappa(\Delta)$ consisting of functions f for which $c_0 = 0$ at all the cusps.

For a positive integer N put

$$\Gamma_0(N) = \left\{ \begin{pmatrix} a & b \\ c & d \end{pmatrix} \in SL_2(\mathbb{Z}) : c \equiv 0 \pmod{N} \right\},$$

$$\Gamma_1(N) = \left\{ \begin{pmatrix} a & b \\ c & d \end{pmatrix} \in \Gamma_0(N) : a \equiv d \equiv 1 \pmod{N} \right\}.$$

We recall that if $\vartheta(z)$ is the theta function and $\gamma \in \Gamma_0(4)$, then the Mordell automorphy factor ([73], page 363),

$$j(\gamma, z) = \frac{\vartheta(\gamma(z))}{\vartheta(z)}$$

satisfies the relation

$$j\left(\begin{pmatrix} a & b \\ c & d \end{pmatrix}, z\right) = \varepsilon_d^{-1}\left(\frac{c}{d}\right)(cz+d)^{1/2},$$

where $\varepsilon_d = 1$ or i according as $d \equiv 1$ or $3 \pmod 4$. By appropriately writing the theta multiplier system ([63], page 51, Theorem 3) one easily shows that

$$j\left(\begin{pmatrix} a & b \\ c & d \end{pmatrix}, z\right)^2 = \left(\frac{-1}{d}\right)(cz+d),$$

where $(-1/d)$ is the nonprincipal character attached to the quadratic extension $\mathbb{Q}((-1)^{1/2})$. We now define an isomorphism from $\Gamma_0(4)$ into $\tilde{G}$ by sending $\gamma \in \Gamma_0(4)$ to $\gamma^* = (\gamma, j(\gamma, z))$ in $\tilde{G}$. If N is divisible by 4, we denote by $\Delta_0(N)$

and $\Delta_1(N)$ the images of $\Gamma_0(N)$ and $\Gamma_1(N)$ in $\tilde{G}$ under this isomorphism and observe that these are Fuchsian subgroups of $\tilde{G}$. Let N be a multiple of 4 and χ be a character modulo N. Denote by $S_\kappa(N, \chi)$ (respectively, $G_\kappa(N, \chi)$) the set of all elements f of $S_\kappa(\Delta_1(N))$ (respectively, $G_\kappa(\Delta_1(N))$) that satisfy

$$f|[\gamma]_\kappa = \chi(d)f \quad \text{for all } \gamma = \begin{pmatrix} a & b \\ c & d \end{pmatrix} \in \Gamma_0(N).$$

If p is a prime, we define a Hecke operator $T^N_{\kappa,\chi}(p^2)$ acting on the functions $f(z)$ of the space $G_\kappa(N, \chi)$ by the prescription: if

$$f(z) = \sum_{n=0}^\infty a(n)q^n,$$

then

$$\left(f|T^N_{\kappa,\chi}(p^2)\right)(z) = \sum_{n=0}^\infty b(n)q^n,$$

where

$$b(n) = a(p^2 n) + \chi_1(p)\left(\frac{n}{p}\right)p^{\lambda-1}a(n) + \chi(p^2)p^{\kappa-2}a\left(\frac{n}{p^2}\right),$$

with $\lambda = (\kappa - 1)/2$, and χ_1 is a character modulo N defined by $\chi_1(n) = \chi(n)(-1/n)^\lambda$. Here we understand that $a(x) = 0$ if x is not an integer and $\chi_1(p)(n/p) = 0$ if $p = 2$.

We recall that the algebra generated by the operators $T^N_{\kappa,\chi}(p^2)$ for all primes p is commutative and therefore we can find μ ($= \dim G_\kappa(N, \chi)$) basis elements $f_1, \ldots, f_\mu$ of $G_\kappa(N, \chi)$, such that

$$f_v|T^N_{\kappa,\chi}(p^2) = \omega_v(p)f_v, \qquad v = 1, \ldots, \mu,$$

for all rational primes p, that is, a basis consisting of common eigenfunctions of the $T^N_{\kappa,\chi}(p^2)$ for all primes p. It is shown in ([102], Remark on page 457) that $\vartheta(z)$ is an element of $G_1(4, \chi_0)$, with χ_0 trivial. One can then easily see that if s is odd, $\vartheta(z)^s$ is an element of $G_s(4, \chi_0)$, and furthermore Hardy's estimate $r_s(n) - \rho_s(n) = O(n^{s/4})$ implies that $\vartheta(z)^s - E_s(z)$ is an element in $S_s(4, \chi_0)$. We can therefore write it as a linear combination of eigenfunctions for the Hecke operators $T^4_{s,\chi_0}(p^2)$, that is,

$$\vartheta(z)^s - E_s(z) = \sum_{v=1}^\mu C_v f_v(z), \qquad \mu = \dim S_s(4, \chi_0). \tag{6.51}$$

We now recall the fundamental results of Shimura.

THEOREM 6.10 Shimura's Theorem [102], page 453
   Let $f(z) = \sum_{n=0}^\infty a(n)q^n$ be an element of $G_\kappa(N, \chi)$, that is, a common eigenfunction of $T^N_{\kappa,\chi}(p^2)$ for all primes p, and let $f|T^N_{\kappa,\chi}(p^2) = \omega(p)f$. Further

let t be a positive integer which has no square factor, other than 1, prime to N. Then the formal Dirichlet series $\sum_{n=1}^{\infty} a(tn^2)n^{-s}$ has the Euler product,

$$\sum_{n=1}^{\infty} a(tn^2)n^{-s} = a(t) \prod_p \left(1 - \chi_1(p) \left(\frac{t}{p}\right) p^{\lambda-1-s}\right) \times$$

$$\left(1 - \omega(p)p^{-s} + \chi(p)^2 p^{\kappa-2-2s}\right)^{-1},$$

where the product is taken over all primes p, and λ and χ_1 are defined as in the definition of the Hecke operators given above.

Shimura's main result is this theorem.

THEOREM 6.11 Shimura's Theorem [102], page 458

Let κ be an odd integer, N a positive integer divisible by 4, χ a character modulo N, and let $f(z) = \sum_{n=1}^{\infty} a(n)q^n$ be an element of $S_\kappa(N, \chi)$. Suppose $\kappa \geq 3$ and put $\lambda = (\kappa - 1)/2$. Further let t be a square-free positive integer, and let χ_t denote the character modulo tN defined by

$$\chi_t(m) = \chi(m) \left(\frac{-1}{m}\right)^\lambda \left(\frac{t}{m}\right).$$

Define a function $F_t(z)$ on $\mathcal{H}$ by

$$F_t(z) = \sum_{n=1}^{\infty} A_t(n)q^n,$$

$$\sum_{n=1}^{\infty} A_t(n)n^{-s} = \left(\sum_{m=1}^{\infty} \chi_t(m)m^{\lambda-1-s}\right) \left(\sum_{m=1}^{\infty} a(tm^2)m^{-s}\right).$$

Suppose that f is a common eigenfunction of $T_{\kappa,\chi}^N(p^2)$ for all prime factors p of N not dividing the conductor of χ_t. Then F_t belongs to $M(\Gamma_0(N_t), \kappa - 1, \chi^2)$ for a certain positive integer N_t. Moreover, if $\kappa \geq 5$, then F_t is a cusp form. Furthermore, suppose that $f|T_{\kappa,\chi}^N(p^2) = \omega(p)f$ and define a function F on $\mathcal{H}$ by

$$F(z) = \sum_{n=1}^{\infty} A_n q^n,$$

$$\sum_{n=1}^{\infty} A_n n^{-s} = \prod_p \left(1 - \omega(p)p^{-s} + \chi(p)^2 p^{\kappa-2-2s}\right)^{-1},$$

where the product is taken over all primes p. Let N_0 be the greatest common divisor of the integers N_t for all square-free t, such that $a(t) = 0$. Then F belongs to $M(\Gamma_0(N_0), \kappa-1, \chi^2)$ if $\kappa \geq 3$ and to $M^0(\Gamma_0(N_0), \kappa-1, \chi^2)$ if $\kappa \geq 5$.

We now observe that Theorem 6.9 is true for $s \leq 7$ by the results of Hardy [48]; so in the following we assume $s \geq 9$. Let us then consider a cusp form $f(z) = \sum_{n=1}^{\infty} a(n)q^n$ in $G_s(\Gamma_0(4), \chi_0)$ which is an eigenfunction of the Hecke operators $T_{s,\chi_0}^4(p^2)$ and suppose $f|T_{s,\chi_0}^4(p^2) = \omega(p)f$. Then from the identity

$$\sum_{n=1}^{\infty} a(tn^2)n^{-s} = a(t) \prod_p \left(1 - \chi_1(p)\left(\frac{t}{p}\right)p^{\lambda-1-s}\right) \times$$

$$\left(1 - \omega(p)p^{-s} + \chi_0(p)p^{\lambda-2-2s}\right)^{-1}$$

$$= \frac{a(t)}{L(s+1-\lambda, \chi_t)} \sum_{n=1}^{\infty} A_n n^{-s},$$

we get by identifying coefficients

$$a(tm^2) = a(t) \sum_{d|m} \mu(d)d^{\lambda-1}\chi_t(d)A\left(\frac{m}{d}\right),$$

where $\chi_t(p) = (t/p)(-1/p)^{\lambda}$. In particular, if $m = p^{\ell}$ and

$$1 - \omega(p)T + p^{\lambda-2}T^2 = (1 - \pi T)(1 - {}^{\sigma}\pi T),$$

then

$$a(tp^{2\ell}) = a(t)\left(\frac{\pi^{\ell+1} - {}^{\sigma}\pi^{\ell+1}}{\pi - {}^{\sigma}\pi} - p^{\lambda-1}\chi_t(p) \cdot \frac{\pi^{\ell} - {}^{\sigma}\pi^{\ell}}{\pi - {}^{\sigma}\pi}\right).$$

Again by Deligne's proof of the Ramanujan-Petersson conjecture ([23], page 302) we have that

$$\pi = p^{(\lambda-2)/2+i\phi_p} \quad \text{with } \phi_p \text{ real.}$$

Now from the identity (6.51) we get by comparing coefficients in the q-expansion that

$$r_s(n) = \rho_s(n) + \sum_{v=1}^{\mu} C_v a_v(n).$$

If t is a positive square-free odd integer, then we can apply the above considerations to obtain that

$$r_s(tp^{2\ell}) = \rho_s(tp^{2\ell}) + \sum_{v=1}^{\mu} C_v a_v(t)\left(\frac{\pi_v^{\ell+1} - {}^{\sigma}\pi_v^{\ell+1}}{\pi_v - {}^{\sigma}\pi_v} - \frac{\pi_v^{\ell} - {}^{\sigma}\pi_v^{\ell}}{\pi_v - {}^{\sigma}\pi_v} \cdot p^{(s-3)/2}\chi_t(p)\right),$$

where

$$\pi_v = p^{s/2-1+i\phi_p} \quad \text{with } \phi_p \text{ real.}$$

This completes the proof of Theorem 6.9. ∎

6.6 Examples: Quadratic Forms and Sums of Squares

This section consists of two examples. Our first example deals with quadratic forms.

In the beginning modular forms were created to deal with problems connected with the representation of integers by sums of squares, and more generally by quadratic forms. For example, the quadratic form in eight variables

$$Q(x_1, \ldots, x_8) = \frac{1}{2} \sum_{i=1}^{8} x_i^2 + \frac{1}{2} \left(\sum_{i=1}^{8} x_i \right)^2 - x_1 x_2 - x_2 x_8$$

has associated with it the theta series

$$\vartheta(z, Q) = 1 + \sum_{n=1}^{\infty} a_n(Q) q^n, \qquad q = \exp(2\pi i z),$$

where $a_n(Q)$ is the number of representations of n by the quadratic form Q. Applying the Poisson summation formula to $\vartheta(z, Q)$ we obtain the identity

$$\vartheta(z, Q) = 240 E_4(z),$$

where $E_4(z)$ is the seminormalized Eisenstein series of weight 4. By comparing coefficients we get

$$a_n(Q) = 240 \sigma_3(n),$$

a result that is also immediate by observing that

$$M_4(\Gamma) = \mathbb{C} \vartheta(z, Q).$$

Even though the main subject of this chapter is the study of the coefficients of modular forms on Γ such as $a_n(Q)$ above, we give below an example of a modular form on $\Gamma_0(11)$ arising from a quadratic form in four variables and of discriminant different from 1.

Let Q be defined by

$$Q(x_1, x_2, x_3, x_4) = x_1^2 + 4(x_2^2 + x_3^2 + x_4^2) + x_1 x_3 + 4x_2 x_3 + 3x_2 x_4 + 7x_3 x_4,$$

and denote by $\vartheta(z, Q)$ the corresponding theta series

$$\vartheta(z, Q) = 1 + \sum_{n=1}^{\infty} a_n(Q) q^n,$$

where again $a_n(Q)$ is the number of solutions of the diophantine equation

$$n = Q(x_1, x_2, x_3, x_4).$$

A simple computation done by Hecke gives the decomposition

$$\vartheta(z, Q) = \frac{12}{5} E_{\chi_0}(z) + \frac{18}{5} f(z),$$

where

$$E_{\chi_0}(z) = \frac{5}{12} + \sum_{n=1}^{\infty} \frac{\chi_0(n) q^n}{1 - q^n},$$

and $\chi_0(n) = 1$ if $\gcd(n, 11) = 1$ and 0 otherwise, and

$$f(z) = q \prod_{n=1}^{\infty} (1 - q^n)^2 (1 - q^{11n})^2 = \sum_{n=1}^{\infty} a_n q^n.$$

The functions $E_{\chi_0}(z)$ and $f(z)$ are respectively an Eisenstein series and a cusp form of weight 2 associated with the subgroup $\Gamma_0(11)$. The Dirichlet series associated with $E_{\chi_0}(z)$ has the relatively simple form

$$\zeta(s)\zeta(s+1)(1 - 11^{1-s}).$$

The Dirichlet series associated with the cusp form $f(z)$ has the Euler product

$$\phi_f(s) = \sum_{n=1}^{\infty} a_n n^{-s} = (1 - 11^{-s}) \prod_{p \nmid 11} (1 - a_p p^{-s} + p^{1-2s})^{-1}.$$

In fact, since the compactification $X_0(11)$ of $\mathcal{H}/\Gamma_0(11)$ is a Riemann surface of genus 1, $f(z)\, dq$ is the unique differential of the first kind on the elliptic curve $X_0(11)$. Thus, by results of Eichler [27] and Shimura [100] the above Dirichlet series is the global zeta function of the elliptic curve $X_0(11)$ of which an affine model is given by the Tate curve

$$y^2 - y = x^3 - x^2.$$

This is equivalent to saying that the coefficients a_p can be defined by the counting function

$$N_p = p + 1 - a_p,$$

where N_p is the number of points on the Tate curve with coordinates in $\mathbb{F}_p$. This determination of a_p gives a quick way of computing the coefficients $a_n(Q)$.

Now the theory of Hecke operators and the Riemann hypothesis for elliptic function fields imply that for $p \neq 11$

$$a_{p^{\ell-1}}(Q) = \frac{12}{5} \cdot \frac{p^{\ell} - 1}{p - 1} + \frac{18}{5} \cdot \frac{\pi^{\ell} - {}^{\sigma}\pi^{\ell}}{\pi - {}^{\sigma}\pi},$$

where $|\pi| = p^{1/2}$ and

$$N_p = p + 1 - (\pi + {}^{\sigma}\pi)$$

is the number of points on the elliptic curve

$$y^2 - y = x^3 - x^2$$

over the finite field $\mathbb{F}_p$ of p elements. This shows how a reciprocity law can be used to completely determine the arithmetical function $a_{p^\ell}(Q)$.

A similar decomposition for the space $M_2(\Gamma_0(N))$ can also be given in the cases

$$N = 14, \ 15, \ 17, \ 19, \ 20, \ 21, \ 24, \ 27, \ 32, \ 36 \text{ and } 49.$$

(See Shimura [100], page 182.) Andrianov [3] has given a detailed exposition of the cases $N = 27, \ 32, \ 36$, and $M_4(\Gamma_0(9))$, where the relevant elliptic curves have complex multiplication, thus making it possible to express the traces of Frobenius in terms of Hecke grössen characters.

Our second example deals with sums of squares.

First we consider the case of s even. We have

$$r_{10}(p^{\ell-1}) = \rho_{10}(p^{\ell-1}) + \frac{32}{5} \cdot \frac{\pi^\ell - {}^\sigma\pi^\ell}{\pi - {}^\sigma\pi},$$

where

$$1 - a(p)T + p^4 T^2 = (1 - \pi T)(1 - {}^\sigma\pi T)$$

and

$$\sum_{n=1}^{\infty} a(n)q^n = \frac{1}{4} \sum_{x=-\infty}^{\infty} \sum_{y=-\infty}^{\infty} (x^4 - 6x^2 y^2 + y^4)q^{x^2 + y^2}$$

is an eigenform of weight 5. Also,

$$r_{12}(p^{\ell-1}) = \rho_{12}(p^{\ell-1}) + 16 \cdot \frac{\pi^\ell - {}^\sigma\pi^\ell}{\pi - {}^\sigma\pi},$$

where

$$1 - a(p)T + p^5 T^2 = (1 - \pi T)(1 - {}^\sigma\pi T)$$

and

$$\sum_{n=1}^{\infty} a(n)q^n = \left(q^{1/4} - 3q^{9/4} + 5q^{25/4} - 7q^{49/4} + \cdots \right)^4$$

is an eigenform of weight 6.

These examples were suggested by computations in Mordell ([74], pages 100 and 103). Other examples can be derived from the tables of Glaisher ([38], page 480):

$$r_{14}(p^{\ell-1}) = \rho_{14}(p^{\ell-1}) + \frac{1456}{61} \cdot \frac{\pi^\ell - {}^\sigma\pi^\ell}{\pi - {}^\sigma\pi},$$

$$r_{16}(p^{\ell-1}) = \rho_{16}(p^{\ell-1}) + \frac{512}{17} \cdot \frac{\pi^\ell - {}^\sigma\pi^\ell}{\pi - {}^\sigma\pi}.$$

In the last two examples the π can also be given as in the early cases, but the generating functions do not throw any light on their meaning.

In the odd cases we have the example of $s = 9$:

$$r_9(p^{2\ell}) = \rho_9(p^{2\ell}) + \frac{64}{17} \cdot \left(\frac{\pi^{\ell+1} - {}^\sigma\pi^{\ell+1}}{\pi - {}^\sigma\pi} - p^3 \cdot \frac{\pi^\ell - {}^\sigma\pi^\ell}{\pi - {}^\sigma\pi} \right),$$

where

$$1 - (a(p^2) + p^3)T + p^7 T^2 = (1 - \pi T)(1 - {}^\sigma\pi T)$$

and

$$f(z) = \sum_{n=1}^\infty a(n)q^n = \frac{\eta(2z)^{12}}{\vartheta(z)^3}$$

is the unique cusp form of weight $9/2$ with respect to $\Delta_0(4)$ ([102], page 477). In this example $\eta(z)$ is the distinguished 24-th root of the Ramanujan modular form $\Delta(z)$. Examples similar to this one can be given for $s = 11$, 13, and 15.

In all of the above examples we have assumed that p is an odd prime. We leave open the problem of how to modify Theorems 6.8 and 6.9 so as to include the case $p = 2$.

In the main theorems and in the examples we have written

$$\pi = p^{(s-2)/4 + i\phi_p}, \qquad \pi = p^{(s-2)/2 + i\phi_p},$$

in the even and odd cases, respectively, but we have said nothing about the nature of the angles ϕ_p. If we decompose $\vartheta(z)^s - E_s(z) = \sum A_v f_v(z)$ as a linear combinations of eigenfunctions $f_v(z)$ of the Hecke operators, then it may happen that all the f_v's have "complex multiplication" and in this case the angles ϕ_p have a regular distribution with respect to the standard Lebesgue measure $d\vartheta$ on the unit circle. In fact the numbers π_v can be given an arithmetical interpretation with the framework of abelian class field theory. This seems to happen in the case of 10, 12 and 16 squares, and it would appear that the cases $s = 9$, 11 and 13 also fall under this category. The Ramanujan example for $s = 24$ is clearly one in which the associated modular form $\vartheta(z)^{24} - E_{24}(z)$ does not have "complex multiplication." Actually, in this case one expects that the angles ϕ_p are distributed according to the Sato-Tate measure $(2/\pi)\sin^2\phi \, d\phi$. In still other cases it may turn out that both types of modular forms are present. A possible example is $\vartheta(z)^{18} - E_{18}(z)$.

6.7 *Liouville's Methods and Elliptic Modular Forms*

In this section we call the reader's attention to the close connection between the elementary results about sums of squares, in particular of the First and Second Fundamental Identities, Theorems 2.1 and 2.2, in Chapter 2, and the analytic methods presented in Chapter 6. The elucidation of the precise relation, even if the discussion were confined only to historical aspects, would take us too far afield and would probably require a separate monograph in its own right. It is our intention here to discuss a few important examples to help the reader amplify his awareness of the deep relation between Liouville's elementary methods and the original ideas of Jacobi, as first exposed in his *Fundamenta Nova*. We encourage the reader to consult the classical sources including some of the articles by Jacobi. Uspensky [112] refers to earlier papers in which he explains the connection between Liouville's elementary methods and elliptic functions. An excellent modern source is Chapter 4 of Fine's monograph [31]. This book contains some of the most beautiful results of the theory.

6.7.1 The Basic Elliptic Modular Forms

In the applications of Liouville's methods to the representation of an integer as a sum of s squares, especially in the cases where $1 \leq s \leq 8$, the integer 12 as well as the characters of the multiplicative group $(\mathbb{Z}/12\mathbb{Z})^{\times} = \{1, 5, 7, 11\}$ of reduced residues modulo 12, have played a crucial role, one which has received very little notice. We now describe, with the hindsight gained from Hecke's brief attempt to develop a theory of modular forms of half integral weight, an arithmetic reason for their unique presence. The four Dirichlet characters we will need are the ones that appeared in the examples in Section 3.4.2. They are the principal character $\chi_0(n)$, the Gauss character

$$
\chi_g(n) = \begin{cases} (-1)^{(n-1)/2} & \text{if } n \text{ is odd,} \\ 0 & \text{if } n \text{ is even,} \end{cases}
$$

the Eisenstein character

$$
\chi_e(n) = \left(\frac{n}{3}\right) = \begin{cases} 0 & \text{if } n \equiv 0 \ (\mathrm{mod}\ 3), \\ 1 & \text{if } n \equiv 1 \ (\mathrm{mod}\ 3), \\ -1 & \text{if } n \equiv 2 \ (\mathrm{mod}\ 3), \end{cases}
$$

and the product character $\chi_{ge}(n) = \chi_g(n)\chi_e(n)$.

As functions on $\mathbb{Z}$, $\chi_g(-1) = \chi_e(-1) = -1$. Therefore, functions like $f(x) = \chi_g(x) \cdot x$, $g(x) = \chi_e(x)\sin(nx)$ and $\chi_{ge}(n) = \chi_g(n)\chi_e(n)$, all ubiquitous components of the Liouville identities, are even functions, while functions like $h(x) = \chi_{ge}(x)\sin(nx)$ are odd functions. These four characters are intimately connected with the simplest cyclotomic extension of degree 4 over the field of rational numbers, $K = \mathbb{Q}(\zeta_{12})$, where ζ_{12} is a primitive twelfth root of 1. K is a biquadratic field whose Galois group is isomorphic to $(\mathbb{Z}/12\mathbb{Z})^{\times}$. It contains the Gaussian field $k_G = \mathbb{Q}(i)$, where $i = \sqrt{-1}$, the Eisenstein field $k_E = \mathbb{Q}(\rho)$, where ρ is a cube root of 1, and the real quadratic field $k_{GE} = \mathbb{Q}(\sqrt{12})$.

The Dirichlet L-function associated to each of these characters is related to a unique modular form of half integral weight. We make this connection precise in the following four examples.

Example 6.3

The principal character χ_0. The uniqueness of the theta function

$$\vartheta(z) = \sum_{n=-\infty}^{\infty} e^{\pi i n^2 z}$$

is a consequence of the integral representation

$$\Lambda(s, \chi_0) = \pi^{-s} \Gamma(s) \prod_p \frac{1}{1 - p^{-2s}}$$

$$= \frac{1}{2s(s-1)} + \frac{1}{2} \int_1^{\infty} \left(y^s + y^{(1/2)-s} \right) (\vartheta(iy) - 1) \frac{dy}{y} \,,$$

and of the uniqueness of $\zeta(s)$ already established in Hamburger's Theorem 4.9 in Chapter 4.

Example 6.4

The Gauss character χ_g. The uniqueness of Jacobi's identity

$$\prod_{n=1}^{\infty} (1 - q^n)^3 = \sum_{m=0}^{\infty} (-1)^m (2m+1) q^{m(m+1)/2} \,,$$

which we can also write in the more suggestive form

$$\eta(z)^3 = \sum_{n=1}^{\infty} \chi_g(n) n \cdot q^{n^2/8}, \qquad q = e^{2\pi i z} \,,$$

follows from the integral representation

$$\Lambda(s, \chi_g) = \pi^{-s} \Gamma(s) (4)^{s-(1/2)} \prod_p \frac{1}{1 - \chi_g(p) p^{1-2s}}$$

$$= \int_1^{\infty} \left(y^s + y^{(3/2)-s} \right) \eta(iy)^3 \frac{dy}{y} \,,$$

and Hecke's generalization [53] of Hamburger's uniqueness result, Theorem 4.9.

Example 6.5

The product character χ_{ge}. The uniqueness of Euler's identity

$$\prod_{n=1}^{\infty} (1 - q^n) = \sum_{n=-\infty}^{\infty} (-1)^n q^{n(3n+1)/2}, \qquad q = e^{2\pi i z} \,,$$

which we can also write in the equivalent form

$$\eta(z) = \sum_{n=1}^{\infty} \chi_{ge}(n) q^{n^2/24}, \tag{6.52}$$

also follows, using the same line of argument as in Hecke [53], from the integral representation for the Euler product

$$\Lambda(s, \chi_{ge}) = \pi^{-s} \Gamma(s)(12)^s \prod_p \frac{1}{1 - \chi_{ge}(p)p^{-2s}}$$

$$= \int_1^\infty \left(y^s + y^{(1/2)-s} \right) \eta(iy) \frac{dy}{y} .$$

Example 6.6

The Eisenstein character χ_e. Finally, we note that this character is associated with the modular form

$$\frac{\eta(z)^4}{\vartheta(z)} = \sum_{n=1}^\infty \chi_e(n) \cdot n \cdot q^{n^2/6} .$$

Its uniqueness, also noted by Hecke in [53], is equivalent to the uniqueness of the Euler product

$$\Lambda(s, \chi_e) = \pi^{-s} \Gamma(s)(3)^{-s} \prod_p \frac{1}{1 - \chi_e(p)p^{1-2s}}$$

$$= \int_1^\infty \left(y^s + y^{(3/2)-s} \right) \cdot \frac{\eta(iy)^4}{\vartheta(iy)} \cdot \frac{dy}{y} .$$

It is worth recalling that in the terminology of Hecke the modular forms mentioned in the four examples above are all of half integral weight:

$$\vartheta(z) \quad \text{weight } \tfrac{1}{2} \quad \text{Nebentype } \chi_0$$
$$\eta(z)^3 \quad \text{weight } \tfrac{3}{2} \quad \text{Nebentype } \chi_g$$
$$\eta(z) \quad \text{weight } \tfrac{1}{2} \quad \text{Nebentype } \chi_{ge}$$
$$\tfrac{\eta(z)^4}{\vartheta(z)} \quad \text{weight } \tfrac{3}{2} \quad \text{Nebentype } \chi_e .$$

REMARK 6.20   For a deeper study of the modular forms $\eta(z)$, $\eta(z)^3$, $\vartheta(z)$, $\eta(z)^4/\vartheta(z)$, and other related theta functions, the reader can consult the encyclopedic monograph of Petersson [85].   ∎

As a typical case of a Liouville-type relation we now give a simple example which involves the character χ_{ge}.

THEOREM 6.12 Example of a Liouville-type Relation
Let χ_{ge} denote the product of the Gauss and Eisenstein characters. Let $\sigma(m)$ denote the sum of the positive divisors of m when m is a positive integer, and let $\sigma(0) = -1/24$. We then have

$$\chi_{ge}(n)n^2 = -24 \sum_{k,h}{}' \sigma(k)\chi_{ge}(h),$$

where the summation $\sum'_{k,\,h}$ is taken over all pairs h, k of integers satisfying the equation

$$n^2 = 24k + h^2,$$

with $k \geq 0$ and $h \geq 1$.

PROOF   Starting with the product

$$\eta(z) = q^{1/24} \prod_{n=1}^{\infty} (1 - q^n), \qquad q = e^{2\pi i z},$$

and denoting by $\eta'(z)$ the derivative of $\eta(z)$ with respect to z, we obtain this formula for the logarithmic derivative

$$
\frac{\eta'(z)}{\eta(z)} = 2\pi i \left\{ \frac{1}{24} - \sum_{n=1}^{\infty} n \cdot \frac{q^n}{1 - q^n} \right\}
$$

$$
= 2\pi i \left\{ \frac{1}{24} - \sum_{n=1}^{\infty} \sigma(n) q^n \right\}.
$$

Multiplying both sides by $\eta(z)$, dividing by $2\pi i$ and using the formula (6.52) for $\eta(z)$, we find

$$
\sum_{n=1}^{\infty} \chi_{ge}(n) \cdot \frac{n^2}{24} \cdot q^{n^2/24} = \left\{ \frac{1}{24} - \sum_{m=1}^{\infty} \sigma(m) q^m \right\} \sum_{h=1}^{\infty} \chi_{ge}(h) q^{h^2/24}.
$$

Multiply the two q-series on the right side formally and equate the coefficients of equal powers of q on both sides to obtain the equation claimed in the theorem statement.   ∎

Another example that can be proved along the same lines is the following identity due to Gauss:

$$
\frac{\eta(2z)^2}{\eta(z)} = \sum_{n=0}^{\infty} q^{(n+(1/2))^2/2}.
$$

In its most elementary form this identity reflects the fact that the following two partition problems have the same answer.

Problem 1. In how many ways may a given positive integer N be expressed as a monotonically decreasing series of odd positive integers?

Example 6.7

$$7 = 1 + 1 + 1 + 1 + 1 + 1 + 1$$
$$= 3 + 1 + 1 + 1 + 1$$
$$= 3 + 3 + 1$$
$$= 5 + 1 + 1$$
$$= 7$$

Problem 2. In how many ways may a given positive integer N be expressed as an ordered sum whose first term is of the form $n(n + 1)/2$, while the remaining terms form a monotonically increasing sequence of even positive integers?

Example 6.8

$$7 = 1 + 2 + 2 + 2$$
$$= 1 + 2 + 4$$
$$= 1 + 6$$
$$= 3 + 2 + 2$$
$$= 3 + 4$$

If M denotes the answer to both problems, then its values are given in this table:

N	1	2	3	4	5	6	7	8	9	10	11	12	13	...
M	1	1	2	2	3	4	5	6	8	10	12	14	17	...

This example is taken from the lucid article [110] by Jacques Tits on the nature of symmetry. There it is revealed that neither an understanding of the significance of the Gauss identity nor knowledge of the logic of an elementary proof of the dual nature of Problem 1 and Problem 2 are sufficient to penetrate the deep symmetries embodied in this identity, a fact that requires that it be viewed as an analytic expression for the "trace" of an infinite dimensional representation of a Kac-Moody Lie algebra of type A_1, a subject with many connections in common with the theory of modular forms presented in this book.

6.7.2 Jacobi's Identity: The Origin of Liouville's Methods

Jacobi's original proof of the identity

$$\eta(z)^3 = \sum_{n=1}^{\infty} \chi_g(n) n \cdot q^{n^2/8}, \qquad q = e^{2\pi i z}, \tag{6.53}$$

was based on his analytic derivation of the equality

$$\vartheta_1' = \pi \vartheta_2 \cdot \vartheta_3 \cdot \vartheta_4 \, ,$$

where

$$\vartheta_1(v, z) = -i \sum_{n=-\infty}^{\infty} (-1)^n q^{(n+(1/2))^2} e^{(2n+1)\pi i v} ,$$

$$\vartheta_2(v, z) = \sum_{n=-\infty}^{\infty} q^{(n+(1/2))^2} e^{(2n+1)\pi i v} ,$$

$$\vartheta_3(v, z) = \sum_{n=-\infty}^{\infty} q^{n^2} e^{2\pi i n v} ,$$

$$\vartheta_4(v, z) = \sum_{n=-\infty}^{\infty} (-1)^n q^{n^2} e^{2\pi i n v} ,$$

and

$$\vartheta_1' = \frac{d}{dv} \vartheta_1(v, z)\Big|_{v=0}, \qquad \vartheta_\lambda = \vartheta_\lambda(0, z) \text{ for } \lambda = 2, 3, 4.$$

Jacobi's alternative proof [57] of his identity was closer to the elementary ideas already used by Euler and Gauss in proofs of similar results.

The starting point is, *mirabile dictu*, the identity we wish to prove:

$$\left\{ \sum_{n=1}^{\infty} \chi_{ge}(n) q^{n^2/24} \right\}^3 = \sum_{m=1}^{\infty} \chi_g(m) m \cdot q^{m^2/8}.$$

Replacing q by q^{24} and applying the operator $q \cdot \frac{d}{dq} \log\{\cdot\}$, we arrive at

$$\frac{3 \sum_{n=1}^{\infty} \chi_{ge}(n) n^2 \cdot q^{n^2}}{\sum_{n=1}^{\infty} \chi_{ge}(n) \cdot q^{n^2}} = \frac{\sum_{m=1}^{\infty} \chi_g(m) m \cdot 3m^2 q^{3m^2}}{\sum_{m=1}^{\infty} \chi_g(m) m \cdot q^{3m^2}}.$$

Cross multiplication yields

$$\sum_{m=1}^{\infty} \sum_{n=1}^{\infty} \chi_{ge}(n) \chi_g(m) m \left\{ m^2 - n^2 \right\} q^{3m^2 + n^2} = 0.$$

By equating each coefficient of $q^{3m^2 + n^2}$ to zero, we obtain that for all integers $N \geq 1$, we have

$$\sum_{m \geq 1} \sum_{n \geq 1} \chi_{ge}(n) \chi_g(m) m \left\{ m^2 - n^2 \right\} = 0, \qquad (6.54)$$

where the summations extend over all solutions to the equation

$$N = 3m^2 + n^2$$

in positive integers. We should note that $\chi_g(m) = 0$ unless m is odd and that $\chi_{ge}(n) = 0$ unless n is relatively prime to 12. Let us now insist that $n \equiv 1 \pmod 6$, but allow n to be a negative integer. This replaces a solution to $N = 3m^2 + n^2$ having $n > 0$, $n \equiv 5 \pmod 6$ by a solution $N = 3m^2 + (-n)^2$ having $-n < 0$, $-n \equiv 1 \pmod 6$. Also, $\chi_{ge}(-n) = \chi_{ge}(n)$ because χ_{ge} is an even function. Indeed, $\chi_{ge}(k) = 0$ if $\gcd(k, 12) > 1$, and

$$\chi_{ge}(k) = \begin{cases} 1 & \text{if } k \equiv 1 \text{ or } 11 \pmod{12}, \\ -1 & \text{if } k \equiv 5 \text{ or } 7 \pmod{12}. \end{cases}$$

Now when $n \equiv 1 \pmod 6$ we have $\chi_{ge}(n) = (-1)^{(n-1)/2}$. Also, for odd m we have $\chi_g(m) = (-1)^{(m-1)/2}$. Therefore, Equation (6.54) is equivalent to

$$\sum_m \sum_n (-1)^{(m-n)/2} m \left\{ m^2 - n^2 \right\} = 0, \tag{6.55}$$

with m and n running through all integers satisfying $N = 3m^2 + n^2$, m odd, m positive and $n \equiv 1 \pmod 6$. Note that n is allowed to be positive or negative.

Example 6.9

If $N = 52$, the solutions to $N = 3m^2 + n^2$ satisfying these conditions are $(m, n) = (1, 7)$ and $(3, -5)$, so Equation (6.55) becomes

$$(-1)^{(1-7)/2}(1)(1^2 - 7^2) + (-1)^{(3-(-5))/2}(3)(3^2 - (-5)^2) = (-1)(-48) + (3)(-16) = 0.$$

Jacobi then notes that since all the steps are reversible, a proof of (6.55) for all N will result in the proof of the original q-series identity (6.53). Jacobi's goal is then to prove (6.55) by elementary means. From the modern point of view his key idea is to use the fact that the integers of the form $N = 3m^2 + n^2$ are precisely the norms of the numbers in the ring

$$\mathbb{Z}[\sqrt{-3}] \stackrel{\text{def}}{=} \{A + B\sqrt{-3} : A, B \in \mathbb{Z}\}.$$

Since $\mathbb{Z}[\sqrt{-3}] = \mathbb{Z} + 2 \cdot \mathbb{Z}[\rho]$, where ρ is a cube root of unity, Jacobi's idea is to use the hexagonal symmetry of the ring $\mathbb{Z}[\rho]$ of Eisenstein integers in order to obtain a partition of the solutions of the equation $N = 3m^2 + n^2$ into two equinumerous families $\{(m, n)\}$ and $\{(m', n')\}$, such that

$$m(-1)^{(m-n)/2} \left(m^2 - n^2 \right) = -m'(-1)^{(m'-n')/2} \left(m'^2 - n'^2 \right).$$

It is not unlikely that these ideas of Jacobi, together with the elementary combinatorial arguments of Euler and Gauss concerning the partition function, served as catalytic elements in Liouville's development of his methods.

It is nevertheless interesting to note that Liouville does not once mention the name of Jacobi in any of his first twelve essays regarding the proof of the fundamental identities.

In this section we show how to derive a Liouville-type identity from an elliptic modular form identity.

Formal manipulation of the Jacobi triple product identity,

$$\prod_{n=1}^{\infty} \{(1-q^{2n})(1-zq^{2n-1})(1-z^{-1}q^{2n-1})\} = \sum_{n=-\infty}^{\infty} z^n \cdot q^{n^2},$$

with $q = e^{2\pi i z}$, provides the Fourier expansion of the function $\vartheta_1'(0,z)/\vartheta_1(u,z)$:

$$\prod_{n \geq 1} \frac{(1-q^n)^2}{1-2q^n \cos 2u + q^{2n}} = 1 - \sin u \sum_{N \geq 1} q^N \sum_{d|N} \sin\left(\frac{2N}{d} - d\right) u,$$

where we have put $z = e^{iu}$ and u is a real variable. This can also be written

$$\prod_{n=1}^{\infty} (1-q^n)^3 \cdot \left\{ \sum_{n \geq 0} (-1)^n \frac{\sin(2n+1)u}{\sin u} \cdot q^{(n^2+n)/2} \right\}^{-1} =$$

$$1 - 4 \sin u \sum_{N \geq 1} q^N \sum_{d|N} \sin\left(\frac{2N}{d}\ d\right) u.$$

After making the substitution $q \mapsto q^8$, we can rewrite this equation in the form

$$\frac{\displaystyle\sum_{n=1}^{\infty} \chi_g(n) \cdot n \cdot \frac{\sin nu}{\sin u} \cdot q^{n^2}}{\displaystyle\sum_{n=1}^{\infty} \chi_g(n) \cdot \frac{\sin nu}{\sin u} \cdot q^{n^2}} = 1 - 4 \sin u \sum_{N=1}^{\infty} q^{8N} \sum_{d|N} \sin\left(\frac{2N}{d} - d\right) u.$$

(For a detailed derivation, starting from Jacobi's triple product formula, the reader may consult the book by Fine [31].)

By using cross-multiplication we can rewrite the last identity in the form

$$\sum_{n=1}^{\infty} \chi_g(n) \{\sin nu - n \sin u\} q^{n^2}$$

$$= 4 \sin u \sum_{m=1}^{\infty} \chi_g(m) \sin mu \cdot q^{m^2} \cdot \sum_{N=1}^{\infty} q^{8N} \sum_{d|N} \sin\left(\frac{2N}{d} - d\right) u$$

$$= \sum_{M=1}^{\infty} q^M \cdot C_M(u),$$

where

$$C_M(u) = 4 \sum_{m,d} \chi_g(m) \sin u \cdot \sin mu \cdot \sin\left(\frac{2N}{d} - d\right) u,$$

the summation being extended over all values of m, d, N satisfying

$$m \geq 1, \quad N \geq 1, \quad d \text{ odd and } > 0, \quad d|N, \quad m^2 + 8N = M.$$

Letting δ be the complementary divisor of d, that is, $\delta \cdot d = N$, and using the addition formula for the sine function, we arrive at the following form for $C_M(u)$:

$$C_M(u) = \sum_{m,d,\delta} \chi_g(m) \{\sin(m + d - 2\delta + 1)u + \sin(m - d + 2\delta - 1)u$$
$$- \sin(m + d - 2\delta - 1)u - \sin(m - d + 2\delta + 1)u\}$$

We thus have

$$\sum_{n=1}^{\infty} \chi_g(n) q^{n^2} \{\sin nu - n \sin u\} = \sum_{\substack{M \geq 1 \\ M \equiv 1 \ (\mathrm{mod} \ 8)}} q^M \cdot C_M(u).$$

By equating the coefficients of equal powers of q we obtain that $C_M(u) = 0$ unless $M = r^2$, in which case we have

$$C_{r^2}(u) = \chi_g(r) \{\sin ru - r \sin u\}.$$

This then provides a family of identities of the form

$$\sum_j A_j^{(M)} \sin ju = 0, \quad \text{for } M = 1, \ 9, \ 17, \ 25, \ \ldots.$$

Since $A_{-j}^{(M)} = A_j^{(M)}$ for all M and j, we may in fact replace $\sin ju$ by an arbitrary odd function of the integer variable j. We thus have proved the following Liouville-type result.

THEOREM 6.13 *Another Liouville-Type Example*
Let M be any positive integer of the form $8n + 1$, let $f(j)$ be any odd function of j, and let S_M denote the sum

$$S_M = \sum_{m,d,\delta} \chi_g(m) \{f(m + d - 2\delta + 1) + f(m - d + 2\delta - 1)$$
$$- f(m + d - 2\delta - 1) - f(m - d + 2\delta + 1)\},$$

extended over all solutions (m, d, δ) of

$$m^2 + 8d\delta = M,$$

where m and d are odd positive integers and δ is a positive integer. Then

$$S_M = \begin{cases} 0 & \text{if } M \text{ is not the square of an integer} \\ \chi_g(r)\{f(r) - rf(1)\} & \text{if } M = r^2. \end{cases}$$

Many other examples of Liouville-type identities can be derived by analogous reasoning. See for example Sections 28–30 on pages 67–72 of Fine [31]. As indicated earlier, the examples we have described only touch the surface of a vast and deep territory relating elliptic modular forms and the representation theory of Lie superalgebras. The interested reader would do well to consult the work of Borcherds [10] concerning the proof of the "moonshine conjecture."

This concludes our discussion of the relation between elliptic modular functions and the Liouville identities.

6.8   *Exercises*

1. By direct calculation of the first 100 terms in the expression $\vartheta(q)^5 =$

$$\{1 + 2q + 2q^4 + 2q^9 + 2q^{16} + 2q^{25} + 2q^{36} + 2q^{49} + 2q^{64} + 2q^{81} + 2q^{100} + \cdots\}^5,$$

verify the following table of values of $r_5(10a + b)$.

					b				
a	0	1	2	3	4	5	6	7	8	9
0	1	10	40	80	90	112	240	320	200	250
1	560	560	400	560	800	960	730	480	1240	1520
2	752	1120	1840	1600	1200	1210	2000	2240	1600	1680
3	2720	3200	1480	1440	3680	3040	2250	2800	3280	4160
4	2800	1920	4320	5040	2800	3472	5920	4480	2960	3370
5	5240	6240	3760	3920	6720	7360	4000	3360	7920	6800
6	4800	6160	6720	8000	5850	3840	8960	9840	4320	6720
7	10720	9280	6200	5280	9840	10480	7600	6720	11040	13440
8	5872	6730	12960	10320	7520	10080	12400	12480	9200	6240
9	14000	16480	8000	10080	16960	13760	8880	8160	13480	17360

2. Prove that if n is a positive integer, then

$$r_5(n) \equiv \begin{cases} 2 \pmod{10} & \text{if } n = 5t^2 \text{ for some positive integer } t, \\ 0 \pmod{10} & \text{otherwise.} \end{cases}$$

3. Let q be a prime $\equiv 3 \pmod 4$. Use Theorem 6.4 and the class number formula

$$h(-q) = \text{ class number of } \mathbb{Q}(\sqrt{-q}) = -\sum_{a=1}^{q-1} \left(\frac{a}{q}\right) \frac{a}{q}$$

to prove Gauss' theorem relating $r_3(q)$ with $h(-q)$.

4. Let p be a prime and $\mathbb{Q}_p$ the p-adic field. Let the sequence of positive integers $\{k_j\}$, $j = 1, 2, 3, \ldots$ converge p-adically to the number $k \in \mathbb{Q}_p$, that is to say, $\mathrm{ord}_p(k_j - k_{j'}) \to \infty$ as $\min(j, j') \to \infty$. Consider

$$\vartheta(q) = \sum_{n=-\infty}^{\infty} q^{n^2}$$

as a formal power series in q. Prove that as a formal power series

$$\vartheta(q)^k = \lim_{j \to \infty} \vartheta(q)^{k_j} = 1 + \sum_{n=1}^{\infty} \rho_k(n) q^n$$

makes sense and represents a well-defined series in $\mathbb{Z}_p[q]$. We call $\rho_k(n)$ the p-adic singular series in the variable k.

5. Verify that the constant $c = 2$ that appears in the second summation formula (6.6) is correct by using the formula

$$r_{12}(n) = \rho_{12}(n) + c\sum (a + ib + jc + kd)^4,$$

where the summation is taken over all integral quaternions $a+ib+jc+kd$ of norm n. Apply the above formula to $n = 1$ and note that $r_{12}(1) = 24$, $\rho_{12}(1) = 8$ and $\sum_{N(u)=1} u^4 = 8$. Hint: Use the well-known fact that in addition to the eight fundamental quaternion units

$$\pm 1, \quad \pm i, \quad \pm j, \quad \pm k,$$

there are sixteen other integral quaternion units given by the expressions

$$u = \frac{\pm 1 \pm i \pm j \pm k}{2},$$

all of which satisfy the simple relation $u^4 = -u$. (The noncommutative multiplication of quaternions $a+ib+jc+kd$ is the usual one: $i^2 = j^2 = k^2 = -1$, and $i \cdot j = k$.

6. Prove Equation (6.55) and Theorems 6.12 and 6.13 by the elementary methods of Chapter 2.

7. Prove that for each integer $n \geq 1$ we have

$$r_{12}(2n + 1) = 8\sigma_5(2n + 1) + 16\left(\sigma_1(2n + 1) \right.$$

$$+ 16 \sum_{j=1}^{n} (-1)^j j^3 \sum_{\substack{k \geq 1 \\ 2n-2jk+1>0}} \sigma_1(2n - 2jk + 1) \Bigg),$$

$$r_{12}(2n + 2) = 8\left(\sigma_5(2n + 2) - 64\sigma_5((n + 1)/2) \right).$$

(This exercise and the next one use the convention that $\sigma_i(x) = 0$ if x is not a positive integer.)

8. Show that if $\mathrm{odd}(m)$ means the largest odd divisor of the positive integer m, then for all integers $n \geq 1$ we have

$$
r_{16}(n) = \frac{32}{17} \left\{ \sigma_7(n) - 2\sigma_7(n/2) + 256\,\sigma_7(n/4) \right.
$$

$$
+ (-1)^{n-1} 16 \left(2^{3\,\mathrm{ord}_2(n)} \sigma_3(\mathrm{odd}(n)) \right)
$$

$$
\left. + 16 \sum_{j=1}^{n-1} (-1)^j j^3 \sum_{\substack{k \geq 1 \\ n-jk>0}} 2^{3\,\mathrm{ord}_2(n-jk)} \sigma_3(\mathrm{odd}(n-jk)) \right) \right\}.
$$

9. Use Equation (6.4) to prove that $f_8(n) \overset{\mathrm{def}}{=} r_8(n)/r_8(1)$ is a multiplicative function.

Chapter 7

Arithmetic Progressions

Except for the last section, this chapter has little to do with sums of squares of integers. It proves a famous theorem of Szemerédi that says that every subset of the positive integers with positive (upper) asymptotic density contains arbitrarily long arithmetic progressions. Three squares x^2, y^2, z^2 form an arithmetic progression if and only if $y^2 - x^2 = z^2 - y^2$, that is, if and only if $x^2 + z^2 = 2y^2$. Therefore, an arithmetic progression of three squares is equivalent to the representation of twice a square as the sum of two squares. Exercises at the end of this chapter make additional connections between arithmetic progressions and sums of squares. A further connection between this chapter and the rest of the book is a proof technique (in Section 7.3) which was also used in Section 6.2.

7.1 Introduction

In this chapter the notation $|A|$ means the cardinality of A if A is a set and the absolute value of A if A is a real number. The meaning will always be clear from the context.

We repeat Definition 3.5 from Chapter 3. For integers a and b, let $[a, b] = \{n \in \mathbb{Z} : a \leq n \leq b\}$. Likewise, let $[a, b) = \{n \in \mathbb{Z} : a \leq n < b\}$. Define $(a, b]$ and (a, b) similarly. For a set A of integers and a positive integer n, let $A(n) = |A \cap [1, n]|$, the number of positive integers $\leq n$ in A. The *upper* and *lower asymptotic densities* of A are $\overline{d}(A) = \limsup_{n \to \infty} A(n)/n$ and $\underline{d}(A) = \liminf_{n \to \infty} A(n)/n$. Thus, $0 \leq \underline{d}(A) \leq \overline{d}(A) \leq 1$. If $\underline{d}(A) = \overline{d}(A)$, we say that A has *(asymptotic) density* $d(A) = \underline{d}(A)$.

Example 7.1

The odd integers have asymptotic density $1/2$, as do the even integers. Both the set of primes and the set of squares have asymptotic density 0. Exercise 19 of Chapter 2 shows that the asymptotic density of the set A of integers representable as a sum of three squares is $5/6$, since $A(n) = n - N(n)$.

An *arithmetic progression of length k*, or *k-progression*, is a set of the form $\{a + xd : x \in [0, k-1]\}$, where a and d are positive integers. In 1926, van der Waerden proved that if the set of positive integers is partitioned into a finite number of sets, then at least one set must contain arbitrarily long arithmetic progressions. However, for any integer $r > 1$, one can partition the positive integers into r sets, none of which contains an infinite arithmetic progression $\{a + xd : x = 0, 1, \ldots\}$ with a and d positive integers. In Section 7.2 we will prove van der Waerden's theorem as well as a version of it for large finite sets of integers.

If the set of positive integers is partitioned into a finite number of sets, then one might expect that the "largest" one contains arbitrarily long arithmetic progressions. It is easy to see that at least one of the sets must have positive upper density. Erdős and Turán [30] conjectured that every set of integers with positive upper density contains arbitrarily long arithmetic progressions. (From this statement van der Waerden's theorem is an immediate corollary.)

Let $r_k(n)$ be the cardinality of the largest subset of $[1, n]$ that contains no arithmetic progression of length k. (This notation conflicts with the meaning of $r_k(n)$ throughout the rest of the book, but both notations have been standard for many decades.)

It is easy to prove that $\tau_k = \lim_{n \to \infty} r_k(n)/n$ exists, $0 \le \tau_k \le 1$, and also that $\tau_k = 0$ if and only if every set of integers with positive upper density contains an arithmetic progression of length k. Hence, the Erdős-Turán conjecture is that all $\tau_k = 0$. Behrend [9] proved that either the conjecture holds or else $\lim_{k \to \infty} \tau_k = 1$. Salem and Spencer [94] gave the lower bound

$$r_3(n) > n^{1 - c/\log\log n}$$

for some $c > 0$. Later Rankin [89] proved a similar lower bound for $r_k(n)$. In 1952, Roth [92] proved that $\tau_3 = 0$. We present his (analytic) argument in Section 7.3. Using elementary (nonanalytic) methods, Szemerédi proved [108] in 1967 that $\tau_4 = 0$, and in 1973 [109] that every $\tau_k = 0$. In Section 7.4 we give an elementary proof of Roth's theorem ($\tau_3 = 0$) using ideas from Szemerédi's beautiful proof [109] of the great conjecture of Erdős and Turán [30], which occupies most of the remainder of this chapter.

After Szemerédi's proof was published, Furstenberg [33] gave a completely different proof based on ergodic theory of the same result. Recently, Gowers [39] has given a new proof of Szemerédi's theorem which allows one to compute the number $n_k(\varepsilon)$ of Szemerédi's Theorem 7.5 explicitly.

Very recently, Green and Tao [41] proved the long-standing conjecture that there are arbitrarily long arithmetic progressions of prime numbers. This fact does not follow from Szemerédi's theorem because the set of primes has density 0. However, Chowla [17] proved in 1944 that there are infinitely many 3-progressions of primes.

We will frequently use the *box principle* (*pigeonhole principle, drawer principle*), which says that if you put $kn + 1$ objects into n boxes (pigeonholes, drawers), then at least one box must contain at least $k + 1$ objects.

7.2 Van der Waerden's Theorem

The short, lucid proof given in this section is due to Graham and Rothschild. The same argument is used (in less clear form) in all other proofs of van der Waerden's theorem.

THEOREM 7.1 Finite Form of van der Waerden's Theorem
For all positive integers ℓ and r there is an $n_0 = n_0(\ell, r)$, such that if $n \geq n_0$ is an integer and $[1, n]$ is partitioned into r sets, then (at least) one set contains an ℓ-progression.

DEFINITION 7.1   *Two m-tuples $(x_1, \ldots, x_m)$ and $(y_1, \ldots, y_m)$ in $[0, \ell]^m$ are ℓ-equivalent if they agree up through their last occurrences of ℓ. If ℓ does not occur in either m-tuple, then they are ℓ-equivalent.*

Example 7.2

If $\ell = 5$, then $(3, 5, 5, 0)$ and $(3, 5, 5, 3)$ are 5-equivalent, while $(4, 5, 5, 3)$ and $(0, 5, 5, 3)$ are not 5-equivalent,

It is clear that ℓ-equivalence is an equivalence relation. With $m = 1$, the ℓ-equivalence classes of $[0, \ell]^1$ are $\{(0), (1), \ldots, (\ell - 1)\}$ and $\{(\ell)\}$.

The following notation will shorten the proof of van der Waerden's theorem. For positive integers ℓ and m, let $S(\ell, m)$ denote this statement:

> For any positive integer r there exists a positive integer $N(\ell, m, r)$, such that for any function $f : [1, N(\ell, m, r)] \longrightarrow [1, r]$ there exist positive integers $a, d_1, \ldots, d_m$, such that $f(a + \sum_{i=1}^{m} x_i d_i)$ is constant on each ℓ-equivalence class of $[0, \ell]^m$.

We now give the proof of Graham and Rothschild [40] for Theorem 7.1.

PROOF   Theorem 7.1 follows immediately from these four statements:
(a) $S(\ell, 1)$ implies Theorem 7.1 for that ℓ and all r.
(b) $S(1, 1)$ is true.
(c) For all positive integers ℓ and m, $S(\ell, m)$ implies $S(\ell, m + 1)$.
(d) For each positive integer ℓ, if $S(\ell, m)$ holds for every positive integer m, then $S(\ell + 1, 1)$ holds.   ∎

Note that Statements (b), (c) and (d) give an inductive proof that $S(\ell, m)$ is true for all positive integers ℓ and m.

We now prove the four statements.

(a) $S(\ell, 1)$ implies Theorem 7.1 for that ℓ and all r.

PROOF   Given ℓ and r, let $n_0 = n_0(\ell, r) = N(\ell, 1, r)$. Let $n \geq n_0$ and suppose $[1, n]$ is partitioned into r sets $A_1, \ldots, A_r$. Define $f : [1, n_0] \longrightarrow [1, r]$ by $f(x) = i$ if $x \in A_i$. By $S(\ell, 1)$ and the definition of ℓ-equivalence, there exist positive integers a and d, such that $f(a + xd)$ is constant for $x \in [0, \ell - 1]$. This means that the arithmetic progression $\{a + xd : x \in [0, \ell - 1]\}$ is wholly contained in one A_i, which is van der Waerden's theorem.   ∎

(b) $S(1, 1)$ is true.

PROOF   The 1-equivalence classes of $[0, 1]$ are $\{(0)\}$ and $\{(1)\}$. Any function is constant on a singleton set.   ∎

(c) For all positive integers ℓ and m, $S(\ell, m)$ implies $S(\ell, m + 1)$.

PROOF   Assume $S(\ell, m)$. For fixed r, let $M = N(\ell, m, r)$, $M' = N(\ell, 1, r^M)$ and $N(\ell, m + 1, r) = MM'$. Suppose $f : [1, MM'] \longrightarrow [1, r]$ is given. Define $g : [1, M'] \longrightarrow [1, r^M]$ by

$$g(k) = 1 + \sum_{j=0}^{M-1} (f(kM - j) - 1)r^j,$$

so that $g(k) = g(k')$ if and only if $f(kM - j) = f(k'M - j)$ for every $j \in [0, M - 1]$. (Any function g with this property will work.) By induction there exist positive integers a' and d', such that $g(a' + xd')$ is constant for $x \in [0, \ell - 1]$. Since $S(\ell, m)$ can apply to the interval $[a'M + 1, (a' + 1)M]$ of length M because arithmetic progressions are invariant under translation, then by the choice of $M = N(\ell, m, r)$, there exist positive integers $a, d_1, \ldots, d_m$ with all sums $a + \sum_{i=1}^m x_i d_i$ in $[a'M + 1, (a' + 1)M]$ for $x_i \in [0, \ell]$ and with $f(a + \sum_{i=1}^m x_i d_i)$ constant on ℓ-equivalence classes of $[0, \ell]^m$. Set $d_i' = d_i$ for $i \in [1, m]$ and $d_{m+1}' = d'M$. Then $f\left(a + \sum_{i=1}^{m+1} x_i d_i'\right)$ is constant on ℓ-equivalence classes of $[0, \ell]^{m+1}$, which proves $S(\ell, m + 1)$.   ∎

(d) For each positive integer ℓ, if $S(\ell, m)$ holds for every positive integer m, then $S(\ell + 1, 1)$ holds.

PROOF   Let $N(\ell + 1, 1, r) = 2N(\ell, r, r)$. For fixed r, let $f : [1, 2N(\ell, r, r)] \longrightarrow [1, r]$ be given. By definition of $N(\ell, r, r)$ there exist positive integers $a, d_1, \ldots, d_r$, such that

$$a + \sum_{i=1}^r x_i d_i \in [1, N(\ell, r, r)]$$

for $x_i \in [0, \ell]$ and $f\left(a + \sum_{i=1}^{r} x_i d_i\right)$ is constant on ℓ-equivalence classes of $[0, \ell]^r$. By the box principle there exist integers $0 \le u < v \le r$, such that

$$f\left(a + \sum_{i=1}^{u} \ell d_i\right) = f\left(a + \sum_{i=1}^{v} \ell d_i\right). \tag{7.1}$$

Therefore

$$h(x) = f\left(\left(a + \sum_{i=1}^{u} \ell d_i\right) + x\left(\sum_{i=u+1}^{v} d_i\right)\right)$$

is constant for $x \in [0, \ell]$: We have $h(x) = h(0)$ for $x \in [0, \ell - 1]$ because

$$\{(\underbrace{\ell, \ldots, \ell}_{u}, \underbrace{x, \ldots, x}_{v-u}, \underbrace{0, \ldots, 0}_{r-v}) : x \in [0, \ell - 1]\}$$

is one equivalence class, and $h(\ell) = h(0)$ by Equation (7.1). This proves $S(\ell + 1, 1)$. ∎

This corollary is also due to van der Waerden.

COROLLARY 7.1
If the set of all positive integers is partitioned into a finite number of sets, then (at least) one set contains arbitrarily long arithmetic progressions.

PROOF   Suppose the corollary were false. Then there would exist a positive integer r and a partition of the positive integers into disjoint sets sets $A_1, \ldots, A_r$, such that there is a uniform bound, say $\ell - 1$, on the length of the arithmetic progressions contained in any A_i. Let $n = n_0(\ell, r)$ and $B_i = A_i \cap [1, n]$. By Theorem 7.1, some B_i contains an ℓ-progression. But this progression is too long to be contained in A_i and this contradiction proves the corollary. ∎

7.3   Roth's Theorem $\tau_3 = 0$

In this section we give Roth's [92] analytic proof that $\tau_3 = 0$. We will use the circle method of Section 6.2. Both are based on Theorem 7.2 of Hardy and Littlewood.

For integer k consider the statement C_k:

If a set A of positive integers contains no k-progression, then $d(A) = 0$.

Clearly C_1 and C_2 are true, and C_{k+1} implies C_k for any positive integer k.

REMARK 7.1   If a set A of positive integers contains an infinite arithmetic progression $\{a + xd : x = 1, 2, \ldots\}$, then clearly its lower density is $\underline{d}(A) \geq 1/d > 0$. However, it is possible to have $\underline{d}(A) > 0$ while A contains no infinite arithmetic progression. ∎

The standard technique for proving Statement C_k for $k \geq 3$ uses the function $r_k(n)$.

DEFINITION 7.2   *In this chapter only, $r_k(n)$ denotes the size of the largest subset of $[1, n]$ that contains no k-progression.*

Note that if A is a set of positive integers containing no k-progression, then $A(n) \leq r_k(n)$. This shows that Statement C_k follows from $\lim_{n \to \infty} r_k(n)/n = 0$, and this is how we will prove C_k.

The following properties of $r_k(N)$ will be useful. (Their proofs follow.)

1. If P is any n-progression, then $r_k(n)$ is the size of the largest subset of P containing no k-progression.

2. For all positive integers k, m and n, $r_k(m + n) \leq r_k(m) + r_k(n)$.

3. If k is a positive integer, then $\tau_k \overset{\text{def}}{=} \lim_{n \to \infty} r_k(n)/n$ exists.

4. For all positive integers n and k, $r_k(n) \geq n\tau_k$.

Property 1 holds because k-progressions are invariant under any nondegenerate linear transformation. Property 2 is immediate from Property 1. Properties 3 and 4 follow from Property 2 and the next proposition with $f(n) = r_k(n)$, which is due to Behrend [9].

PROPOSITION 7.1

If $f(n)$ is a nonnegative real-valued function defined on positive integers n and satisfying $f(m+n) \leq f(m)+f(n)$, then $\lim_{n \to \infty} f(n)/n$ exists and equals $\inf_{n \geq 1} f(n)/n$.

PROOF   Since $f(n) \leq nf(1)$, which is immediate from the triangle inequality, we see that $f(n)/n$ is bounded. As $f(0)$ does not appear in the statement of the proposition, we may define $f(0) = 0$. Let m and n be positive integers. Then $0 \leq m - n\lfloor m/n \rfloor < n$. Using $f(m + n) \leq f(m) + f(n)$ and induction on m we find

$$f(m) \leq \left\lfloor \frac{m}{n} \right\rfloor f(n) + f\left(m - n \left\lfloor \frac{m}{n} \right\rfloor\right)$$
$$\leq \frac{m}{n} f(n) + nf(1).$$

Therefore,

$$\frac{f(m)}{m} \le \frac{f(n)}{n} + \frac{n}{m} f(1).$$

Letting m go to infinity we find

$$\limsup_{m\to\infty} \frac{f(m)}{m} \le \frac{f(n)}{n} \qquad (7.2)$$

for every positive integer n. Now let n go to infinity. We get

$$\limsup_{m\to\infty} \frac{f(m)}{m} \le \liminf_{n\to\infty} \frac{f(n)}{n}.$$

Hence the limit exists, and by Inequality (7.2), it equals $\inf_{n\ge 1} f(n)/n$. ∎

The usual method of proving C_k is to show $\tau_k = 0$. In 1952, Roth [92] proved that $\tau_3 = 0$. His proof is an unusual application of the Hardy-Littlewood circle technique we have already seen in Section 6.2. We give it here because it is very different from the method Szemerédi [109] used to prove $\tau_k = 0$ for all k.

For real α let $e(\alpha) = e^{2\pi i\alpha}$. Then $e(x+y) = e(x)e(y)$, $e(-x) = \overline{e(x)}$ (the complex conjugate), $e(x)$ is periodic with period 1, and for integers n we have $e(n) = 1$ and

$$\int_0^1 e(n\alpha)\, d\alpha = \begin{cases} 1 & \text{if } n = 0 \\ 0 & \text{if } n \ne 0. \end{cases}$$

The following fact is an easy exercise.

THEOREM 7.2 Hardy-Littlewood Circle Method Principle
Let $A_1, \ldots, A_r$ be finite sets of integers. Let $S_i(x) = \sum_{a\in A_i} e(ax)$. Then

$$\int_0^1 \prod_{j=1}^r S_j(\alpha)\, d\alpha = \sum_{a_1\in A_1} \cdots \sum_{a_r\in A_r} \int_0^1 e\left(\left(\sum_{j=1}^r a_j\right)\alpha\right) d\alpha$$

$= $ *the number of solutions of* $\sum_{j=1}^r a_j = 0$ *with* $a_j \in A_j$.

THEOREM 7.3 Roth's Theorem [92]
We have $\tau_3 = 0$.

PROOF   We assume that $\tau = \tau_3 > 0$ and obtain a contradiction. Let N be a large integer, $r = r_3(N)$, and $A = \{a_1, \ldots, a_r\}$ be a (maximal) subset of $[1, N]$ containing no 3-progression. Define $k(n) = 1$ if $n \in A$ and $k(n) = 0$ if

$n \notin A$. Let m be a positive integer to be chosen later which depends only on N. Let

$$S(\alpha) = \sum_{i=1}^{r} e(a_i\alpha) = \sum_{n=1}^{N} k(n)e(n\alpha)$$

and

$$S^*(\alpha) = \frac{r_3(m)}{m} \sum_{n=1}^{N} e(n\alpha).$$

Let $E(\alpha) = S^*(\alpha) - S(\alpha)$. We will show that $E(\alpha)$ is small compared to N. Let $d_n = r_3(m)/m - k(n)$ so that $|d_n| \leq 1$ and $E(\alpha) = \sum_{n=1}^{N} d_n e(n\alpha)$. Let $F(\alpha) = \sum_{\nu=0}^{m-1} e(-\nu\alpha)$, so that $F(0) = m$. Let q be a positive integer and for $n = 1, 2, \ldots, N - mq$ define

$$\sigma(n) = \sigma(m, q, n) = \sum_{\nu=0}^{m-1} d_{n+\nu q}.$$

(In this section $\sigma(n)$ is *not* the divisor function.)

The proof uses five lemmas to show that $E(\alpha)$ is small.

LEMMA 7.1

$F(q\alpha)E(\alpha) = \sum_{n=1}^{N-mq} \sigma(n)e(n\alpha) + R(\alpha)$, where $|R(\alpha)| \leq 2m^2 q$. In particular, with $\alpha = 0$, we have $mE(0) = \sum_{n=1}^{N-mq} \sigma(n) + R$, where $|R| \leq 2m^2 q$.

PROOF   We have

$$F(q\alpha)E(\alpha) = \sum_{\nu=0}^{m-1} e(-\nu q\alpha) \sum_{n=1}^{N} d_n e(n\alpha)$$

$$= \sum_{\nu=0}^{m-1} \sum_{n=1}^{N} d_n e((-\nu q + n)\alpha)$$

$$= \sum_{\nu=0}^{m-1} \sum_{n=1-\nu q}^{N-\nu q} d_{n+\nu q} e(n\alpha)$$

$$= \sum_{n=1}^{N-mq} \left(\sum_{\nu=0}^{m-1} d_{n+\nu q} \right) e(n\alpha) + R(\alpha)$$

$$= \sum_{n=1}^{N-mq} \sigma(n)e(n\alpha) + R(\alpha),$$

where $R(\alpha)$ contains at most $2m^2 q$ terms, each ≤ 1 in absolute value. ∎

LEMMA 7.2
For every positive integer n, $\sigma(n) \geq 0$.

PROOF By definition,

$$\sigma(n) = \sum_{\nu=0}^{m-1} \left(\frac{r_3(m)}{m} - k(n + \nu q) \right)$$

$$= r_3(m) - \sum_{\nu=0}^{m-1} k(n + \nu q)$$

$$= r_3(m) - |\{n + \nu q : \nu \in [0, m-1], \ n + \nu q \in A\}| \geq 0$$

because the set is a subset of an m-progression and contains no 3-progression, and can have no more than $r_3(m)$ elements. $\blacksquare$

LEMMA 7.3
Let α be a real number and M be a positive integer. Then there exist integers p and q, such that $|p - q\alpha| < 1/M$ and $1 \leq q \leq M$.

PROOF Let $\{\beta\} = \beta - \lfloor \beta \rfloor$ denote the fractional part of β and $\|\beta\| = \min(\{\beta\}, 1 - \{\beta\}) = \min_{n \in \mathbb{Z}} |\beta - n|$ denote the distance from β to the nearest integer. Apply the box principle to the $M + 1$ objects $\{g\alpha\}$, $g \in [0, M]$ and the M boxes $[j/M, (j+1)/M)$, $j \in [0, M-1]$. By the box principle there must exist two integers $0 \leq g < g' \leq M$, such that $|\{\alpha g\} - \{\alpha g'\}| < 1/M$. Let $q = g' - g$. Let p be the nearest integer to $q\alpha$. Then

$$|p - q\alpha| = \|q\alpha\| = \|g\alpha - g'\alpha\| \leq |\{\alpha g\} - \{\alpha g'\}| < \frac{1}{M},$$

which proves the lemma. $\blacksquare$

LEMMA 7.4
For every real number α there is a positive integer q, such that $1 \leq q \leq 2m$ and $|F(q\alpha)| \geq 2m/\pi$.

PROOF Apply Lemma 7.3 with $\alpha = \alpha$ and $M = 2m$. There exist integers p and q, such that $1 \leq q \leq 2m$ and $|p - q\alpha| < 1/(2m)$. Let $\beta = p - \alpha q$. Then $|m\beta| < 1/2$ and

$$|F(q\alpha)| = |F(-\beta)| = \left| \sum_{\nu=0}^{m-1} e(\nu\beta) \right| = \left| \frac{1 - e(m\beta)}{1 - e(\beta)} \right|$$

$$= \left| \frac{\sin \pi m\beta}{\sin \pi\beta} \right| = m \cdot \left| \frac{\sin \pi m\beta}{\pi m\beta} \right| \cdot \left| \frac{\pi\beta}{\sin \pi\beta} \right| \geq m \cdot \frac{2}{\pi} \cdot 1.$$

∎

LEMMA 7.5

For all real α we have

$$|E(\alpha)| < \frac{\pi}{2}N\left(\frac{r_3(m)}{m} - \frac{r}{N}\right) + 4\pi m^2.$$

PROOF   Let α be given. Choose q by Lemma 7.4. By Lemmas 7.1 and 7.2 we have

$$|F(q\alpha)E(\alpha)| \le \sum_{n=1}^{N-mq} \sigma(n) + |R(\alpha)|$$
$$\le mE(0) + |R| + |R(\alpha)|$$
$$\le mE(0) + 4m^2q.$$

By Lemma 7.4 we have $q \le 2m$ and

$$|E(\alpha)| \le |F(q\alpha)|^{-1}\left(mE(0) + 8m^3\right)$$
$$\le \frac{\pi}{2m}\left(mE(0) + 8m^3\right)$$
$$\le \frac{\pi}{2}E(0) + 4m^2.$$

Finally, $E(0) = S^*(0) - S(0) = r_3(m)N/m - r$.   ∎

We now resume the proof of Roth's theorem.

Let $I = \int_0^1 S^2(\alpha)S(-2\alpha)\,d\alpha$ and $I^* = \int_0^1 S^2(\alpha)S^*(-2\alpha)\,d\alpha$. We will obtain the desired contradiction by showing that I and $|I^* - I|$ are small but I^* is large. Now I is the number of solutions to $a_i + a_j - 2a_k = 0$. Since A contains no 3-progression the only solutions are $a_i = a_j = a_k$ and so $I = r$. Similarly, I^* is $r_3(m)/m$ times the number of solutions to $a_i + a_j - 2n = 0$. This equation has one solution for each pair (a_i, a_j) in which a_i and a_j are elements of A with the same parity. Let V be the number of even numbers in A and let W be the number of odd numbers in A. Then $V + W = r$ and the number of solutions to $a_i + a_j = 2n$ is $V^2 + W^2 \ge (V+W)^2/2 = r^2/2$ by the Cauchy-Schwartz inequality. Hence,

$$I^* \ge \frac{r_3(m)}{m} \cdot \frac{r^2}{2} \ge \frac{\tau}{2}(\tau N)^2 = \frac{\tau^3 N^2}{2}.$$

We may assume that N is so large that $r/N < 2\tau$. Then

$$|I^* - I| \ge \frac{\tau^3 N^2}{2} - r > \frac{\tau^3 N^2}{2} - 2\tau N.$$

On the other hand we have by Lemma 7.5,

$$|I^* - I| = \left| \int_0^1 S^2(\alpha)(S^*(-2\alpha) - S(-2\alpha)) \, d\alpha \right|$$

$$\leq \left(\sup_{0 \leq \alpha \leq 1} |S^*(-2\alpha) - S(-2\alpha)| \right) \int_0^1 |S^2(\alpha)| \, d\alpha$$

$$= \sup_{0 \leq \alpha \leq 1} |E(\alpha)| \int_0^1 S(\alpha)S(-\alpha) \, d\alpha$$

$$\leq \left(\frac{\pi}{2} N \left(\frac{r_3(m)}{m} - \frac{r}{N} \right) + 4\pi m^2 \right) r$$

because $\int_0^1 S(\alpha)S(-\alpha) \, d\alpha$ is the number of solutions to $a_i - a_j = 0$ with $a_i, a_j \in A$, which is r. Now choose $m = \lfloor N^{1/4} \rfloor$. Since $r < 2\tau N$ we have

$$|I^* - I| \leq \left(\frac{\pi}{2} \left(\frac{r_3(m)}{m} - \frac{r}{N} \right) + 4\pi N^{-1/2} \right) 2\tau N^2,$$

or $|I^* - I| \leq N^2 \varepsilon(N)$, where $\varepsilon(N) \to 0$ as $N \to \infty$. But this contradicts our lower bound on $|I^* - I|$, and the proof of Roth's theorem is complete. ∎

REMARK 7.2   A more careful analysis of the proof shows that $r_3(n) < cn/\log \log n$ for some constant $c > 0$. ∎

7.4   *Szemerédi's Proof of Roth's Theorem*

Roth's beautiful proof [92] that $\tau_3 = 0$ uses analytic methods (integrals, complex numbers) far removed from sets of integers and arithmetic progressions of integers. In contrast, Szemerédi's equally beautiful proof of the same theorem avoids analysis and sticks to the integers.

If $A_1, \ldots, A_r$ are sets of integers, let $A_1 + \cdots + A_r = \{a_1 + \cdots + a_r : a_i \in A_i\}$. Recall that the half-open interval $[a, b)$ is the set of all integers n with $a \leq n < b$.

LEMMA 7.6
If ℓ is a positive integer and $0 < \delta < 1$, then there is a number $p(\delta, \ell)$, such that if $n \geq p(\delta, \ell)$, $C \subseteq [0, n)$ and $|C| \geq \delta n$, then there is a set $S_\ell \subseteq C$ of the form

$$S_\ell = \{y\} + \{0, x_1\} + \cdots + \{0, x_\ell\},$$

where y is a nonnegative integer and each x_i is a positive integer.

PROOF   Let $p(\delta, \ell) = (4/\delta)^{2^\ell}/2$. Thus $n \geq p(\delta, \ell)$ implies

$$4\left(\frac{\delta}{4}\right)^{2^k} n \geq 2 \quad \text{for } k \in [0, \ell]. \tag{7.3}$$

With $k = 0$ we have $|C| \geq \delta n \geq 2$, so that $|C| - 1 \geq |C|/2$. The number of (unordered) pairs of elements of C is

$$\sum_{a < a' \in C} 1 = \binom{|C|}{2} = \frac{|C|(|C| - 1)}{2} \geq \frac{|C|^2}{4} \geq \frac{\delta^2 n^2}{4}.$$

For a positive integer d define $D(d)$ to be the number of $a \in C$ for which $a + d$ is also in C. Since $C \subseteq [0, n)$ we have

$$\sum_{d=1}^{n-1} D(d) = \sum_{a < a' \in C} 1 \geq \frac{\delta^2 n}{4} n.$$

By the box principle there exists a positive integer x_1, such that $D(x_1) \geq \delta^2 n/4$. Let C_1 be the set of all $a \in C$ for which $a + x_1 \in C$. It follows that $C_1 + \{0, x_1\} \subseteq C$ and $|C_1| = D(x_1) \geq \delta^2 n/4 \geq 2$ by Inequality (7.3) with $k = 1$.

Repeat this process with C replaced by C_1. We find C_2 and x_2, such that C_2 is the set of all $a \subset C_1$ for which $a \mid x_2 \subset C_1$. We have $C_2 \mid \{0, x_2\} \subseteq C_1$ and $|C_2| \geq 4(\delta^2/4)^{2^2} n \geq 2$ by Inequality (7.3) with $k = 2$. Continue this process. Finally we reach C_ℓ and x_ℓ, such that $C_\ell + \{0, x_\ell\} \subseteq C_{\ell-1}$ and $|C_\ell| \geq 2$. Let $y \in C_\ell$. Then $\{y\} + \{0, x_\ell\} \subseteq C_{\ell-1}$ and $S_\ell \subseteq C$. $\blacksquare$

DEFINITION 7.3   For $\varepsilon > 0$ let $n_3(\varepsilon)$ be the least positive integer, such that $n \geq n_3(\varepsilon)$ implies $r_3(n)/n < \tau_3 + \varepsilon$. This number exists because $\lim_{n \to \infty} r_3(n)/n = \tau_3$.

We now give Szemerédi's elementary proof that $\tau_3 = 0$.

PROOF   Suppose that $\tau = \tau_3 > 0$. Let $\varepsilon = \tau^2/32$. Then $\varepsilon < \tau/6$ because $\tau < 1$. Let $m = n_3(\varepsilon)$, $\delta = \tau/8$ and ℓ be an integer, such that $\ell \geq 32m/\tau$. Let n be an integer so large that $n \geq p(\delta, \ell)$, $n \geq 4n_3(\varepsilon)$ and $4|n$. Since $r_3(n) \geq n\tau$, there exists a set $A \subseteq [0, n)$, such that $|A| \geq n\tau$ and A contains no 3-progression. Let $B = A \cap [0, n/4)$, $C = A \cap [n/4, n/2)$ and $D = A \cap [n/2, n)$. Since $n/4 \geq n_3(\varepsilon)$ we have $|B|, |C| \leq (\tau + \varepsilon)n/4$ and $|D| \leq (\tau + \varepsilon)n/2$. Since $|A| \geq n\tau$ and $\varepsilon < \tau/6$ we must have $|B|, |C| \geq (\tau - 3\varepsilon)n/4 \geq \tau n/8$.

Apply Lemma 7.6 with δ, ℓ and C as above to get a set $S_\ell \subseteq C$. With the parameters $y, x_1, \ldots, x_\ell$ of S_ℓ, define $S_0 = \{y\}$ and $S_i = \{y\} + \{0, x_1\} + \cdots + \{0, x_i\}$ for $i \in [1, \ell]$. Let $D_i = \{2s - b : b \in B, s \in S_i\}$. Hence,

$$D_i = D_{i-1} \cup (D_{i-1} + \{2x_i\}). \tag{7.4}$$

We have

$$S_0 \subseteq S_1 \subseteq \cdots \subseteq S_\ell \subseteq C$$
$$D_0 \subseteq D_1 \subseteq \cdots \subseteq D_\ell \subseteq [0, n).$$

Now $|S_0| = 1$, so $|D_0| = |B| \geq \tau n/8$. Thus, for $i \in [0, \ell]$ we have $\tau n/8 \leq |D_i| \leq n$. By the box principle there is a $j \in [1, \ell]$, such that $|D_j| - |D_{j-1}| \leq n/\ell$. Write $D_{j-1} = \bigcup_{\sigma=1}^r P_\sigma$, where the P_σ are disjoint arithmetic progressions with common difference $2x_j$ and with maximum length in D_{j-1}. Since P_σ is maximal we have $P_\sigma + \{2x_j\} \not\subseteq D_{j-1}$. But of course $P_\sigma + \{2x_j\} \subseteq D_j$ by Equation (7.4). Hence to each P_σ corresponds a unique element of D_j not in D_{j-1}. Therefore, $r = |D_j| - |D_{j-1}| \leq n/\ell$. Let P'_σ be P_σ with m elements dropped from each end. (If $|P_\sigma| \leq 2m$, then P'_σ is the null set.) Let $E = \bigcup_{\sigma=1}^r P'_\sigma \subseteq D_{j-1}$. Then

$$|E| \geq |D_{j-1}| - 2mr \geq \tau n/8 - 2mn/\ell = (\tau/8 - 2m/\ell)n \geq \tau n/16$$

by our choice of ℓ. By the construction of E we can write $[0, n) - E = \bigcup_{\sigma=1}^t Q_\sigma$, where the Q_σ are disjoint arithmetic progressions with difference $2x_j$, such that $|Q_\sigma| \geq m$. Therefore,

$$|E \cap A| = |A| - \sum_{\sigma=1}^t |A \cap Q_\sigma|$$

$$\geq \tau n - \sum_{\sigma=1}^t (\tau + \varepsilon)|Q_\sigma| \quad \text{(since } m = n_3(\varepsilon))$$

$$= \tau n - (\tau + \varepsilon)(n - |E|)$$

$$\geq \tau |E| - \varepsilon n \geq \left(\frac{\tau^2}{16} - \varepsilon\right) n \geq \frac{\tau^2 n}{32}$$

since $\varepsilon < \tau^2/32$. Hence $|E \cap A| > 0$ and there exists an integer $d \in E \cap A$. Therefore $d \in D_{j-1}$, so $d = 2c - b$ for some $b \in B$ and $c \in S_{j-1} \subseteq C$. Thus, the integers $b, c, d \in A$ form an arithmetic progression, and this result contradicts the choice of A. Therefore, $\tau = \tau_3 = 0$. ∎

REMARK 7.3   Some of the ideas of this proof will appear in the proof (in the following sections) that all $\tau_k = 0$. In particular, the set S_ℓ is an example of a configuration of order ℓ, which we will define in Section 7.6. ∎

7.5   Bipartite Graphs

In this section we begin the proof of the great theorem of Szemerédi [109] that all $\tau_k = 0$. In this proof lowercase Greek letters will denote real numbers between 0 and 1. Lowercase Roman letters usually denote nonnegative integers and uppercase Roman letters denote sets (or sometimes large integers).

DEFINITION 7.4   *Let A and B be disjoint sets with m and n elements, respectively. For $X \subseteq A$ and $Y \subseteq B$, let $[X, Y] = \{\{x, y\} : x \in X, y \in Y\}$. A bipartite graph is a subset of $[A, B]$. If I is a bipartite graph, $X \subseteq A$ and $Y \subseteq B$, let $k_I(X, Y) = k(X, Y) = |[X, Y] \cap I|$ and $\beta(X, Y) = k(X, Y)|X|^{-1}|Y|^{-1}$. For $u \in A \cup B$ let $k_I(u) = \{v \in A \cup B : \{u, v\} \in I\}$.*

Note that $0 \leq \beta(X, Y) \leq 1$. Note also that $k_I(u)$ is the set of all vertices directly connected to u by an edge in I.

REMARK 7.4   When we say, "Let I be a bipartite graph," we always mean that I is a subset of $[A, B]$ for sets A and B of sizes m and n, respectively.
∎

The main result of this section is Theorem 7.4, which says roughly that any large bipartite graph can be decomposed into nearly regular subgraphs. We also prove several lemmas in this section.

LEMMA 7.7
For all real numbers ε_1, ε_2 and δ strictly between 0 and 1, and for all nonempty bipartite graphs I, there exist subsets $\overline{X} \subseteq A$, $\overline{Y} \subseteq B$ and a positive integer r, such that:

(a$_1$) $r \leq 1/\delta$,

(b$_1$) $|\overline{X}| > \varepsilon_1^r |A|$ *and* $|\overline{Y}| > \varepsilon_2^r |B|$, *and*

(c$_1$) *for all $S \subseteq \overline{X}$ and $T \subseteq \overline{Y}$ with $|S| > \varepsilon_1 |\overline{X}|$ and $|T| > \varepsilon_2 |\overline{Y}|$, we have $\beta(S, T) > \beta(\overline{X}, \overline{Y}) - \delta$.*

PROOF   Since $0 < \delta < 1$, there is a positive integer r satisfying (a$_1$). Also, $\overline{X} = A$, $\overline{Y} = B$ satisfy (b$_1$). If $X \subseteq A$ and $Y \subseteq B$ do not satisfy (c$_1$), then there exist $X' \subseteq X$ and $Y' \subseteq Y$ with $|X'| > \varepsilon_1 |X|$ and $|Y'| > \varepsilon_2 |Y|$, but $\beta(X', Y') \leq \beta(X, Y) - \delta$. For a nonnegative integer t define X_t and Y_t by $X_0 = A$, $Y_0 = B$, and for $t \geq 0$, $X_{t+1} = X_t'$, $Y_{t+1} = Y_t'$, provided $X_t \subseteq A$ and $Y_t \subseteq B$ and they do not satisfy (c$_1$), Thus,

$$|X_t| > \varepsilon_1 |X_{t-1}| > \cdots > \varepsilon_1^t |X_0| = \varepsilon_1^t |A|$$
$$|Y_t| > \varepsilon_2 |Y_{t-1}| > \cdots > \varepsilon_2^t |Y_0| = \varepsilon_2^t |B|.$$

Also,

$$0 \leq \beta(X_t, Y_t) \leq \beta(X_{t-1}, Y_{t-1}) - \delta \leq \cdots \leq \beta(X_0, Y_0) - t\delta$$
$$= \beta(A, B) - t\delta \leq 1 - t\delta.$$

Thus $t \leq 1/\delta$. If for all $s \leq 1/\delta$, X_s and Y_s are defined and do not satisfy (c_1), then X_{s+1} and Y_{s+1} are also defined, and this contradicts $t \leq 1/\delta$. Hence, for some $r \leq 1/\delta$, $\overline{X} = X_r$ and $\overline{Y} = Y_r$ are defined and satisfy (c_1). ∎

LEMMA 7.8

For all real numbers ε_1, ε_2 and δ strictly between 0 and 1, and with $\varepsilon_2 < 1/2$, there exist positive integers M and N, such that for all bipartite graphs I with $|A| = m > M$ and $|B| = n > M$, there exist subsets $X \subseteq A$, $Y \subseteq B$ and a positive integer r, such that

(a_2) $r \leq 1/\delta$,

(b_2) $|X| > (1/2)\varepsilon_1^r|A|$ and $|Y| > \varepsilon_2^r|B|$,

(c_2) *for all $S \subseteq X$ and $T \subseteq Y$ with $|S| > 2\varepsilon_1|X|$ and $|T| > \varepsilon_2|Y|$, we have $\beta(S,T) > \beta(X,Y) - \delta$, and*

(d_2) *for all $x \in X$, $|k(x) \cap Y| \leq (\beta(X,Y) + 2\delta)|Y|$.*

PROOF By Lemma 7.7, there exist subsets $\overline{X} \subseteq A$, $\overline{Y} \subseteq B$ and a positive integer r satisfying (a_1), (b_1) and (c_1). Let $\beta = \beta(\overline{X}, \overline{Y})$ and

$$Z = \left\{ x \in \overline{X} \, : \, |k(x) \cap \overline{Y}| > (\beta + \delta)|\overline{Y}| \right\}.$$

We claim $|Z| \leq |\overline{X}|/2$. If not, then there exists $Z' \subseteq Z$, such that $|\overline{X}|/2 < |Z'| \leq |\overline{X}|/2 + 1$. Now

$$k(\overline{X}, \overline{Y}) = k(Z', \overline{Y}) + k(\overline{X} - Z', \overline{Y}),$$

so that $\beta = \alpha\beta(Z', \overline{Y}) + (1 - \alpha)\beta(\overline{X} - Z', \overline{Y})$, where $\alpha = |Z'|/|\overline{X}|$. By definition of Z, $k(Z', \overline{Y}) > (\beta + \delta)|Z'||\overline{Y}|$, and so $\beta(Z', \overline{Y}) > \beta + \delta$. Also, since $|Z'| \leq (1/2)|\overline{X}| + 1$, we have $|\overline{X} - Z'| \geq (1/2)|\overline{X}| - 1 > \varepsilon_1|\overline{X}|$ provided $((1/2) - \varepsilon_1)|\overline{X}| > 1$, that is, provided $|\overline{X}| > ((1/2) - \varepsilon_1)^{-1}$. This condition holds when $M > \left(\varepsilon_1^{1/\delta}((1/2) - \varepsilon_1) \right)^{-1}$. Hence $|\overline{X} - Z'| > \varepsilon_1|\overline{X}|$ and, by (c_1), we have $\beta(\overline{X} - Z', \overline{Y}) > \beta - \delta$. Hence $\beta \geq \alpha(\beta + \delta) + (1 - \alpha)(\beta - \delta) = \beta + (2\alpha - 1)\delta$. But by hypothesis, $\alpha = |Z'|/|\overline{X}| > 1/2$, so $\beta > \beta$, which is impossible. Hence $|Z| \leq |\overline{X}|/2$, as claimed.

Let $X = \overline{X} - Z$ and $Y = \overline{Y}$. Thus $|X| > (1/2)\varepsilon_1^r|A|$ and (a_2) and (b_2) are satisfied. Finally, (d_2) follows from the definition of Z, while (c_1) implies (c_2). ∎

THEOREM 7.4 Decomposition of Bipartite Graphs

For all real numbers ε_1, ε_2, δ, ρ and σ strictly between 0 and 1, there exist integers m_0, n_0, M, N, such that for all bipartite graphs I with $|A| = m > M$ and $|B| = n > N$, there exist disjoint $C_i \subseteq A$, $i < m_0$, and for each $i < m_0$, disjoint $C_{i,j} \subseteq B$, $j < n_0$, such that:

(a) $|A - \bigcup_{i<m_0} C_i| < \rho m$ and $|B - \bigcup_{j<n_0} C_{i,j}| < \sigma n$ for each $i < m_0$;

(b) for all $i < m_0$, $j < n_0$, $S \subseteq C_i$ and $T \subseteq C_{i,j}$ with $|S| > \varepsilon_1 |C_i|$ and $|T| > \varepsilon_2 |C_{i,j}|$, we have $\beta(S,T) \geq \beta(C_i, C_{i,j}) - \delta$; and

(c) for all $i < m_0$, $j < n_0$ and $x \in C_i$, we have

$$|k(x) \cap C_{i,j}| \leq (\beta(C_i, C_{i,j}) + \delta)|C_{i,j}|.$$

PROOF   Let $r = \lfloor 4/\delta \rfloor + 1$. Choose n_0 so that $(1 - \varepsilon_2^r)^{n_0} < \sigma$. Let $\varepsilon(0) = \varepsilon_1$ and

$$\varepsilon(t+1) = \varepsilon_1 \prod_{i=0}^{t} \left(\frac{\varepsilon(i)}{4}\right)^r$$

for $0 \leq t \leq n_0$. Choose m_0 so that $(1 - \varepsilon(n_0 + 1))^{m_0} < \rho$.

At this point we need Lemma 7.9, which is stated and proved below. We now finish this proof quickly using that lemma.

Apply Lemma 7.9 recursively to prove Theorem 7.4. Begin with $A_0 = A$, so $|A_0| = m > \rho m$. By Lemma 7.9, there exists a subset $C_0 \subseteq A_0$ with $|C_0| > \varepsilon(n_0+1)|A_0|$, such that the conclusions of Lemma 7.9 hold. Let $A_1 = A_0 - C_0$. Then $|A_1| < (1 - \varepsilon(n_0 + 1))|A_0|$. If $|A_1| < \rho m$, then we stop; otherwise we continue. Apply Lemma 7.9 to A_1 to obtain C_1 with $|C_1| > \varepsilon(n_0 + 1)|A_1|$. Let $A_2 = A_1 - C_1$. Eventually, we get to $A_{m_0} = A_{m_0-1} - C_{m_0-1}$ with

$$|A_{m_0}| < (1 - \varepsilon(n_0 + 1))^{m_0}|A_0| < \rho m,$$

which is the first part of (a). Lemma 7.9 gives the other conclusions of Theorem 7.4.   ∎

LEMMA 7.9
In the situation of the proof of Theorem 7.4 where this lemma is needed, there exist integers M and N, such that if $m > M$, $n > N$ and $A' \subseteq A$ with $|A'| > \rho m$, then there exist $C \subseteq A'$ and $\overline{C}_j \subseteq B$ for $j < n_0$, such that $|C| > \varepsilon(n_0 + 1)|A'|$ and C and the $\overline{C}_j$ satisfy the conclusions of Theorem 7.4, except the first part of (a), if we choose $C_i = C$ and $C_{i,j} = \overline{C}_j$.

PROOF   Define $\overline{C}_j \subseteq B$ and $Z_j \subseteq A'$ for $j \in [-1, n_0]$ by $Z_{-1} = A'$, $\overline{C}_j = \emptyset$ (the empty set); and if Z_j and $\overline{C}_j$ have been defined there are two possibilities:

(i) If $|B - \bigcup_{\nu<j} \overline{C}_\nu| < \sigma n$, set $Z_j = Z_{j-1}$ and $\overline{C}_j = \emptyset$.

(ii) If $|B - \bigcup_{\nu<j} \overline{C}_\nu| \geq \sigma n$, apply Lemma 7.8 with $\varepsilon_1 = (1/2)\varepsilon(n_0 - j)$, $\varepsilon_2 = \varepsilon_2$, $\delta = \delta/4$, $A = Z_{j-1}$ and $B = B - \bigcup_{\nu<j} \overline{C}_\nu$. There exist

$X \subseteq Z_{j-1}$, $Y \subseteq B - \bigcup_{\nu<j} \overline{C}_\nu$ and an integer $\overline{r} > 0$ satisfying the conclusions of Lemma 7.8. Set $Z_j = X$ and $\overline{C}_j = Y$. By (a$_2$),

$$\overline{r} \le \frac{4}{\delta} < \left\lfloor \frac{4}{\delta} \right\rfloor + 1 = r.$$

Also,

(b$_3$) we have

$$|Z_j| > \frac{1}{2}\left(\frac{1}{2}\varepsilon(n_0-j)\right)^{\overline{r}} |Z_{j-1}| > \left(\frac{\varepsilon(n_0-j)}{4}\right)^r |Z_{j-1}|,$$

$$|\overline{C}_j| > \varepsilon_2^{\overline{r}}\left|B - \bigcup_{\nu<j}\overline{C}_\nu\right| > \varepsilon_2^r\left|B - \bigcup_{\nu<j}\overline{C}_\nu\right|,$$

(c$_3$) for all $S \subseteq Z_j$ and $T \subseteq \overline{C}_j$ with $|S| > \varepsilon(n_0-j)|Z_j|$ and $|T| > \varepsilon_2|\overline{C}_j|$, we have

$$\beta(S,T) > \beta(Z_j,\overline{C}_j) - \frac{\delta}{2},$$

(d$_3$) for all $x \in Z_j$, we have

$$|k(x) \cap \overline{C}_j| \le \left(\beta(Z_j,\overline{C}_j) + \frac{\delta}{2}\right)|\overline{C}_j|.$$

Once we use (i) to define Z_j and $\overline{C}_j$, we always use it until Z_{n_0} and $\overline{C}_{n_0}$ are defined. Let j_0 be the largest index $j < n_0$ for which (ii) is used. By (b$_3$) we have

$$|\overline{C}_0| > \varepsilon_2^r|B|,$$
$$|B - \overline{C}_0| < (1-\varepsilon_2^r)|B|,$$
$$|\overline{C}_1| > \varepsilon_2^r|B - \overline{C}_0|,$$
$$|B - \overline{C}_0 - \overline{C}_1| < (1-\varepsilon_2^r)^2|B|,$$

$$\vdots$$

$$\left|B - \bigcup_{\nu<j_0}\overline{C}_\nu\right| < (1-\varepsilon_2^r)^{j_0}|B|,$$

$$|\overline{C}_{j_0}| > \varepsilon_2^r\left|B - \bigcup_{\nu<j_0}\overline{C}_\nu\right|,$$

$$\left|B - \bigcup_{\nu<j_0+1}\overline{C}_\nu\right| < (1-\varepsilon_2^r)^{j_0+1}|B|.$$

If $j_0 = n_0 - 1$, then

$$\left| B - \bigcup_{\nu < n_0} \overline{C}_\nu \right| < (1 - \varepsilon_2^r)^{n_0} |B| < \sigma n.$$

Otherwise (i) was used. Hence in either case

$$\left| B - \bigcup_{\nu < n_0} \overline{C}_\nu \right| < \sigma n,$$

which is part of (a) of Theorem 7.4.

Also we have

$$|Z_{n_0}| > \left(\frac{\varepsilon(0)}{4} \right) |Z_{n_0 - 1}|$$

$$\vdots$$

$$> \prod_{\nu = j+1}^{n_0} \left(\frac{\varepsilon(n_0 - \nu)}{4} \right)^r |Z_j| = \prod_{i=0}^{n_0 - j - 1} \left(\frac{\varepsilon(i)}{4} \right)^r |Z_j|,$$

by using either (b_3), or $Z_j = Z_{j-1}$ when (i) is used. Therefore, by definition of $\varepsilon(n_0 - j)$,

$$\varepsilon_1 |Z_{n_0}| > \varepsilon(n_0 - j)|Z_j| \quad \text{for } j \in [-1, n_0].$$

Let $C = Z_{n_0}$. Then $|C| = |Z_{n_0}| > \varepsilon(n_0 + 1)|A'|$. Now suppose $S \subseteq C$ and $T \subseteq \overline{C}_j$ for some $j \in [0, n_0)$ with $|S| > \varepsilon_1 |C|$ and $|T| > \varepsilon_2 |\overline{C}_j|$. Then $|S| > \varepsilon_1 |Z_{n_0}| > \varepsilon(n_0 - j)|Z_j|$. Therefore by ($c_3$), we have $\beta(S, T) > \beta(Z_j, \overline{C}_j) - \delta/2$. By ($d_3$), we have $\beta(C, \overline{C}_j) \leq \beta(Z_j, \overline{C}_j) + \delta/2$, so $\beta(S, T) > \beta(C, \overline{C}_j) - \delta$, which is (d) of Theorem 7.4.

Since $|Z_{n_0}| > \varepsilon(n_0 - j)|Z_j|$, it follows by ($c_3$) with $S = Z_{n_0} = C$ and $T = \overline{C}_j$ that we have $\beta(C, \overline{C}_j) > \beta(Z_j, \overline{C}_j) - \delta/2$. Thus by ($d_3$), for $x \in C \subseteq Z_j$, we have

$$|k(x) \cap \overline{C}_j| \leq \left(\beta(Z_j, \overline{C}_j) + \frac{\delta}{2} \right) |\overline{C}_j|$$

$$< (\beta(C, \overline{C}_j) + \delta)|\overline{C}_j|,$$

which is (c) of Theorem 7.4, and Lemma 7.9 is proved.   ∎

If we apply Theorem 7.4 to the bipartite graph $\overline{I} = [A, B] - I$ we obtain this result.

COROLLARY 7.2

For all real numbers ε_1, ε_2, δ, ρ and σ strictly between 0 and 1, there exist integers m_0, n_0, M, N, such that for all bipartite graphs I with $|A| = m > M$ and $|B| = n > N$, there exist disjoint $\overline{C}_i \subseteq A$, $i < m_0$, and for each $i < m_0$, disjoint $\overline{C}_{i,j} \subseteq B$, $j < n_0$, such that:

(ā) $|A - \bigcup_{i<m_0} \overline{C}_i| < \rho m$ and $|B - \bigcup_{j<n_0} \overline{C}_{i,j}| < \sigma n$ for each $i < m_0$;

(b̄) for all $i < m_0$, $j < n_0$, $S \subseteq \overline{C}_i$ and $T \subseteq \overline{C}_{i,j}$ with $|S| > \varepsilon_1 |\overline{C}_i|$ and $|T| > \varepsilon_2 |\overline{C}_{i,j}|$, we have $\beta(S,T) < \beta(\overline{C}_i, \overline{C}_{i,j}) + \delta$; and

(c̄) for all $i < m_0$, $j < n_0$ and $x \in \overline{C}_i$, we have

$$|k(x) \cap \overline{C}_{i,j}| \geq \left(\beta(\overline{C}_i, \overline{C}_{i,j}) - \delta\right) |\overline{C}_{i,j}|.$$

7.6 Configurations

In this section we define configurations, which generalize arithmetic progressions, and prove some facts about them. For positive integers ℓ_1, ..., ℓ_m (where possibly $m = 0$) we define a set $B(\ell_1, \ldots, \ell_m)$ of sets of positive integers called *configurations* as follows.

DEFINITION 7.5   *Let $B(\emptyset) = \{\{n\} : n$ is a positive integer$\}$, the set of all singleton sets $\{n\}$, where n is a positive integer. When m is a positive integer, let $B(\ell_1, \ldots, \ell_m)$ denote the set of all sets X of positive integers of the form $X = \bigcup_{i<\ell_m} X_i$, where for some $Y \in B(\ell_1, \ldots, \ell_{m-1})$ and some positive integer d we have $X_i = Y + \{di\}$ for $i \in [0, \ell_m)$ and $X_0 < X_1 < \cdots < X_{\ell_m-1}$. Here the notation $X < X'$ means $x < x'$ for all $x \in X$, $x' \in X'$. The elements of $B(\ell_1, \ldots, \ell_m)$ are called configurations of order m. If $X \in B(\ell_1, \ldots, \ell_m)$, then $|X| = \ell_1 \cdots \ell_m$. Two configurations $X, Y \in B(\ell_1, \ldots, \ell_m)$ are called congruent if $X = Y + \{d\}$ for some integer d.*

Example 7.3

The elements of $B(\ell_1)$ are just the ℓ_1-progressions of positive integers. The elements of $B(\ell_1, \ell_2)$ are the sets of ℓ_2 equally spaced nonoverlapping ℓ_1-progressions with the same common difference.

Example 7.4

For $X \in B(\ell_1, \ldots, \ell_m)$, the sets X_i, for $0 \leq i < \ell_m$, in the definition of configuration are congruent configurations in $B(\ell_1, \ldots, \ell_{m-1})$.

Now assume that we are given a set R of positive integers with positive upper asymptotic density: $\overline{d}(R) > 0$. Eventually we will show that R contains arbitrarily long arithmetic progressions. At present, however, we will define a sequence $t_1, t_2, \ldots$ of positive integers, certain sets

$$P(t_1, \ldots, t_m) \subseteq S(t_1, \ldots, t_m) \subseteq B(t_1, \ldots, t_m),$$

and other sequences of integers and functions. The elements of $P(t_1, \ldots, t_m)$ and $S(t_1, \ldots, t_m)$ will be called *perfect* and *saturated* configurations of order

m, respectively. Let

$$S(\emptyset) = B(\emptyset) = \{\{n\} \; : \; n \text{ is a positive integer}\},$$
$$P(\emptyset) = \{\{n\} \; : \; n \in R\}.$$

If $X \in B(t_1, \ldots, t_{m-1}, \ell)$ and $S(t_1, \ldots, t_{m-1})$ and $P(t_1, \ldots, t_{m-1})$ have been defined, we let

$$s^m(X) = |\{i < \ell \; : \; X_i \in S(t_1, \ldots, t_{m-1})\}|,$$
$$p^m(X) = |\{i < \ell \; : \; X_i \in P(t_1, \ldots, t_{m-1})\}|.$$

Let $f_1(\ell) = \max\{|X \cap R| \; : \; X \in B(\ell)\} = \max\{p^1(X) \; : \; X \in B(\ell)\}$. Clearly, $f_1(\ell + \ell') \leq f_1(\ell) + f_1(\ell')$, so by Proposition 7.1,

$$\alpha_1 = \lim_{\ell \to \infty} \frac{f_1(\ell)}{\ell}$$

exists. Also, $\alpha_1 > 0$ because $\overline{d}(R) > 0$. Let $\varepsilon_1(\ell) = |\alpha_1 - f_1(\ell)/\ell|$.

Let K be a positive integer. We will prove that R contains an arithmetic progression of $K - 1$ terms, and this will prove Szemerédi's theorem because K is arbitrary. Define a positive integer t_1 to be sufficiently large depending on K so that $\varepsilon_1(t_1)$ is small. We shall explain later precisely what this means.

We have now defined several sets and functions for the case $m = 1$. Next we define these objects and several others for $m = 2$. Later we will define them recursively for general m. Let

$$S(t_1) = \{X \in B(t_1) \; : \; p^1(X) > (\alpha_1 - \varepsilon_1(t_1)^{1/4})t_1\},$$
$$g_2(\ell) = \max\{s^2(X) \; : \; X \in B(t_1, \ell)\},$$
$$\beta_2 = \limsup_{\ell \to \infty} \frac{g_2(\ell)}{\ell}, \text{ and}$$
$$\mu_2(\ell) = \left| \beta_2 - \frac{g_2(\ell)}{\ell} \right|.$$

The corresponding quantities for $m = 1$ are degenerate: $g_1(\ell) = \ell$, $\beta_1 = 1$, $\mu_1(\ell) = 0$. Lemma 7.11 will show that $\mu_2(\ell) \to 0$ and Proposition 7.2 will show that β_2 is close to 1.

DEFINITION 7.6   *Two configurations $X, Y \in B(t_1, \ldots, t_m)$ are called R-equivalent if for any two elements $x \in X$ and $y \in Y$ in corresponding positions we have $x \in R$ if and only if $y \in R$.*

Let $X = \bigcup_{i < \ell} X_i \in B(t_1, \ell)$. The R-equivalence relation partitions the set of $X_i \in S(t_1)$ into equivalence classes, in fact, into at most 2^{t_1} equivalence classes. Let $\overline{p}^2(X)$ be the cardinality of the largest equivalence class. Define

$$\overline{f}_2(\ell) = \max\left\{ \overline{p}^2(X) \; : \; X \in B(t_1, \ell) \quad \text{and} \quad s^2(X) \geq \left(\beta_2 - \sqrt{\mu_2(\ell)} \right) \ell \right\}$$

and $\alpha_2 = \limsup_{\ell \to \infty} \bar{f}_2(\ell)/\ell$. Hence there is a sequence $\{\ell_n\}$, such that $\lim_{n \to \infty} \bar{f}_2(\ell_n)/\ell_n = \alpha_2$. Therefore there exist $X^{(n)} \in B(t_1, \ell_n)$ for which

$$s^2\left(X^{(n)}\right) \geq \left(\beta_2 - \sqrt{\mu_2(\ell_n)}\right)\ell_n$$

and

$$\lim_{n \to \infty} \bar{p}^2\left(X^{(n)}\right)/\ell_n = \alpha_2.$$

Now for infinitely many n, the same R-equivalence class occurs in the definition of $\bar{p}^2\left(X^{(n)}\right)$. Choose one such class and denote it by $P(t_1)$. Clearly, $P(t_1) \subseteq S(t_1)$. Let

$$f_2(\ell) = \max\left\{p^2(X) \;:\; X \in B(t_1, \ell) \quad \text{and} \quad s^2(X) \geq \left(\beta_2 - \sqrt{\mu_2(\ell)}\right)\ell\right\}$$

and $\varepsilon_2(\ell) = |\alpha_2 - f_2(\ell)/\ell|$. Lemmas 7.11 and 7.13 will show

$$\lim_{\ell \to \infty} g_2(\ell)/\ell = \beta_2 \quad \text{and} \quad \lim_{\ell \to \infty} f_2(\ell)/\ell = \alpha_2.$$

Now choose a positive integer t_2 sufficiently large depending on α_2, β_2, K, and t_1 so that $\varepsilon_2(t_2)$ and $\mu_2(t_2)$ are small.

Assume that for some positive integer m we have defined

$$P(t_1, \ldots, t_{m-1}) \subseteq S(t_1, \ldots, t_{m-1}) \subseteq B(t_1, \ldots, t_{m-1}),$$

$g_m(\ell)$, $f_m(\ell)$, α_m, β_m, $\varepsilon_m(\ell)$, $\mu_m(\ell)$ and t_m so that

$$g_m(\ell) = \max\{s^m(X) \;:\; X \in B(t_1, \ldots, t_{m-1}, \ell)\},$$

$$\beta_m = \lim_{\ell \to \infty} \frac{g_m(\ell)}{\ell},$$

$$\mu_m(\ell) = \left|\beta_m - \frac{g_m(\ell)}{\ell}\right|,$$

$$f_m(\ell) = \max\left\{p^m(X) \;:\; X \in B(t_1, \ldots, t_{m-1}, \ell) \text{ and}\right.$$

$$\left. s^m(X) \geq \left(\beta_m - \sqrt{\mu_m(\ell)}\right)\ell\right\},$$

$$\alpha_m = \lim_{\ell \to \infty} \frac{f_m(\ell)}{\ell},$$

$$\varepsilon_m(\ell) = \left|\alpha_m - \frac{f_m(\ell)}{\ell}\right|,$$

and the positive integer t_m is sufficiently large depending on α_m, β_m, K and t_{m-1} to insure that $\varepsilon_m(t_m)$ and $\mu_m(t_m)$ are small. We give the next step in the definition. Let

$$S(t_1, \ldots, t_m) = \left\{X \in B(t_1, \ldots, t_m) \;:\; s^m(X) \geq \left(\beta_m - \sqrt{\mu_m(t_m)}\right)t_m\right.$$

$$\text{and} \quad p^m(X) > \left.\left(\alpha_m - \sqrt{\sqrt{\varepsilon_m(t_m)} + \sqrt{\mu_m(t_m)}}\right)t_m\right\},$$

$$g_{m+1}(\ell) = \max\{s^{m+1}(X) \;:\; X \in B(t_1, \ldots, t_m, \ell)\}.$$

LEMMA 7.10

With the notation above we have

$$g_{m+1}(\ell_1 + \ell_2) \leq g_{m+1}(\ell_1) + g_{m+1}(\ell_2).$$

PROOF   Let $X \in B(t_1, \ldots, t_m, \ell_1 + \ell_2)$ have $s^{m+1}(X) = g_{m+1}(\ell_1 + \ell_2)$. Then $X = Y \cup Z$ where $Y = \bigcup_{i < \ell_1} X_i$ and $Z = \bigcup_{i < \ell_2} X_{\ell_1 + i}$. Clearly, $Y \in B(t_1, \ldots, t_m, \ell_1)$ and $Z \in B(t_1, \ldots, t_m, \ell_2)$, so $s^{m+1}(Y) \leq g_{m+1}(\ell_1)$ and $s^{m+1}(Z) \leq g_{m+1}(\ell_2)$. But $s^{m+1}(X) = s^{m+1}(Y) + s^{m+1}(Z)$. ∎

By Proposition 7.1 we have

LEMMA 7.11

For every positive integer m, $\beta_{m+1} = \lim_{\ell \to \infty} g_{m+1}(\ell)/\ell$ exists and $\beta_{m+1} \leq g_{m+1}(\ell)/\ell$ for every positive integer ℓ.

Let $\mu_{m+1}(\ell) = g_{m+1}(\ell)/\ell - \beta_{m+1}$, so that $\mu_{m+1}(\ell) \geq 0$ for every ℓ.

LEMMA 7.12

We may assume that $\mu_m(\ell) > 0$ for $m \geq 2$ and all positive integers ℓ. In other words, if $m \geq 2$, ℓ is a positive integer and $\mu_m(\ell) = 0$, then R contains arbitrarily long arithmetic progressions.

REMARK 7.5   Note that Lemma 7.12 implies that we may assume $\beta_m < 1$ for $m \geq 2$. ∎

PROOF   We will later choose t_1 so large that $\alpha_1 - \varepsilon_1(t_1)^{1/4} > 0$ and t_m so large that $\beta_m - \sqrt{\mu_m(t_m)} > 0$ (so $\beta_m > 0$). It follows by induction that if $X' \in S(t_1, \ldots, t_{m-1})$, then $X' \cap R \neq \emptyset$.

If $\mu_m(\ell) = 0$, then $g_m(\ell) = \ell\beta_m$. Let ν be a large positive integer. By Lemmas 7.10 and 7.11, $g_m(\ell\nu) = \ell\nu\beta_m$. Hence there is an $X = \bigcup_{i < \ell\nu} X_i \in B(t_1, \ldots, t_{m-1}, \ell\nu)$, such that $X_i \in S(t_1, \ldots, t_{m-1})$ for exactly $\ell\nu\beta_m$ indices $i < \ell\nu$. For $k < \nu$, let $Y_k = \bigcup_{j < \ell} X_{\ell k + j}$. Then $s^m(Y_k) \leq g_m(\ell) = \ell\beta_m$ for $k < \nu$. But $\sum_{k < \nu} s^m(Y_k) = s^m(X) = \ell\nu\beta_m$. Hence $s^m(Y_k) = \ell\beta_m$ for $k < \nu$.

The R-equivalence relation partitions $S(t_1, \ldots, t_{m-1}, \ell)$ into at most $2^{t_1 \cdots t_{m-1}\ell}$ equivalence classes. This induces a partition of $[0, \nu)$ into at most $2^{t_1 \cdots t_{m-1}\ell}$ classes, where i and j are in the same class if and only if Y_i and Y_j are R-equivalent.

Let k be a positive integer. Since ν may be as large as we please, Theorem 7.1 guarantees that some class contains a k-progression $\{a + xd : x \in [0, k)\}$. Hence $Y_{a + xd}$ for $x \in [0, k)$ are R-equivalent. But $\beta_m > 0$ and $s^m(Y_a) = \ell\beta_m >$

0, so Y_a contains an $X_i \in S(t_1, \ldots, t_{m-1})$. Then $X_i \cap R \neq \emptyset$ and $Y_a \cap R \neq \emptyset$, so R contains a k-progression. ∎

Let $X = \bigcup_{i < \ell} X_i \in B(t_1, \ldots, t_m, \ell)$. The R-equivalence relation partitions the set of $X_i \in S(t_1, \ldots, t_m)$ into at most $2^{t_1 \cdots t_m}$ equivalence classes. Let $\overline{p}^{m+1}(X)$ denote the cardinality of the largest equivalence class. Let

$$\overline{f}_{m+1}(\ell) = \max \left\{ \overline{p}^{m+1}(X) \;:\; X \in B(t_1, \ldots, t_m, \ell) \right.$$
$$\left. \text{and} \quad s^{m+1}(X) \geq \left(\beta_{m+1} - \sqrt{\mu_{m+1}(\ell)} \right) \ell \right\},$$
$$\alpha_{m+1} = \limsup_{\ell \to \infty} \frac{\overline{f}_{m+1}(\ell)}{\ell}.$$

Note that $\overline{p}^{m+1}(X) \geq s^{m+1}(X) 2^{-t_1 \cdots t_m}$ for every $X \in B(t_1, \ldots, t_m, \ell)$. Take $Y \in B(t_1, \ldots, t_m, \ell)$ with $s^{m+1}(Y) = g_{m+1}(\ell)$. Then $s^{m+1}(Y) \geq \beta_{m+1} \ell$ by Lemma 7.11, so that

$$\overline{f}_{m+1} \geq \overline{p}^{m+1}(Y) \geq g_{m+1}(\ell) 2^{-t_1 \cdots t_m}$$
$$\text{and} \quad \alpha_{m+1} \geq \beta_{m+1} 2^{-t_1 \cdots t_m}.$$

Just as in the case $m = 1$, there exists an R-equivalence class $P(t_1, \ldots, t_m)$, such that if we define

$$f_{m+1}(\ell) = \max \left\{ p^{m+1}(X) \;:\; X \in B(t_1, \ldots, t_m, \ell) \right.$$
$$\left. \text{and} \quad s^{m+1}(X) \geq \left(\beta_{m+1} - \sqrt{\mu_{m+1}(\ell)} \right) \ell \right\},$$

then $\alpha_{m+1} = \limsup_{\ell \to \infty} f_{m+1}(\ell)/\ell$.

LEMMA 7.13
For every positive integer m we have $\alpha_{m+1} = \lim_{\ell \to \infty} f_{m+1}(\ell)/\ell$.

PROOF Let $\varepsilon > 0$ and ℓ be a positive integer. There is a large integer L, such that $f_{m+1}(L)/L > \alpha_{m+1} - \varepsilon$. Hence there is an $X \in B(t_1, \ldots, t_m, L)$, such that $p^{m+1}(X) > (\alpha_{m+1} - \varepsilon)L$ and $s^{m+1}(X) \geq \left(\beta_{m+1} - \sqrt{\mu_{m+1}(L)} \right) L$. Write $L = \ell \nu + \overline{\ell}$ where $0 \leq \overline{\ell} < \ell$. Let $X = \bigcup_{i < L} X_i$, with $X_i \in B(t_1, \ldots, t_m)$. For $0 \leq i < \nu$, let
$$Y_i = \bigcup_{j < \ell} X_{\ell i + j} \in B(t_1, \ldots, t_m, \ell)$$
and let $X' = \bigcup_{j < \nu \ell} X_j = \bigcup_{i < \nu} Y_i$. Let
$$T = \left\{ j < \nu \;:\; s^{m+1}(Y_j) < \left(\beta_{m+1} - \sqrt{\mu_{m+1}(\ell)} \right) \ell \right\}$$

and let $|T| = \alpha\nu$ (so $0 \le \alpha \le 1$). We have $s^{m+1}(Y_j) \le (\beta_{m+1} + \mu_{m+1}(\ell))\,\ell$ by definition of μ_{m+1}. Hence

$$s^{m+1}(X') = \sum_{j \in T} s^{m+1}(Y_j) + \sum_{j \notin T} s^{m+1}(Y_j)$$

$$\le \left(\beta_{m+1} - \sqrt{\mu_{m+1}(\ell)}\right)\ell\alpha\nu + (\beta_{m+1} + \mu_{m+1}(\ell))\,\ell(1 - \alpha)\nu.$$

But $s^{m+1}(X') \ge s^{m+1}(X) - \ell \ge \left(\beta_{m+1} - \sqrt{\mu_{m+1}(L)}\right)L - \ell$. Hence

$$\beta_{m+1} - \sqrt{\mu_{m+1}(L)} - \frac{1}{\nu} \le \alpha\left(\beta_{m+1} - \sqrt{\mu_{m+1}(\ell)}\right) + (1 - \alpha)\left(\beta_{m+1} + \mu_{m+1}(\ell)\right)$$

$$= \beta_{m+1} + \mu_{m+1}(\ell) - \alpha\left(\sqrt{\mu_{m+1}(\ell)} + \mu_{m+1}(\ell)\right),$$

so that

$$\alpha\left(\sqrt{\mu_{m+1}(\ell)} + \mu_{m+1}(\ell)\right) \le \mu_{m+1}(\ell) + \frac{1}{\nu} + \sqrt{\mu_{m+1}(\ell)}. \qquad (7.5)$$

By Lemma 7.12, we may assume that $\mu_{m+1}(\ell) > 0$, so for sufficiently large L, Inequality (7.5) becomes

$$\alpha\left(\sqrt{\mu_{m+1}(\ell)} + \mu_{m+1}(\ell)\right) < 2\mu_{m+1}(\ell),$$

so that

$$\alpha \le 2\sqrt{\mu_{m+1}(\ell)}. \qquad (7.6)$$

Now suppose that for all $j < \nu$ we have either

$$s^{m+1}(Y_j) < \left(\beta_{m+1} - \sqrt{\mu_{m+1}(\ell)}\right)\ell$$

or

$$p^{m+1}(Y_j) \le (\alpha_{m+1} - 2\varepsilon)\,\ell.$$

The first inequality holds for only $\alpha\nu$ indices j (those in T). Hence at least $(1 - \alpha)\nu$ indices j satisfy the second inequality. But $j \in T$ implies

$$p^{m+1}(Y_j) < \left(\beta_{m+1} - \sqrt{\mu_{m+1}(\ell)}\right)\ell$$

because $P(t_1, \ldots, t_m) \subseteq S(t_1, \ldots, t_m)$. Thus

$$\sum_{j < \nu} p^{m+1}(Y_j) < \alpha\nu\left(\beta_{m+1} - \sqrt{\mu_{m+1}(\ell)}\right)\ell + (1 - \alpha)\nu\,(\alpha_{m+1} - 2\varepsilon)\,\ell.$$

But

$$\sum_{j < \nu} p^{m+1}(Y_j) = p^{m+1}(X') \ge p^{m+1}(X) - \ell > (\alpha_{m+1} - \varepsilon)L - \ell.$$

Therefore

$$(\alpha_{m+1} - \varepsilon)L - \ell < \alpha\nu \left(\beta_{m+1} - \sqrt{\mu_{m+1}(\ell)} \right) \ell + (1 - \alpha)\nu \left(\alpha_{m+1} - 2\varepsilon \right) \ell.$$

and

$$\alpha_{m+1} - \varepsilon - \frac{1}{\nu} < \alpha \left(\beta_{m+1} - \sqrt{\mu_{m+1}(\ell)} \right) + (1 - \alpha) \left(\alpha_{m+1} - 2\varepsilon \right). \quad (7.7)$$

The right side of Inequality (7.7) is a convex combination of $\beta_{m+1} - \sqrt{\mu_{m+1}(\ell)}$ and $\alpha_{m+1} - 2\varepsilon$. By Inequality (7.6), $1 - \alpha$ is arbitrarily close to 1 for ℓ sufficiently large. Since $\beta_{m+1} - \sqrt{\mu_{m+1}(\ell)}$ is bounded, we obtain a contradiction from Inequality (7.7) by choosing ℓ and L sufficiently large.

Hence our assumption was false and there does exist an index $j < \nu$, such that both

$$s^{m+1}(Y_j) \geq \left(\beta_{m+1} - \sqrt{\mu_{m+1}(\ell)} \right) \ell$$

and

$$p^{m+1}(Y_j) > (\alpha_{m+1} - 2\varepsilon)\ell.$$

Therefore, $f_{m+1}(\ell)/\ell > \alpha_{m+1} - 2\varepsilon$ for all sufficiently large ℓ. Since $\varepsilon > 0$ was arbitrary, we have proved Lemma 7.13. ∎

Finally, we choose t_{m+1} sufficiently large depending on α_{m+1}, β_{m+1}, K and t_m so that $\varepsilon_{m+1}(t_{m+1})$ and $\mu_{m+1}(t_{m+1})$ are sufficiently small (to be made precise later). This completes the inductive step in the definition.

The next proposition shows that we may choose β_{m+1} as close to 1 as we like.

PROPOSITION 7.2
For all $m \geq 1$, we have

$$\beta_{m+1} \geq 1 - 2 \left(\sqrt{\mu_m(t_m)} + \sqrt{\sqrt{\varepsilon_m(t_m)} + \sqrt{\mu_m(t_m)}} \right). \quad (7.8)$$

PROOF   Let $\phi(t_m)$ denote the right side of Inequality (7.8). By the definitions of ε_m and α_m, for all positive integers ℓ there exists an $X \in B(t_t, \ldots, t_{m-1}, t_m\ell)$, such that $s^m(X) \geq \left(\beta_m - \sqrt{\mu_m(t_m\ell)} \right) t_m\ell$ and $p^m(X) \geq (\alpha_m - \varepsilon(t_m\ell)) t_m\ell$. Write

$$X = \bigcup_{i < t_m\ell} X_i, \quad X_i \in B(t_1, \ldots, t_{m-1}),$$

and for $j < \ell$, set

$$Y_j = \bigcup_{i < t_m} X_{jt_m + i} \in B(t_1, \ldots, t_m).$$

We write X as Y when it is considered an element of $B(t_1, \ldots, t_m, \ell)$.

We shall show that $s^{m+1}(Y) \geq \phi(t_m)\ell$ for sufficiently large ℓ. This will prove Proposition 7.2 since then $g_{m+1}(\ell) \geq s^{m+1}(Y) \geq \phi(t_m)\ell$ for large ℓ so that $\beta_{m+1} = \lim_{\ell \to \infty} g_{m+1}(\ell)/\ell \geq \phi(t_m)$. The argument is like the proof of Lemma 7.13 used twice, once for s^m and once for p^m.

Let $T_1 = \left\{ j < \ell : s^m(Y_j) < \left(\beta_m - \sqrt{\mu_m(t_m)} \right) t_m \right\}$ and $|T_1| = a\ell$. By definition of μ_m,

$$s^m(Y_y) \leq (\beta_m + \mu_m(t_m)) t_m.$$

Therefore,

$$\sum_{j < \ell} s^m(Y_j) \leq \left(\beta_m - \sqrt{\mu_m(t_m)} \right) t_m \, a\ell + (\beta_m + \mu_m(t_m)) t_m (1 - a)\ell.$$

But $\sum_{j < \ell} s^m(Y_j) = s^m(X) \geq \left(\beta_m - \sqrt{\mu_m(t_m\ell)} \right) t_m \ell$, so

$$\beta_m - \sqrt{\mu_m(t_m\ell)} \leq \beta_m - a\sqrt{\mu_m(t_m)} + (1 - a)\mu_m(t_m)$$

and

$$a \left(\sqrt{\mu_m(t_m)} + \mu_m(t_m) \right) \leq \mu_m(t_m) + \sqrt{\mu_m(t_m\ell)}. \tag{7.9}$$

If $m = 1$, then $\mu_1(\ell') = 0$ for all ℓ', $s^1(Y_j) = t_1$ for all $j < \ell$, and hence $T_1 = \emptyset$. Therefore $a = 0$ and

$$a \leq 2\sqrt{\mu_m(t_m)}. \tag{7.10}$$

Now suppose $m > 1$. By Lemma 7.12 we may assume $\mu_m(t_m) > 0$. Then for ℓ sufficiently large, we have from Inequality (7.9)

$$a \leq \frac{2\mu_m(t_m)}{\sqrt{\mu_m(t_m)} + \mu_m(t_m)} < 2\sqrt{\mu_m(t_m)}.$$

Hence Inequality (7.10) holds in all cases.

Now let

$$T_2 = \left\{ j < \ell : s^m(Y_j) \geq \left(\beta_m - \sqrt{\mu_m(t_m)} \right) t_m \quad \text{and} \right.$$

$$\left. p^m(Y_j) < \left(\alpha_m - \sqrt{\sqrt{\varepsilon_m(t_m)} + \sqrt{\mu_m(t_m)}} \right) t_m \right\},$$

and $|T_2| = b\ell$ (so $0 \le b \le 1$). For $j \in T_1$ we have $p^m(Y_j) \le (\alpha_m + \varepsilon_m(t_m))t_m$. Therefore,

$$\sum_{j<\ell} p^m(Y_j) = \sum_{j\in T_1} p^m(Y_j) + \sum_{j\in T_2} p^m(Y_j) + \sum_{j\notin T_1\cup T_2} p^m(Y_j)$$

$$\le a\ell t_m + \left(\alpha_m - \sqrt{\sqrt{\varepsilon_m(t_m)} + \sqrt{\mu_m(t_m)}}\right)t_m b\ell$$

$$+ (1 - a - b)\ell(\alpha_m + \varepsilon_m(t_m))t_m$$

$$\le 2t_m\ell\sqrt{\mu_m(t_m)} + \left(\alpha_m - \sqrt{\sqrt{\varepsilon_m(t_m)} + \sqrt{\mu_m(t_m)}}\right)t_m b\ell$$

$$+ (1 - b)(\alpha_m + \varepsilon_m(t_m))t_m\ell$$

when ℓ is sufficiently large that Inequality (7.10) holds. But

$$\sum_{j<\ell} p^m(Y_j) = p^m(X) \ge (\alpha_m - \varepsilon_m(t_m\ell))t_m\ell.$$

Thus,

$$\alpha_m - \varepsilon_m(t_m\ell) \le 2\sqrt{\mu_m(t_m)} + \alpha_m + \varepsilon_m(t_m)$$

$$- b\left(\sqrt{\sqrt{\varepsilon_m(t_m)} + \sqrt{\mu_m(t_m)}} + \varepsilon_m(t_m)\right),$$

and so

$$b\left(\sqrt{\sqrt{\varepsilon_m(t_m)} + \sqrt{\mu_m(t_m)}} + \varepsilon_m(t_m)\right) \le \qquad (7.11)$$

$$2\sqrt{\mu_m(t_m)} + \varepsilon_m(t_m) + \varepsilon_m(t_m\ell).$$

We may assume that $\varepsilon_1(t_1) > 0$. (The proof is like that of Lemma 7.12.) For $m > 1$ we have $\mu_m(t_m) > 0$ by Lemma 7.12. Hence, $\sqrt{\varepsilon_m(t_m)} + \sqrt{\mu_m(t_m)} > 0$ for every positive integer m. If $\varepsilon_m(\ell') = 0$ for all sufficiently large ℓ', then we may choose ℓ so large that

$$\varepsilon_m(t_m\ell) \le \varepsilon_m(t_m) \le \sqrt{\varepsilon_m(t_m)}.$$

Otherwise, we may choose a large t_m so that $\varepsilon_m(t_m) > 0$, and then the preceding inequalities also hold for sufficiently large ℓ. Hence by Inequality (7.11) we have $b \le 2\sqrt{\sqrt{\varepsilon_m(t_m)} + \sqrt{\mu_m(t_m)}}$. But for $j \notin T_1 \cup T_2$, we have $Y_j \in S(t_1, \ldots, t_m)$. Therefore $s^{m+1}(Y) \ge (1 - a - b)\ell \ge \phi(t_m)\ell$ and Proposition 7.2 is proved. ∎

Define

$$C(t_1, \ldots, t_m, \ell) = \{X \in B(t_1, \ldots, t_m, \ell) : s^{m+1}(X) = \ell\}.$$

Thus, if $X = \bigcup_{i<\ell} X_i \in C(t_1, \ldots, t_m, \ell)$, then for all $i < \ell$ we have $X_i \in S(t_1, \ldots, t_m)$.

LEMMA 7.14

If $\beta_{m+1} > 1 - 1/\ell$, then $C(t_1, \ldots, t_m, \ell) \neq \emptyset$.

PROOF   By Lemma 7.11, if $\beta_{m+1} > 1 - 1/\ell$, then

$$g_{m+1}(\ell) \geq \ell\beta_{m+1} > \ell(1 - 1/\ell) = \ell - 1,$$

so that $g_{m+1}(\ell) = \ell$. Thus there exists an $X \in B(t_1, \ldots, t_m)$ with $s^{m+1}(X) = \ell$ so that $X \in C(t_1, \ldots, t_m, \ell) \neq \emptyset$.   ∎

By Proposition 7.2 and the fact that $\alpha_{m+1} \geq \beta_{m+1} 2^{-t_1 \cdots t_m}$, we have that $S(t_1, \ldots, t_m)$ and $P(t_1, \ldots, t_m)$ are both nonempty for all m (provided t_m is suitably chosen).

For $X = \bigcup_{i<\ell} X_i \in B(t_1, \ldots, t_m, \ell)$ and $C \subseteq [0, \ell)$, define

$$s^m(X, C) = |\{i < \ell : i \in C \text{ and } X_i \in S(t_1, \ldots, t_{m-1})\}|,$$
$$p^m(X, C) = |\{i < \ell : i \in C \text{ and } X_i \in P(t_1, \ldots, t_{m-1})\}|.$$

PROPOSITION 7.3

For all δ and τ there exists a positive integer ℓ, such that if $X = \bigcup_{i<\ell} X_i \in C(t_1, \ldots, t_m, \ell)$, $C \subseteq [0, t_m)$, $|C| \geq \tau t_m$, and t_m is sufficiently large depending of ℓ, then there exists an integer $i < \ell$, such that $p^m(X_i, C) > (\alpha_m - \delta)|C|$ and an integer $i' < \ell$, such that $p^m(X_i, C) < (\alpha_m + \delta)|C|$.

PROOF   The proof is very similar to that of Proposition 7.2. Write $X = \bigcup_{j<t_m} X_{i,j}$ for $i < \ell$ and let $Y_j = \bigcup_{i<\ell} X_{i,j}$ for $j < t_m$. Thus $Y_j \in B(t_1, \ldots, t_{m-1}, \ell)$. Since $X \in C(t_1, \ldots, t_m, \ell)$ we have $X_i \in S(t_1, \ldots, t_m)$ for all $i < \ell$. Therefore

$$s^m(X_i) \geq \left(\beta_m - \sqrt{\mu_m(t_m)} \right) t_m$$

$$\text{and} \quad p^m(X_i) > \left(\alpha_m - \sqrt{\sqrt{\varepsilon_m(t_m)} + \sqrt{\mu_m(t_m)}} \right) t_m$$

for $i < \ell$. Thus,

$$\sum_{i<\ell} s^m(X_i) = \sum_{j<t_m} s^m(Y_j) \geq \left(\beta_m - \sqrt{\mu_m(t_m)} \right) t_m \ell.$$

Let $T_1 = \left\{ j < t_m \; : \; s^m(Y_j) < \left(\beta_m - \sqrt{\mu_m(\ell)} \right) \ell \right\}$ and $|T_1| = at_m$. By definition of μ_m, we have $s^m(Y_j) \leq (\beta_m + \mu_m(\ell))\ell$ for all $j < t_m$. Thus,

$$\sum_{j < t_m} s^m(Y_j) = \sum_{j \in T_1} s^m(Y_j) + \sum_{j \notin T_1} s^m(Y_j)$$

$$\leq \left(\beta_m - \sqrt{\mu_m(t_m)} \right) t_m a \ell + (\beta_m + \mu_m(\ell))(1 - a)t_m \ell.$$

Hence

$$\left(\beta_m - \sqrt{\mu_m(t_m)} \right) t_m \ell \leq \left(\beta_m - \sqrt{\mu_m(t_m)} \right) t_m a \ell + (\beta_m + \mu_m(\ell))(1 - a)t_m \ell$$

and

$$a \left(\mu_m(\ell) + \sqrt{\mu_m(\ell)} \right) \leq \mu_m(\ell) + \sqrt{\mu_m(t_m)}.$$

Since $\mu_m(\ell) > 0$ by Lemma 7.12, we have $a \leq 2\sqrt{\mu_m(\ell)}$ provided t_m is sufficiently large depending on ℓ.

Let $T_2 = \left\{ j < t_m \; : \; s^m(Y_j) \geq \left(\beta_m - \sqrt{\mu_m(\ell)} \right) \ell \text{ and } p^m(Y_j) < \left(\alpha_m - \sqrt{\sqrt{\varepsilon_m(\ell)} + \sqrt{\mu_m(\ell)}} \right) \ell \right\}$, and $|T_2| = bt_m$. For $j \notin T_1$ we have by definition of ε_m,

$$p_m(Y_j) \leq f_m(\ell) \leq (\alpha_m + \varepsilon_m(\ell))\ell.$$

Therefore, since $T_1 \cap T_2 = \emptyset$,

$$\sum_{j < t_m} p^m(Y_j) = \sum_{j \in T_1} p^m(Y_j) + \sum_{j \in T_2} p^m(Y_j) + \sum_{j \notin T_1 \cup T_2} p^m(Y_j)$$

$$\leq at_m \ell + \left(\alpha_m - \sqrt{\sqrt{\varepsilon_m(\ell)} + \sqrt{\mu_m(\ell)}} \right) bt_m \ell$$

$$+ (1 - a - b)(\alpha_m + \varepsilon_m(\ell))t_m \ell$$

$$\leq 2t_m \ell \sqrt{\mu_m(\ell)} + \left(\alpha_m - \sqrt{\sqrt{\varepsilon_m(\ell)} + \sqrt{\mu_m(\ell)}} \right) bt_m \ell$$

$$+ (1 - a - b)(\alpha_m - \varepsilon_m(\ell))t_m \ell$$

for t_m sufficiently large depending on ℓ. But

$$\sum_{j < t_m} p^m(Y_j) = \sum_{i < \ell} p^m(X_i) \geq \left(\alpha_m - \sqrt{\sqrt{\varepsilon_m(t_m)} + \sqrt{\mu_m(t_m)}} \right) t_m \ell.$$

It follows that

$$b \left(\sqrt{\sqrt{\varepsilon_m(\ell)} + \sqrt{\mu_m(\ell)}} + \varepsilon_m(\ell) \right) \leq$$

$$2\sqrt{\mu_m(\ell)} + \varepsilon_m(\ell) + \sqrt{\sqrt{\varepsilon_m(t_m)} + \sqrt{\mu_m(t_m)}}.$$

As in Proposition 7.2 we may assume $\varepsilon_1(t_1) > 0$. If $m > 1$ and $\varepsilon_m(\ell') = 0$ for all sufficiently large ℓ', then we may assume $\varepsilon_m(\ell) = 0$. But $\mu_m(\ell) > 0$ and we have

$$b \leq 4\sqrt{\sqrt{\varepsilon_m(\ell)} + \sqrt{\mu_m(\ell)}}$$

for sufficiently large t_m. Otherwise we may choose ℓ so that $\varepsilon_m(\ell) > 0$ and we obtain

$$b \leq 2\sqrt{\sqrt{\varepsilon_m(\ell)} + \sqrt{\mu_m(\ell)}}$$

for sufficiently large t_m depending on ℓ. Hence in either case

$$\left| [0, t_m) - T_1 - T_2 \right| = \left| \left\{ j < t_m \,:\, s^m(Y_j) \geq \left(\beta_m - \sqrt{\mu_m(\ell)} \right) \ell \text{ and } \right. \right.$$
$$\left. \left. p^m(Y_j) \geq \left(\alpha_m - \sqrt{\sqrt{\varepsilon_m(\ell)} + \sqrt{\mu_m(\ell)}} \right) \ell \right\} \right| \quad (7.12)$$
$$\geq (1 - a - b)t_m$$
$$\geq \left(1 - 2\left(\sqrt{\mu_m(\ell)} + 2\sqrt{\sqrt{\varepsilon_m(\ell)} + \sqrt{\mu_m(\ell)}} \right) \right) t_m.$$

By the definition of ε_m, we have for $j \in [0, t_m) - T_1$,

$$p^m(Y_j) \leq (\alpha_m + \varepsilon_m(\ell))\ell. \quad (7.13)$$

Hence for ℓ sufficiently large depending on δ and τ, and m sufficiently large depending on ℓ, we have by Inequalities (7.12) and (7.13),

$$\left| \{ j < t_m \,:\, (\alpha_m - \delta/2)\ell < p^m(Y_j) < (\alpha_m + \delta/2)\ell \} \right| > \left(1 - \frac{\tau\delta}{4} \right) t_m.$$

Now suppose $p^m(X_i, C) \leq (\alpha_m - \delta)|C|$ for all $i < \ell$. Then

$$\sum_{i < \ell} p^m(X_i, C) = \sum_{j \in C} p^m(Y_j) \leq (\alpha_m - \delta)|C|\ell.$$

On the other hand,

$$\sum_{j \in C} p^m(Y_j) \geq \sum_{j \in C - (T_1 \cup T_2)} p^m(Y_j)$$
$$\geq \left(\alpha_m - \frac{\delta}{2} \right) \ell |C - (T_1 \cup T_2)|$$
$$\geq \left(\alpha_m - \frac{\delta}{2} \right) \ell \left(|C| - \frac{\tau\delta}{4} t_m \right).$$

Thus

$$\left(\alpha_m - \frac{\delta}{2} \right) \left(|C| - \frac{\tau\delta}{4} t_m \right) \leq (\alpha_m - \delta)|C|,$$

and so

$$\frac{\delta}{2}|C| \leq \left(\alpha_m - \frac{\delta}{2}\right)\frac{\tau\delta}{4}t_m < \frac{\tau\delta}{4}t_m,$$

which contradicts the assumption that $|C| \geq t_m$. This proves the first assertion of the proposition.

Suppose $p^m(X_i, C) \geq (\alpha_m + \delta)|C|$ for all $i < \ell$. Then

$$\sum_{i<\ell} p^m(X_i, C) = \sum_{j\in C} p^m(Y_j) \geq (\alpha_m + \delta)|C|\ell.$$

But by Inequality (7.13)

$$|\{j < t_m \ : \ p^m(Y_j) > (\alpha_m + \varepsilon_m(\ell))\ell\}| \leq |T_1| = at_m \leq 2\sqrt{\mu_m(\ell)}t_m.$$

Thus

$$\sum_{j\in C} p^m(Y_j) = \sum_{j\in C\cap T_1} p^m(Y_j) + \sum_{j\in C-T_1} p^m(Y_j)$$
$$\leq 2\sqrt{\mu_m(\ell)}t_m\ell + (\alpha_m + \varepsilon_m(\ell))|C|\ell.$$

Comparing the two inequalities for $\sum_{j\in C} p^m(Y_j)$, we have

$$(\alpha_m + \delta)|C|\ell \leq 2\sqrt{\mu_m(\ell)}t_m\ell + (\alpha_m + \varepsilon_m(\ell))|C|\ell,$$

and

$$|C| \leq \frac{2\sqrt{\mu_m(\ell)}}{\delta - \varepsilon_m(\ell)}t_m,$$

which is impossible for sufficiently large ℓ. This proves the second assertion and the proof of the proposition is complete. ∎

7.7   More Definitions

For $i < K$ we define the two sets

$$D^i(t_1, \ldots, t_m, K) = \{X \in C(t_1, \ldots, t_m, K) \ : \ X = \bigcup_{j<K} X_j \text{ and}$$
$$X_j \in P(t_1, \ldots, t_m) \text{ for all } j < i\},$$
$$D^{*i}(t_1, \ldots, t_m, K) = \{X \in B(t_1, \ldots, t_m, K) \ : \ X = \bigcup_{j<K} X_j \text{ and}$$
$$X_j \in P(t_1, \ldots, t_m) \text{ for all } j < i\}.$$

Note that $D^i \subseteq D^{*i}$ for all $i < K$, and

$$D^0(t_1, \ldots, t_m, K) = C(t_1, \ldots, t_m, K),$$
$$D^{*0}(t_1, \ldots, t_m, K) = B(t_1, \ldots, t_m, K).$$

We will prove the theorem, in Section 7.10, by fixing K and showing by induction on i that $D^i(t_1, \ldots, t_m, K) \neq \emptyset$ provided the t_k satisfy certain conditions. With $m = 0$ and $i = K - 1$ we will then have $D^{K-1}(K) \neq \emptyset$, so that there exists $X = \bigcup_{j<K} X_j$ with $X_j \in P(\emptyset)$ for $j < K - 1$, that is, X is an arithmetic progression of $K - 1$ terms contained in R.

For a positive integer t define $E(t, K) = \{(j_0, \ldots, j_{K-1}) : j_i < t \text{ for all } i < K \text{ and } j_0, \ldots, j_{K-1} \text{ forms a } K\text{-progression}\}$. We allow $j_0 = j_1 = \cdots = j_{K-1}$ and $j_0 > j_1 > \cdots > j_{K-1}$ as well as $j_0 < j_1 < \cdots < j_{K-1}$.

LEMMA 7.15
Let $X = \bigcup_{i<K} X_i \in B(t_1, \ldots, t_m, K)$ and write $X_i = \bigcup_{j<t_m} X_{i,j}$ as usual. Then $(j_0, \ldots, j_{K-1}) \in E(t_m, K)$ if and only if $\bigcup_{i<K} X_{i,j_i} \in B(t_1, \ldots, t_{m-1}, K)$.

PROOF   If $(j_0, \ldots, j_{K-1}) \in E(t_m, K)$, then clearly

$$\bigcup_{i<K} X_{i,j_i} \in B(t_1, \ldots, t_{m-1}, K).$$

If $\bigcup_{i<K} X_{i,j_i} \in B(t_1, \ldots, t_{m-1}, K)$, then $j_0, \ldots, j_{K-1}$ must form an arithmetic progression of length K with $j_i < t_m$ for all $i < K$.   ∎

For $i < K$ and $j < t$ define

$$E(t, K, j, i) = \{(j_0, \ldots, j_{K-1}) \in E(t_K) : j_i = j\}$$

and $e(t, K, j, i) = |E(t, K, j, i)|$.

LEMMA 7.16
We have

$$e(t, K, j, i) \leq t, \tag{7.14}$$

and for $t/4 < j < 3t/4$, $K \geq 2$, and $t \geq 4$,

$$e(t, K, j, i) \geq \frac{t}{K^2}. \tag{7.15}$$

PROOF   If $(j_0, \ldots, j_{K-1}) \in E(t, K, j, i)$, then all $j_s < t$ and $j_i = j$. Thus, if $K > 1$, then the choice of j_{i+1} (or j_{i-1} if $i = K - 1$) determines the rest of the progression. But there are at most t choices for j_{i+1} (or j_{i-1}), so that $e(t, K, j, i) \leq t$. If $K = 1$, then $e(t, K, j, i) = 1 \leq t$. This proves Inequality (7.14).

To prove Inequality (7.15), we note that since we allow both increasing and decreasing arithmetic progressions, the worst case occurs when $j = \lfloor t/2 \rfloor$ and $i = K - 1$. Clearly

$$(a, a + d, \ldots, a + (K-1)d) \in E(t, K, \lfloor t/2 \rfloor, K-1)$$

whenever $a \geq 0$, $d \geq 0$ and $a + (K-1)d = \lfloor t/2 \rfloor$, that is, whenever $0 \leq d \leq \lfloor t/2 \rfloor / (K-1)$. Since

$$\frac{1}{K-1} \left\lfloor \frac{t}{2} \right\rfloor \geq \frac{t}{K^2}$$

for $K \geq 2$ and $t \geq 4$, we have proved Inequality (7.15). ∎

LEMMA 7.17

If $\ell > 1$, $t \geq \ell^2 - \ell$ and $L \subseteq [0, t)$ with $|L| > (1 - 1/\ell)t$, then L contains an ℓ-progression.

PROOF Suppose L contains no ℓ-progression. Then each interval $I_j = [j\ell, (j+1)\ell)$, for $j \geq 0$, can contain at most $\ell - 1$ numbers from L. Write $t = q\ell + r$ with $0 \leq r < \ell$. We have $q \geq \ell - 1$ because $t \geq \ell^2 - \ell$. If some I_j with $j \in [0, q)$ contains fewer than $\ell - 1$ numbers of L, then

$$t - \frac{t}{\ell} < |L| \leq q(\ell-1) + r - 1 = t - q - 1 = t - \left\lfloor \frac{t}{\ell} \right\rfloor - 1 < t - \frac{t}{\ell}, \quad (7.16)$$

which is a contradiction. Hence $|I_j \cap L| = \ell - 1$ for $j \in [0, q)$. If ℓ divides t, then $r = 0$ and $|L| = (\ell - 1)q = (1 - 1/\ell)t$, a contradiction. Therefore $r \neq 0$ and $t > \ell^2 - \ell$. Furthermore, $[q\ell, t) \subseteq L$ or else we obtain the contradiction (7.16). Let

$$A = [q\ell - (\ell^2 - \ell), q\ell] \cap \left(L + \{q\ell + \ell - \ell^2\} \right).$$

Then A contains no ℓ-progression, $A \subseteq [0, \ell^2 - \ell]$, $|A| = \ell^2 - \ell + 1$ and $\ell^2 - \ell \in A$. Just as for L, we may assume that $|A \cap I_j| = \ell - 1$ for $j \in [0, \ell-1)$. For $j \in [0, \ell-1)$, let $k_j + j\ell$ be the unique element of I_j not in A. Then $k_{j+1} \leq k_j$ or else A contains the progression $[k_j + j\ell + 1, k_{j+1} + (j+1)\ell)$ of length not less than ℓ. However, $k_j = 0$ for some $j \in [0, \ell-1)$ or else A contains the ℓ-progression $0, \ell, 2\ell, \ldots, \ell^2 - \ell$. Hence, $k_{\ell-2} = 0$ and A contains the ℓ-progression $[\ell^2 - 2\ell + 1, \ell^2 - \ell]$. As we have arrived at a contradiction in every case, L must contain an ℓ-progression. ∎

REMARK 7.6 Lemma 7.17 is best possible when ℓ is prime in that

$$r_\ell(\ell^2 - \ell - 1) = (\ell - 1)^2 - 1 > (1 - 1/\ell)(\ell^2 - \ell - 1)$$

for prime ℓ. When $\ell > 1$ is composite, Lemma 7.17 holds with $\ell^2 - \ell$ replaced by a smaller number (but not much smaller). ∎

We now define another subclass of $C(t_1, \ldots, t_m, K)$.

DEFINITION 7.7   For $i \le s < K$ and $X \in C(t_1, \ldots, t_m, K)$, let

$$F(X, j, i, s) = \left\{ (j_0, \ldots, j_{K-1}) \in E(t_m, K, j, s) : \right.$$

$$\left. \bigcup_{i' < K} X_{i', j_{i'}} \in D^{*i}(t_1, \ldots, t_{m-1}, K) \right\},$$

and let $f(X, j, i, s) = |F(X, j, i, s)|$. A configuration $X \in C(t_1, \ldots, t_m, K)$ is called a *homogeneous K-tuple of type* (m, i) if for every s with $i \le s < K$ the condition

$$f(X, j, i, s) \le 2^i \alpha_m^i t_m$$

fails for at most $2i\alpha_m^K(1 - \beta_m)t_m$ indices $j < t_m$, and the condition

$$f(X, j, i, s) \ge \frac{1}{K^2} \frac{1}{2^i} \alpha_m^i t_m$$

fails for at most $2i\alpha_m^K(1 - \beta_m)t_m$ indices j such that $t_m/4 < j < 3t_m/4$. Let $G^i(t_1, \ldots, t_m, K)$ be the set of all homogeneous K-tuples of type (m, i). The elements of $D^i(t_1, \ldots, t_m, K)$ will be called simply K-tuples of type (m, i).

We will use induction to prove the existence of K-tuples of type (m, i). Note that by Lemma 7.16,

$$G^0(t_1, \ldots, t_m, K) = C(t_1, \ldots, t_m, K).$$

Now let $X \in B(t_1, \ldots, t_m, K)$ and let $0 \le i \le s < K$. We shall define a bipartite graph $I(X, i, s)$ which will eventually connect Theorem 7.4 with the rest of the proof. Although the vertex sets A and B are supposed to be disjoint, we shall use the set $[0, t_m)$ for both A and B. This will cause no confusion. The pair $\{j, j'\} \in [A, B]$ is an element of the graph $I(X, i, s)$ if $(j_0, \ldots, j_{K-1}) \in F(X, j, i, s)$ and $j_i = j'$.

The graph $I(X, i, s)$ is crucial to the remainder of the proof. We may picture $I(X, i, s)$ as follows. Write $X \in B(t_1, \ldots, t_m, K)$ as $X = \bigcup_{i < K} X_i$, with $X_i \in B(t_1, \ldots, t_m)$ and $X_i = \bigcup_{j < t_m} X_{i,j}$ for $i < K$ as usual. Display the $X_{i,j}$ in the following way:

		$\underline{B}$		$\underline{A}$		
$X_{0,0}$	$\cdots$	$X_{i,0}$	$\cdots$	$X_{s,0}$	$\cdots$	$X_{K-1,0}$
$X_{0,1}$	$\cdots$	$X_{i,1}$	$\cdots$	$X_{s,1}$	$\cdots$	$X_{K-1,1}$
$\vdots$		$\vdots$		$\vdots$		$\vdots$
$\vdots$		$\vdots$		$X_{s,j}$		$\vdots$
$\vdots$		$\vdots$		$\vdots$		$\vdots$
$\vdots$		$X_{i,j'}$		$\vdots$		$\vdots$
$\vdots$		$\vdots$		$\vdots$		$\vdots$
X_{0,t_m-1}	$\cdots$	X_{i,t_m-1}	$\cdots$	X_{s,t_m-1}	$\cdots$	X_{K-1,t_m-1}

The set A corresponds to the set of indices j of the $X_{s,j}$ with $j < t_m$. The set B corresponds to the set of indices j' of the $X_{i,j'}$ with $j' < t_m$. The pair $\{j, j'\}$ is in the bipartite graph $I(X, i, s)$ if and only if the arithmetic progression $j_0, \ldots, j_{K-1}$ defined by letting $j_i = j'$ and $j_s = j$ has all its terms in $[0, t_m)$ and $X_{i',j_{i'}} \in P(t_1, \ldots, t_{m-1})$ for all $i' < i$.

REMARK 7.7   Note that if $X \in G^i(t_1, \ldots, t_m, K)$, then the graph $I(X, i, s)$ must satisfy strong valency constraints.   ∎

7.8   The Choice of t_m

In this section we specify how t_m should be chosen. Recall that we began with a set R of positive integers having positive upper density and an integer $K \geq 2$ and we want to show that R contains a $(K-1)$-progression.

Assume that in the inductive definitions in Section 7.6, for some $m \geq 1$ we have already defined $P(t_1, \ldots, t_{m-1}) \subseteq S(t_1, \ldots, t_{m-1})$, $g_m(\ell)$, $f_m(\ell)$, α_m, β_m, $\varepsilon_m(\ell)$ and $\mu_m(\ell)$.

1. Define constants $\varepsilon_1^{(m)}$, $\varepsilon_2^{(m)}$, $\delta^{(m)}$, $\rho^{(m)}$, $\sigma^{(m)}$ by

   (i) $\varepsilon_1^{(m)} = \alpha_m^K(1 - \beta_m)$,

   (ii) $\varepsilon_2^{(m)} = \alpha_m/2$,

   (iii) $\delta^{(m)} = 10^{-1}2^{-K}K^{-2}\alpha_m^K$,

   (iv) $\rho^{(m)} = \alpha_m^K(1 - \beta_m)$,

   (v) $\sigma^{(m)} = 10^{-1}2^{-K}K^{-2}\alpha_m^K$.

2. Denote by $m_0^{(m)}$, $n_0^{(m)}$, $M^{(m)}$, $N^{(m)}$ the numbers given by Theorem 7.4 for the corresponding constants chosen in 1.

   (A) Let $t_m > 2M^{(m)}$ and $t_m > N^{(m)}$.

3. Let $\tau^{(m)} = \sigma^{(m)}/n_0^{(m)}$ and let ℓ'_m be a large number for which the statement of Proposition 7.3 is satisfied with $\delta = \delta^{(m)}$ and $\tau = \tau^{(m)}$ provided t_m is sufficiently large depending on ℓ'_m. Also choose $\ell'_m > t_{m-1}$.

   (B) Let t_m be so large that 3 is valid.

4. By Theorem 7.1, for positive integers i and j there exists a least integer $w(i, j)$, such that if $[0, w(i, j))$ is partitioned into j classes, then at least one class contains an i-progression. Choose ℓ_m so that $\ell_m^{2^{-K}}$ is an integer exceeding $w(\ell'_m, 4Km_0^{(m)}n_0^{(m)})$. Thus, by Proposition 7.2 we have

$$2K^3 2^{2K}\ell_m < \frac{1}{\sqrt{1 - \beta_{m+1}}}$$

provided t_m is sufficiently large.

(C) Let t_m be so large that 4 holds. (This also guarantees that $\beta_m > 0$ for $m > 1$.)

(D) Let $t_m \geq 4$, so that Lemma 7.16 holds with $t = t_m$.

(E) Let $t_m > 4(\ell_{m-1}^2 - \ell_{m-1})$ for $m \geq 2$. (This insures that Lemma 7.17 holds with $\ell = \ell_{m-1}$ and $t \geq t_m/4$.)

(F) Let t_m be so large that $\sqrt{\mu_m(t_m)} < \min(\beta_m, 1 - \beta_m)$, which is possible since we may assume that $0 < \beta_m < 1$.

5. If $\varepsilon_m(\ell) = 0$ for all sufficiently large ℓ, omit this condition. Otherwise, we can choose t_m so large that $\varepsilon_m(t_m) > 0$. (This insures that Proposition 7.2 is valid.)

(G) Choose t_m so that $\varepsilon_1(t_1) > 0$ and 5 holds for $m > 1$.

In the following, t_m denotes a fixed number satisfying all of the conditions (A) – (G).

7.9   Well-Saturated K-tuples

Suppose $X \in G^i(t_1, \ldots, t_m, K)$ with $0 \leq i < K - 1$. For $i + 1 \leq s < K$, let the sets $C_\mu(X, s)$, for $\mu < m_0^{(m)}$, and $C_{\mu,\nu}(X, s)$, for $\mu < m_0^{(m)}$ and $\nu < n_0^{(m)}$, satisfy Proposition 7.4 with the choice of constants given in 1 of Section 7.8, where the vertex sets A and B of Proposition 7.4 are defined by

$$A = A^{(m)} = \{j : t_m/4 < j < 3t_m/4\},$$
$$B = B^{(m)} = [0, t_m),$$

and the bipartite graph is $I = I(X, i, s)$.

Similarly, for $i + 1 \leq s < K$, let the sets $\overline{C}_\mu(X, s)$, for $\mu < m_0^{(m)}$, and $\overline{C}_{\mu,\nu}(X, s)$, for $\mu < m_0^{(m)}$ and $\nu < n_0^{(m)}$, satisfy Corollary 7.2 with the choice of constants given in 1 of Section 7.8, where the vertex sets A and B of Corollary 7.2 are defined by

$$A = \overline{A}^{(m)} = [0, t_m), \qquad B = \overline{B}^{(m)} = [0, t_m),$$

and the bipartite graph is $I = I(X, i, s)$. The common indexing of the disjoint vertex sets A and B should cause no confusion.

Assume $m \geq 1$ and write X in the usual way as

$$X = \bigcup_{i' < K} X_{i'} \in G^i(t_1, \ldots, t_m, K),$$
$$X_{i'} = \bigcup_{j < t_m} X_{i',j} \in B(t_1, \ldots, t_m), \quad i' < K.$$

DEFINITION 7.8   We shall say that X is *well saturated* if for all s with $i + 1 \leq s < K$ we have

$$\left|p^m(X_i, C_{\mu,\nu}(X, s)) - \alpha_m |C_{\mu,\nu}(X, s)|\right| \leq \delta^{(m)} |C_{\mu,\nu}(X, s)|$$

and

$$\left|p^m(X_i, \overline{C}_{\mu,\nu}(X, s)) - \alpha_m |\overline{C}_{\mu,\nu}(X, s)|\right| \leq \delta^{(m)} |\overline{C}_{\mu,\nu}(X, s)|$$

whenever

$$|C_{\mu,\nu}(X, s)| \geq \tau^{(m)} t_m \quad \text{and} \quad |\overline{C}_{\mu,\nu}(X, s)| \geq \tau^{(m)} t_m,$$

where $\mu < m_0^{(m)}$ and $\nu < n_0^{(m)}$.

PROPOSITION 7.4

If $i + 1 < K$ and $X \in G^i(t_1, \ldots, t_m, K)$ is well saturated, then $X \in G^{i+1}(t_1, \ldots, t_m, K)$.

PROOF   Choose a fixed s with $i + 1 \leq s < K$. We must show that

(a) $f(X, j, i + 1, s) \leq 2^{i+1} \alpha_m^{i+1} t_m$ fails for at most $2(i + 1)\alpha_m^K(1 - \beta_m)t_m$ indices $j < t_m$, and

(b) $f(X, j, i + 1, s) \geq K^{-2} 2^{-(i+1)} \alpha_m^{i+1} t_m$ fails for at most $2(i + 1)\alpha_m^K \cdot (1 - \beta_m)t_m$ indices j with $t_m/4 < j < 3t_m/4$.

We first prove (b). To be concise, set $I = I(X, i, s)$ and

$$C_\mu = C_\mu(X, s), \qquad C_{\mu,\nu} = C_{\mu,\nu}(X, s)$$
$$\overline{C}_\mu = \overline{C}_\mu(X, s), \qquad \overline{C}_{\mu,\nu} = \overline{C}_{\mu,\nu}(X, s).$$

Define $Z \subseteq A^{(m)}$ by

$$Z = \left\{ j : \frac{t_m}{4} < j < \frac{3t_m}{4} \text{ and } j \in \bigcup_{\mu < m_0^{(m)}} C_\mu \text{ and} \right.$$

$$\left. f(X, j, i + 1, s) \leq \left(\frac{\alpha_m}{2}\right)^{i+1} \frac{t_m}{K^2} \text{ and } f(X, j, i, s) \geq \left(\frac{\alpha_m}{2}\right)^i \frac{t_m}{K^2} \right\}.$$

Case (i). Assume $|Z| \leq \alpha_m^K(1 - \beta_m)t_m$. By Theorem 7.4, which applies by (A) in Section 7.8, we have

$$\left| A^{(m)} - \bigcup_{\mu < m_0^{(m)}} C_\mu \right| \leq \rho^{(m)} |A^{(m)}| \leq \rho^{(m)} t_m = \alpha_m^K(1 - \beta_m)t_m.$$

Also, by definition of $G^i(t_1, \ldots, t_m, K)$, there are at most $2i\alpha_m^K(1 - \beta_m)t_m$ indices j with $f(X, j, i, s) < (\alpha_m/2)^i t_m/K^2$. Thus

$$\left| \left\{ j : \frac{t_m}{4} < j < \frac{3t_m}{4} \text{ and } f(X, j, i+1, s) \leq \left(\frac{\alpha_m}{2}\right)^{i+1} \frac{t_m}{K^2} \right\} \right|$$
$$\leq \alpha_m^K(1 - \beta_m)t_m + \alpha_m^K(1 - \beta_m)t_m + 2i\alpha_m^K(1 - \beta_m)t_m$$
$$= 2(i+1)\alpha_m^K(1 - \beta_m)t_m,$$

which is just (b).

Case (ii). Assume $|Z| > \alpha_m^K(1 - \beta_m)t_m$. If $|Z \cap C_\mu| \leq \alpha_m^K(1 - \beta_m)|C_\mu|$ for all $\mu < m_0^{(m)}$, then, since the C_μ are disjoint,

$$|Z| = \left| Z \cap \bigcup_{\mu < m_0^{(m)}} C_\mu \right| = \sum_{\mu < m_0^{(m)}} |Z \cap C_\mu|$$
$$\leq \alpha_m^K(1 - \beta_m) \sum_{\mu < m_0^{(m)}} |C_\mu| \leq \alpha_m^K(1 - \beta_m)|A^{(m)}| \leq \alpha_m^K(1 - \beta_m)t_m,$$

which is a contradiction. Thus, for some $\mu < m_0^{(m)}$,

$$|Z \cap C_\mu| > \alpha_m^K(1 - \beta_m)|C_\mu| = \varepsilon_1^{(m)}|C_\mu|.$$

Let $Z' = Z \cap C_\mu$. We first show that $k(Z', B^{(m)})$ is not very large.

Let $X = B^{(m)} - \bigcup_{\nu < n_0^{(m)}} C_{\mu,\nu} = [0, t_m) - \bigcup_{\nu < n_0^{(m)}} C_{\mu,\nu}$. By Theorem 7.4, $|X| < \sigma^{(m)}t_m$. Also by Theorem 7.4, for $x \in C_\mu$ we have

$$|k(x) \cap C_{\mu,\nu}| \leq \left(\beta(C_\mu, C_{\mu,\nu}) + \delta^{(m)}\right)|C_{\mu,\nu}|,$$

and so $k(Z', C_{\mu,\nu}) \leq \left(\beta(C_\mu, C_{\mu,\nu}) + \delta^{(m)}\right)|C_{\mu,\nu}||Z'|$. Therefore,

$$k(Z', B^{(m)}) = \sum_{\nu < n_0^{(m)}} k(Z', C_{\mu,\nu}) + k(Z', X)$$
$$\leq \sum_{\nu < n_0^{(m)}} \left(\beta(C_\mu, C_{\mu,\nu}) + \delta^{(m)}\right)|C_{\mu,\nu}||Z'| + |Z'||X| \quad (7.17)$$
$$\leq \sum_{\nu < n_0^{(m)}} \beta(C_\mu, C_{\mu,\nu})|C_{\mu,\nu}||Z'| + \left(\delta^{(m)} + \sigma^{(m)}\right)|Z'|t_m.$$

Next let $\overline{L} = \{j < t_m : X_{i,j} \in P(t_1, \ldots, t_{m-1})\}$. Thus for all $j \in A^{(m)}$,

$$|k(j) \cap \overline{L}| = f(X, j, i+1, s) \quad (7.18)$$

since any edge from j to a vertex $j' \in B^{(m)}$ already has $j_s = j$ and $j_i = j'$ for some arithmetic progression $j_0, j_1, \ldots, j_{K-1}$, such that the sets $X_{0,j_0}, \ldots,$

$X_{i-1,j_{i-1}}$ all belong to $P(t_1,\ldots,t_{m-1})$. Consequently, if $j' \in \overline{L}$, then $X_{i,j'} = X_{i,j_i}$ also belongs to $P(t_1,\ldots,t_{m-1})$ and so $(j_0,\ldots,j_{K-1}) \in F(X,j,i+1,s)$.

By hypothesis, X is well saturated, so for $|C_{\mu,\nu}| \geq \tau^{(m)}t_m$ we have

$$
\begin{aligned}
|\overline{L} \cap C_{\mu,\nu}| &\geq (\alpha_m - \delta^{(m)})|C_{\mu,\nu}| \\
&> \frac{\alpha_m}{2}|C_{\mu,\nu}| \text{ since } \delta^{(m)} = \frac{1}{10}\left(\frac{\alpha_m}{2}\right)^K \frac{1}{K^2} < \frac{\alpha_m}{2} \\
&= \varepsilon_2^{(m)}|C_{\mu,\nu}|.
\end{aligned}
$$

But we have already seen that $|Z'| > \varepsilon_1^{(m)}|C_\mu|$. Thus we may apply Theorem 7.4 and obtain

$$\beta(Z',\overline{L} \cap C_{\mu,\nu}) \geq \beta(C_\mu, C_{\mu,\nu}) - \delta^{(m)} \tag{7.19}$$

for those ν, such that $|C_{\mu,\nu}| \geq \tau^{(m)}t_m$. Therefore,

$$
\begin{aligned}
k(Z',\overline{L}) &\geq \sum_{\nu < n_0^{(m)}} k(Z', C_{\mu,\nu} \cap \overline{L}) \\
&= \sum_{\nu < n_0^{(m)}} \beta\left(Z', \overline{L} \cap C_{\mu,\nu}\right) |\overline{L} \cap C_{\mu,\nu}| |Z'|.
\end{aligned}
$$

Now break the sum into two parts according as $|C_{\mu,\nu}| \geq \tau^{(m)}t_m$ or $|C_{\mu,\nu}| < \tau^{(m)}t_m$. By Inequality (7.19) we find

$$
\begin{aligned}
k(Z',\overline{L}) &\geq \sum_{\nu < n_0^{(m)}} \left(\beta(C_\mu, C_{\mu,\nu}) - \delta^{(m)}\right) |\overline{L} \cap C_{\mu,\nu}| |Z'| \\
&\quad - \sum_{\substack{\nu < n_0^{(m)} \\ |C_{\mu,\nu}| < \tau^{(m)}t_m}} \beta\left(Z', \overline{L} \cap C_{\mu,\nu}\right) |\overline{L} \cap C_{\mu,\nu}| |Z'| \\
&\geq \sum_{\nu < n_0^{(m)}} \beta(C_\mu, C_{\mu,\nu}) |\overline{L} \cap C_{\mu,\nu}| |Z'| \\
&\quad - \delta^{(m)} |Z'| \sum_{\nu < n_0^{(m)}} |\overline{L} \cap C_{\mu,\nu}| \\
&\quad - \sum_{\substack{\nu < n_0^{(m)} \\ |C_{\mu,\nu}| < \tau^{(m)}t_m}} 1 \cdot |\overline{L} \cap C_{\mu,\nu}| |Z'| \\
&\geq \sum_{\nu < n_0^{(m)}} \beta(C_\mu, C_{\mu,\nu}) |\overline{L} \cap C_{\mu,\nu}| |Z'| \\
&\quad - \delta^{(m)}|Z'|t_m - n_0^{(m)}\tau^{(m)}|Z'|t_m \\
&\geq \sum_{\nu < n_0^{(m)}} \beta(C_\mu, C_{\mu,\nu}) |\overline{L} \cap C_{\mu,\nu}| |Z'| - 2\delta^{(m)}|Z'|t_m
\end{aligned}
$$

since $\tau^{(m)} = \sigma^{(m)}/\nu < n_0^{(m)}$. But

$$|\overline{L} \cap C_{\mu,\nu}| = p^m(X_i, C_{\mu,\nu}),$$

so that the fact that X is well saturated implies

$$|\overline{L} \cap C_{\mu,\nu}| \geq \alpha_m |C_{\mu,\nu}| - \delta^{(m)} |C_{\mu,\nu}|$$

provided $|C_{\mu,\nu}| \geq \tau^{(m)} t_m$. Thus

$$\sum_{\nu < n_0^{(m)}} \beta(C_\mu, C_{\mu,\nu}) |\overline{L} \cap C_{\mu,\nu}| \, |Z'|$$

$$\geq \sum_{\substack{\nu < n_0^{(m)} \\ |C_{\mu,\nu}| \geq \tau^{(m)} t_m}} \beta(C_\mu, C_{\mu,\nu}) |\overline{L} \cap C_{\mu,\nu}| \, |Z'|$$

$$\geq \sum_{\nu < n_0^{(m)}} \beta(C_\mu, C_{\mu,\nu})(\alpha_m - \delta^{(m)}) |C_{\mu,\nu}| \, |Z'|$$

$$- \sum_{\substack{\nu < n_0^{(m)} \\ |C_{\mu,\nu}| < \tau^{(m)} t_m}} \beta(C_\mu, C_{\mu,\nu})(\alpha_m - \delta^{(m)}) |C_{\mu,\nu}| \, |Z'|$$

$$\geq \alpha_m \sum_{\nu < n_0^{(m)}} \beta(C_\mu, C_{\mu,\nu}) |C_{\mu,\nu}| \, |Z'| - 2\delta^{(m)} |Z'| t_m,$$

as in the preceding inequality. Substituting this inequality for the corresponding term in the preceding inequality for $k(Z', \overline{L})$ we obtain

$$k(Z', \overline{L}) \geq \alpha_m \sum_{\nu < n_0^{(m)}} \beta(C_\mu, C_{\mu,\nu}) |C_{\mu,\nu}| \, |Z'| - 4\delta^{(m)} |Z'| t_m. \qquad (7.20)$$

Since $\sigma^{(m)} = \delta^{(m)}$, we may multiply Inequality (7.17) by α_m to obtain

$$\alpha_m k(Z', B^{(m)}) \leq \alpha_m \sum_{\nu < n_0^{(m)}} \beta(C_\mu, C_{\mu,\nu}) |C_{\mu,\nu}| \, |Z'| + 2\delta^{(m)} \alpha_m |Z'| t_m.$$

Thus by (7.20) we have

$$k(Z', \overline{L}) \geq \alpha_m k(Z', B^{(m)}) - 6\delta^{(m)} |Z'| t_m.$$

But by the definition of Z', for each $j \in Z'$ we have

$$f(X, j, i, s) \geq \frac{1}{K^2} \left(\frac{\alpha_m}{2} \right)^i t_m$$

and

$$f(X, j, i + 1, s) \leq \frac{1}{K^2} \left(\frac{\alpha_m}{2} \right)^{i+1} t_m.$$

Hence by Equation (7.18) we see

$$
\frac{1}{K^2} \left(\frac{\alpha_m}{2}\right)^{i+1} |Z'| t_m \geq \sum_{j \in Z'} f(X, j, i+1, s) = k(Z', \overline{L})
$$

$$
\geq \alpha_m k(Z', B^{(m)}) - 6\delta^{(m)} |Z'| t_m
$$

$$
= \alpha_m \sum_{j \in Z'} f(X, j, i, s) - 6\delta^{(m)} |Z'| t_m
$$

$$
\geq \frac{2}{K^2} \left(\frac{\alpha_m}{2}\right)^{i+1} |Z'| t_m - 6\delta^{(m)} |Z'| t_m.
$$

Therefore,

$$
\frac{1}{K^2} \left(\frac{\alpha_m}{2}\right)^{i+1} \leq 6\delta^{(m)},
$$

which contradicts the definition of $\delta^{(m)}$ since by 7.8.1(iii),

$$
\delta^{(m)} = \frac{1}{10} \cdot \frac{1}{K^2} \left(\frac{\alpha_m}{2}\right)^K < \frac{1}{6} \cdot \frac{1}{K^2} \left(\frac{\alpha_m}{2}\right)^{i+1}
$$

for $i + 1 = K$. This proves (b).

The proof of (a) is similar to that of (b). Set

$$
\overline{Z} = \Big\{ j \in \bigcup_{\mu < m_0^{(m)}} \overline{C}_\mu \ : \ f(X, j, i+1, s) \geq 2^{i+1} \alpha_m^{i+1} t_m
$$

$$
\text{and } f(X, j, i, s) \leq 2^i \alpha_m^i t_m \Big\}.
$$

As before, if $|\overline{Z}| \leq \alpha_m^K (1 - \beta_m) t_m$, then we would be done. Hence we may assume $|\overline{Z}| > \alpha_m^K (1 - \beta_m) t_m$. Again, as before, for some $\mu < m_0^{(m)}$, we have

$$
|\overline{Z} \cap \overline{C}_\mu| > \alpha_m^K (1 - \beta_m) |\overline{C}_\mu| = \varepsilon_1^{(m)} |\overline{C}_\mu|.
$$

We let $\overline{Z}' = \overline{Z} \cap \overline{C}_\mu$, so that $|\overline{Z}'| > \varepsilon_1^{(m)} |\overline{C}_\mu|$. By Corollary 7.2, for each $x \in \overline{Z}'$ we have

$$
|k(x) \cap \overline{C}_{\mu,\nu}| \geq \left(\beta(\overline{C}_\mu, \overline{C}_{\mu,\nu}) - \delta^{(m)} \right) |\overline{C}_{\mu,\nu}|.
$$

Therefore,

$$
k(\overline{Z}', \overline{C}_{\mu,\nu}) = \sum_{x \in \overline{Z}'} |k(x) \cap \overline{C}_{\mu,\nu}| \geq \left(\beta(\overline{C}_\mu, \overline{C}_{\mu,\nu}) - \delta^{(m)} \right) |\overline{C}_{\mu,\nu}| |\overline{Z}'|,
$$

and so

$$k(\overline{Z}', \overline{B}^{(m)}) \geq \sum_{\nu < n_0^{(m)}} k(\overline{Z}', \overline{C}_{\mu,\nu})$$

$$\geq \sum_{\nu < n_0^{(m)}} \left(\beta(\overline{C}_\mu, \overline{C}_{\mu,\nu}) - \delta^{(m)} \right) |\overline{C}_{\mu,\nu}| \, |\overline{Z}'| \qquad (7.21)$$

$$\geq \sum_{\nu < n_0^{(m)}} \beta(\overline{C}_\mu, \overline{C}_{\mu,\nu}) |\overline{C}_{\mu,\nu}| \, |\overline{Z}'| - \delta^{(m)} |\overline{Z}'| t_m.$$

Also, by Corollary 7.2,

$$k(\overline{Z}', \overline{L}) = \sum_{\nu < n_0^{(m)}} k(\overline{Z}', \overline{C}_{\mu,\nu} \cap \overline{L}) + k\left(\overline{Z}', \left(\overline{B}^{(m)} - \bigcup_{\nu < n_0^{(m)}} \overline{C}_{\mu,\nu}\right) \cap \overline{L}\right)$$

$$\leq \sum_{\nu < n_0^{(m)}} k(\overline{Z}', \overline{C}_{\mu,\nu} \cap \overline{L}) + \sigma^{(m)} |\overline{Z}'| t_m$$

$$= \sum_{\nu < n_0^{(m)}} \beta(\overline{Z}', \overline{C}_{\mu,\nu} \cap \overline{L}) |\overline{C}_{\mu,\nu} \cap \overline{L}| \, |\overline{Z}'| + \sigma^{(m)} |\overline{Z}'| t_m. \qquad (7.22)$$

Now break the sum into two parts according as $|\overline{C}_{\mu,\nu}| \geq \tau^{(m)} t_m$ or $|\overline{C}_{\mu,\nu}| < \tau^{(m)} t_m$. By hypothesis, X is well saturated, so for $|\overline{C}_{\mu,\nu}| > \tau^{(m)} t_m$ we have

$$(\alpha_m - \delta^{(m)}) |\overline{C}_{\mu,\nu}| \leq |\overline{L} \cap \overline{C}_{\mu,\nu}| \leq (\alpha_m + \delta^{(m)}) |\overline{C}_{\mu,\nu}|.$$

Since $\alpha_m - \delta^{(m)} > \alpha_m/2 = \varepsilon_2^{(m)}$, we can apply Corollary 7.2 to the first summand in Inequality (7.22) (after the split) and obtain

$$k(\overline{Z}', \overline{L}) \leq \sum_{\nu < n_0^{(m)}} \left(\beta(\overline{C}_\mu, \overline{C}_{\mu,\nu}) + \delta^{(m)} \right) |\overline{L} \cap \overline{C}_{\mu,\nu}| \, |\overline{Z}'|$$

$$+ \tau^{(m)} t_m |\overline{Z}'| n_0^{(m)} + \sigma^{(m)} |\overline{Z}'| t_m$$

$$\leq \sum_{\substack{\nu < n_0^{(m)} \\ |\overline{C}_{\mu,\nu}| \geq \tau^{(m)} t_m}} \beta(\overline{C}_\mu, \overline{C}_{\mu,\nu}) |\overline{L} \cap \overline{C}_{\mu,\nu}| \, |\overline{Z}'|$$

$$+ \sum_{\substack{\nu < n_0^{(m)} \\ |\overline{C}_{\mu,\nu}| < \tau^{(m)} t_m}} \beta(\overline{C}_\mu, \overline{C}_{\mu,\nu}) |\overline{L} \cap \overline{C}_{\mu,\nu}| \, |\overline{Z}'| + 3\delta^{(m)} |\overline{Z}'| t_m$$

$$\leq \sum_{\nu < n_0^{(m)}} \beta(\overline{C}_\mu, \overline{C}_{\mu,\nu})(\alpha_m + \delta^{(m)}) |\overline{C}_{\mu,\nu}| \, |\overline{Z}'|$$

$$+ n_0^{(m)} \tau^{(m)} |\overline{Z}'| t_m + 3\delta^{(m)} |\overline{Z}'| t_m$$

$$\leq \alpha_m \sum_{\nu < n_0^{(m)}} \beta(\overline{C}_\mu, \overline{C}_{\mu,\nu}) |\overline{C}_{\mu,\nu}| \, |\overline{Z}'| + 5\delta^{(m)} |\overline{Z}'| t_m.$$

Thus by Inequality (7.21) we obtain

$$k(\overline{Z}', \overline{L}) \le \alpha_m k(\overline{Z}', \overline{B}^{(m)}) + 6\delta^{(m)}|\overline{Z}'|t_m.$$

By the definition of $\overline{Z}'$, for each $j \in \overline{Z}'$,

$$f(X, j, i, s) \le 2^i \alpha_m^i t_m$$

and

$$f(X, j, i+1, s) \ge 2^{i+1} \alpha_m^{i+1} t_m.$$

Hence,

$$2^{i+1} \alpha_m^{i+1} |\overline{Z}'| t_m \le \sum_{j \in \overline{Z}'} f(X, j, i+1, s) = k(\overline{Z}', \overline{L})$$

$$\le \alpha_m k(\overline{Z}', \overline{B}^{(m)}) + 6\delta^{(m)} |\overline{Z}'| t_m$$

$$\le \alpha_m 2^i \alpha_m^i |\overline{Z}'| t_m + 6\delta^{(m)} |\overline{Z}'| t_m.$$

Therefore, $2^i \alpha_m^{i+1} \le 6\delta^{(m)}$, which contradicts the definition of $\delta^{(m)}$. This proves (a) and completes the proof of Proposition 7.4. ∎

PROPOSITION 7.5

Suppose, for some i in $0 \le i < K - 1$ that $X^{(\xi)} \in G^i(t_1, \ldots, t_m, K)$ for $\xi < \ell$, where $\ell \ge \ell_m^{2-K}$. Assume that

$$X = \bigcup_{\xi < \ell} X_i^{(\xi)} \in C(t_1, \ldots, t_m, \ell)$$

and that for each $j < i$, $X_j^{(\xi)}$ and $X_j^{(\eta)}$ are R-equivalent for all $\xi, \eta < \ell$. Then there exists a $\pi < \ell$, such that

$$X^{(\pi)} \in G^{i+1}(t_1, \ldots, t_m, K).$$

PROOF   The R-equivalence of $X_j^{(\xi)}$ and $X_j^{(\eta)}$ implies

$$I(X^{(\xi)}, i, s) = I(X^{(\eta)}, i, s) \text{ for } i \le s < K; \; \xi, \eta < \ell,$$

since the edges of $I(X, i, s)$ are completely specified by the R-equivalence classes of the X_j for $j < i$. Hence we may assume

$$C_\mu(X^{(\xi)}, s) = C_\mu(X^{(\eta)}, s) = C_\mu(s),$$
$$C_{\mu,\nu}(X^{(\xi)}, s) = C_{\mu,\nu}(X^{(\eta)}, s) = C_{\mu,\nu}(s),$$

with a similar assumption holding for $\overline{C}_\mu(s)$ and $\overline{C}_{\mu,\nu}(s)$.

By Proposition 7.4, it will be enough to show that for some $\pi < \ell$, $X^{(\pi)}$ is well saturated. Suppose this is not so. Then for each $\xi < \ell$, there exist μ, ν and s such that either

(i) $|C_{\mu,\nu}(s)| \geq \tau^{(m)} t_m$ and

$$\left| p^m(X_i^{(\xi)}, C_{\mu,\nu}(s)) - \alpha_m |C_{\mu,\nu}(s)| \right| > \delta^{(m)} |C_{\mu,\nu}(s)|$$

or

(ii) $|\overline{C}_{\mu,\nu}(s)| \geq \tau^{(m)} t_m$ and

$$\left| p^m(X_i^{(\xi)}, \overline{C}_{\mu,\nu}(s)) - \alpha_m |\overline{C}_{\mu,\nu}(s)| \right| > \delta^{(m)} |\overline{C}_{\mu,\nu}(s)|.$$

There are actually four possibilities here, depending on which way the inequalities go when the absolute value signs are removed. But by the choice of ℓ and ℓ_m (see 7.8.4), we have

$$\ell \geq \ell_m^{2^{-K}} > w(\ell'_m, 4K m_0^{(m)} n_0^{(m)}).$$

Since there are at most $m_0^{(m)}$ choices for μ, at most $n_0^{(m)}$ choices for ν, and at most K choices for s, then by definition of the van der Waerden function w (see 7.8.4), there exists an arithmetic progression of length ℓ'_m in $[0, \ell)$, such that (i) and (ii) involve the same μ, ν and s and so that the inequalities go the same way. Let $\xi_0, \ldots, \xi_{\ell'_m - 1}$ denote this arithmetic progression and suppose to be definite that

$$p^m(X_i^{(\xi_j)}, C_{\mu,\nu}(s)) - \alpha_m |C_{\mu,\nu}(s)| > \delta^{(m)} |C_{\mu,\nu}(s)| \qquad (7.23)$$

for all $j < \ell'_m$. Since

$$\bigcup_{\xi < \ell} X_i^{(\xi)} \in C(t_1, \ldots, t_m, \ell)$$

by hypothesis, we have $X_i^{(\xi)} \in S(t_1, \ldots, t_m)$ for all $\xi < \ell$. But $|C_{\mu,\nu}(s)| \geq \tau^{(m)} t_m$ for the special choice of μ, ν and s above so that we can apply Proposition 7.3 to

$$\bigcup_{j < \ell'_m} X^{(\xi_j)}$$

(by 7.8.3) and obtain for some $j < \ell'_m$

$$p^m(X_i^{(\xi_j)}, C_{\mu,\nu}(s)) < (\alpha_m + \delta^{(m)}) |C_{\mu,\nu}(s)|,$$

which contradicts Inequality (7.23). The other three cases are treated in exactly the same way and Proposition 7.5 is proved. ∎

PROPOSITION 7.6

Let $\ell > 0$, $m \geq 0$ and i with $0 \leq i < K$ be fixed integers and assume that

$$X^{(\xi)} = \bigcup_{i' < K} X_{i'}^{(\xi)} \in G^i(t_1, \ldots, t_{m+1}, K)$$

for $\xi < \ell \leq \ell_m$. Suppose that for all $j < i$ and $\xi, \eta < \ell$, $X_j^{(\xi)}$ and $X_j^{(\eta)}$ are R-equivalent. Write, as usual,

$$X_{i'}^{(\xi)} = \bigcup_{j' < t_{m+1}} X_{i',j'}^{(\xi)}$$

for $i < K$. Then there exists a sequence of ℓ_m arithmetic progressions

$$(j_0^{(\zeta)}, \ldots, j_{K-1}^{(\zeta)}) \in E(t_{m+1}, K), \quad \zeta < \ell_m,$$

such that $j_i^{(0)}, \ldots, j_i^{(\ell_m - 1)}$ forms an arithmetic progression of length ℓ_m and such that for all $\xi < \ell$ and $\zeta < \ell_m$,

$$\bigcup_{i' < K} X_{i', j_{i'}^{(\zeta)}}^{(\xi)} \in D^i(t_1, \ldots, t_m, K).$$

PROOF   By the R-equivalence assumptions we have

$$F(X^{(\xi)}, j, i, s) = F(X^{(\eta)}, j, i, s)$$

for all $\xi, \eta < \ell$, $i \leq s < k$ and $j < t_{m+1}$. To be concise we write $F(X^{(\xi)}, j, i, s) = F(j, i, s)$ and $f(X^{(\xi)}, j, i, s) = f(j, i, s)$. By definition of $G^i(t_1, \ldots, t_{m+1}, K)$, there exist for each s in $i \leq s < K$ sets $Z_s, \overline{Z}_s \subseteq [0, t_{m+1})$, such that $|Z_s| \leq 2K\alpha_{m+1}^K(1 - \beta_{m+1})t_{m+1}$, $|\overline{Z}_s| \leq 2K\alpha_{m+1}^K(1 - \beta_{m+1})t_{m+1}$, and

(a) $f(j, i, s) \leq 2^i \alpha_{m+1}^i t_{m+1}$ if $j \notin Z_s$, $j < t_{m+1}$,

(b) $f(j, i, s) \geq (\alpha_{m+1}/2)^i K^{-2} t_{m+1}$ if $j \notin \overline{Z}_s$, $t_{m+1}/4 < j < 3t_{m+1}/4$.

Define $\overline{Z}$ by $\overline{Z} = \{j : t_{m+1} < j < 3t_{m+1}/4$ and there is no $(j_0, \ldots, j_{K-1}) \in F(j, i, i)$, such that

$$\bigcup_{i' < K} X_{i', j_{i'}}^{(\xi)} \in D^i(t_1, \ldots, t_m, K) \tag{7.24}$$

holds for *all* $\xi < \ell\}$. Set $P = \bigcup_{j \in \overline{Z}} F(j, i, i)$ and $p = |P|$. In other words, $\overline{Z}$ is the set of all indices j with $t_{m+1}/4 < j < 3t_{m+1}/4$, such that for each $(j_0, \ldots, j_{K-1}) \in F(j, i, i)$, there exists at least one $\xi < \ell$ so that (7.24) fails. Note that by the definition of $F(j, i, i)$ we have

$$X_{i', j_{i'}}^{(\xi)} \in P(t_1, \ldots, t_m) \subseteq S(t_1, \ldots, t_m)$$

whenever $i' < i$. Therefore the failure of (7.24) implies

$$X_{i', j_{i'}}^{(\xi)} \notin S(t_1, \ldots, t_m)$$

for some i' with $i \leq i' < K$. But by (b) with $s = i$ we see that for each $j \in \overline{Z} - \overline{Z}_i$,

$$f(j, i, i) = |F(j, i, i)| \geq (\alpha_{m+1}/2)^i K^{-2} t_{m+1}.$$

Since $F(j,i,i)$ and $F(j',i,i)$ are disjoint for $j \neq j'$ and

$$|\overline{Z}_i| \leq 2K\alpha_{m+1}^K(1-\beta_{m+1})t_{m+1},$$

we have

$$p = |P| \geq \left(|\overline{Z}| - 2K(1-\beta_{m+1})\alpha_{m+1}^Kt_{m+1}\right)K^{-2}\left(\frac{\alpha_{m+1}}{2}\right)^i t_{m+1}. \quad (7.25)$$

On the other hand, it is clear from the definition of P that

$$P \subseteq \bigcup_{i \leq s < K} \bigcup_{\xi < \ell} \left(\bigcup_{\substack{j < t_{m+1}, j \in Z_s \\ X_{s,j}^{(\xi)} \notin S(t_1,\ldots,t_m)}} F(j,i,s) \cup \bigcup_{\substack{j < t_{m+1}, j \notin Z_s \\ X_{s,j}^{(\xi)} \notin S(t_1,\ldots,t_m)}} F(j,i,s) \right),$$

so that

$$p = |P| \leq K \cdot \ell \cdot 2K\alpha_{m+1}^K(1-\beta_{m+1})t_{m+1} \cdot t_{m+1}$$
$$+ K \cdot \ell \cdot 2^i\alpha_{m+1}^i t_{m+1} \max_{s,\xi}|\{j : X_{s,j}^{(\xi)} \notin S(t_1,\ldots,t_m) \text{ and } j \notin Z_s\}|$$
$$\leq K\ell 2K\alpha_{m+1}^K(1-\beta_{m+1})t_{m+1}^2 \quad (7.26)$$
$$+ K\ell 2^K\alpha_{m+1}^K t_{m+1} \max_{s,\xi}|\{j : X_{s,j}^{(\xi)} \notin S(t_1,\ldots,t_m)\}|.$$

But by hypothesis, for all $\xi < \ell$,

$$X^{(\xi)} \in G^i(t_1,\ldots,t_{m+1},K) \subseteq C(t_1,\ldots,t_{m+1},K).$$

Thus $X_s^{(\xi)} \in S(t_1,\ldots,t_{m+1})$ for all $s < K$ and $\xi < \ell$. Therefore,

$$s^{m+1}(X_s^{(\xi)}) \geq \left(\beta_{m+1} - \sqrt{\mu_{m+1}(t_{m+1})}\right)t_{m+1},$$

so that

$$\max_{s,\xi}|\{j : X_{s,j}^{(\xi)} \notin S(t_1,\ldots,t_m)\}| \leq \left(1 - \beta_{m+1} + \sqrt{\mu_{m+1}(t_{m+1})}\right)t_{m+1}$$
$$\leq 2(1-\beta_{m+1})t_{m+1}$$

by the choice of t_{m+1} (see (F) in Section 7.8). Hence by Inequality (7.26),

$$p \leq K\ell\alpha_{m+1}^K t_{m+1}^2(1-\beta_{m+1})(2K + 2^{K+1}).$$

Combining this with Inequality (7.25) we obtain

$$|\overline{Z}| \leq 2K(1-\beta_{m+1})\alpha_{m+1}^K t_{m+1} + K^3\ell 2^i\alpha_{m+1}^{K-i}(1-\beta_{m+1})(2K+2^{K+1})t_{m+1}$$
$$\leq (1-\beta_{m+1})2K^3\ell 2^{2K}t_{m+1}$$
$$\leq (1-\beta_{m+1})2K^3\ell_m 2^{2K}t_{m+1}$$
$$\leq \sqrt{1-\beta_{m+1}}t_{m+1},$$

provided

$$K^3 2^{2K} \ell_m \leq \frac{1}{2} (1 - \beta_{m+1})^{-1/2}.$$

But this inequality holds by the choice of t_{m+1} (see 7.8.4).

Now $X_i^{(0)} \in S(t_1, \ldots, t_{m+1})$ because $X^{(0)} \in C(t_1, \ldots, t_{m+1}, K)$. Thus $s^{m+1}(X_i^{(0)}) \geq \left(\beta_{m+1} - \sqrt{\mu_{m+1}(t_{m+1})} \right) t_{m+1}$ and so

$$
\begin{aligned}
& s^{m+1} \left(X_i^{(0)}, \left(\frac{t_{m+1}}{4}, \frac{3t_{m+1}}{4} \right) - \overline{Z} \right) \\
\geq \ & \left(\beta_{m+1} - \sqrt{\mu_{m+1}(t_{m+1})} \right) t_{m+1} - \frac{t_{m+1}}{2} - \sqrt{1 - \beta_{m+1}} t_{m+1} \\
\geq \ & \left(\beta_{m+1} - \frac{1}{2} - 2\sqrt{1 - \beta_{m+1}} \right) t_{m+1} \quad \text{by choice of } t_{m+1} \text{ (see (F))} \\
> \ & \frac{1}{2} \left(1 - \frac{1}{\ell_m} \right) \quad \text{by choice of } t_{m+1} \text{ (see (C))}.
\end{aligned}
$$

By (E) in the choice of t_{m+1} and by Lemma 7.17, $\left(\frac{1}{4} t_{m+1}, \frac{3}{4} t_{m+1} \right) - \overline{Z}$ contains an arithmetic progression $j^{(0)}, \ldots, j^{(\ell_m - 1)}$ of length ℓ_m. Therefore, by the definition of $\overline{Z}$, for each $\zeta < \ell_m$, there exists some arithmetic progression $j_0^{(\zeta)}$, $\ldots, j_{K-1}^{(\zeta)} \in F(j^{(\zeta)}, i, i)$, such that $\bigcup_{i' < K} X_{i', j_i^{(\zeta)}}^{(\xi)} \in D^i(t_1, \ldots, t_m, K)$ for all $\xi < \ell$. This completes the proof of Proposition 7.6. ∎

7.10 Szemerédi's Theorem

We need a little more notation before we prove the main result.

DEFINITION 7.9   *For integers $m \leq m'$ suppose that*

$$X \in B(t_1, \ldots, t_{m'}, K) \quad \text{and} \quad Y \in B(t_1, \ldots, t_m, K).$$

We write $Y \sqsubset X$ if for all $i < K$, Y_i is a subconfiguration of X_i of order m. If $Y \sqsubset X$ and $Y' \sqsubset X'$, we shall say the position of Y in X is the same as the position of Y' in X' if for each $i < K$, there is a j_i, such that Y_i is the j_i-th subconfiguration of X_i and Y_i' is the j_i-th subconfiguration of X_i'.

LEMMA 7.18
For every positive integer m and every integer i with $0 \leq i < K$, there exists a number $h(m, i)$, such that if $m' \geq h(m, i)$ and

$$X^{(\xi)} \in D^i(t_1, \ldots, t_{m'}, K) \quad \text{for } \xi < \ell \leq \ell_m^{2^{-i}},$$

then there exist

$$Y^{(\xi)} \in G^i(t_1, \ldots, t_m, K) \quad \text{for } \xi < \ell,$$

such that $Y^{(\xi)} \sqsubset X^{(\xi)}$ and for all $\xi, \eta < \ell$, the position of $Y^{(\xi)}$ in $X^{(\xi)}$ is the same as the position of $Y^{(\eta)}$ in $X^{(\eta)}$.

PROOF The proof is by induction on i.

Case $i = 0$. We have already seen that

$$D^0(t_1, \ldots, t_m, K) = C(t_1, \ldots, t_m, K)$$
$$\text{and } G^0(t_1, \ldots, t_m, K) = C(t_1, \ldots, t_m, K).$$

Let $h(m, 0) = m$. Let $X^{(\xi)} \in C(t_1, \ldots, t_{m'}, K)$ with $\xi < \ell \le \ell_m$ and $m' \ge m$. We want to show that there exist $Y^{(\xi)} \in C(t_1, \ldots, t_m, K)$ for $\xi < \ell$, such that $Y^{(\xi)} \sqsubset X^{(\xi)}$ and for each $\xi < \ell$, the position of $Y^{(\xi)}$ in $X^{(\xi)}$ is the same.

If $m' = m$, take $Y^{(\xi)} = X^{(\xi)}$ for $\xi < \ell$.

Now assume that $m' > m$ and that the assertion holds for $m' - 1$. The hypothesis of R-equivalence required in Proposition 7.6 is vacuous here since $i = 0$. Thus by Proposition 7.6, there exist

$$Z^{(\xi)} = \bigcup_{i' < K} X^{(\xi)}_{i', j^{(0)}_{i'}} \in C(t_1, \ldots, t_{m'-1}, K).$$

Since $Z^{(\xi)}_{i'} = X^{(\xi)}_{i', j^{(0)}_{i'}}$, the position of each $Z^{(\xi)}$ in $X^{(\xi)}$ is the same, and $Z^{(\xi)} \sqsubset X^{(\xi)}$. Now apply the induction hypothesis to $Z^{(\xi)} \in C(t_1, \ldots, t_{m'-1}, K)$ for $\xi < \ell$. By induction, there exist $Y^{(\xi)} \in C(t_1, \ldots, t_m, K)$, for $\xi < \ell$, such that $Y^{(\xi)} \sqsubset Z^{(\xi)}$ and for each $\xi < \ell$, the position of $Y^{(\xi)}$ in $Z^{(\xi)}$ is the same. Thus $Y^{(\xi)} \sqsubset X^{(\xi)}$ and for each $\xi < \ell$, the position of $Y^{(\xi)}$ in $X^{(\xi)}$ is the same. This completes the case $i = 0$.

Case $i > 0$. Assume for some $i \ge 0$ that Lemma 7.18 holds. We now prove it for $i + 1$. Let $m^* = h(m, i) + 1$ and $h(m, i + 1) = h(m^*, i)$. Suppose $m' \ge h(m, i + 1) = h(m^*, i)$ and

$$X^{(\xi)} \in D^{i+1}(t_1, \ldots, t_{m'}, K) \text{ for } \xi < \ell \le \ell_m^{2^{-(i+1)}}.$$

Thus $X^{(\xi)} \in D^i(t_1, \ldots, t_{m'}, K)$. Since $\ell_m^{2^{-(i+1)}} \le \ell_m^{2^{-i}}$, we can apply the induction hypothesis to obtain configurations $Y^{(\xi)} \in G^i(t_1, \ldots, t_{m^*}, K)$, for $\xi < \ell$, such that $Y^{(\xi)} \sqsubset X^{(\xi)}$ and the position of each $Y^{(\xi)}$ in $X^{(\xi)}$ is the same. Since $X^{(\xi)} \in D^{i+1}(t_1, \ldots, t_{m'}, K)$, we see that $X^{(\xi)}_j \in P(t_1, \ldots, t_{m'})$ for all $j \le i$. By the definition of $P(t_1, \ldots, t_{m'})$, all the $X^{(\xi)}$, for $j \le i$, are R-equivalent. Since the position of each $Y^{(\xi)}$ in $X^{(\xi)}$ is the same, all the $Y^{(\xi)}_j$, for $j \le i$, are R-equivalent. Since $h(m, i) \ge m$, we have $\ell_m = \ell_{h(m.i)}$ and we can apply Proposition 7.6 to the $Y^{(\xi)}$ for $\xi \le \ell$. The conclusion of Proposition 7.6 asserts that there exists a sequence of arithmetic progressions

$$(j_0^{(\zeta)}, \ldots, j_{K-1}^{(\zeta)}) \in E(t_{h(m,i)+1}, K), \quad \zeta < \ell_{h(m,i)},$$

such that for every $\xi < \ell$ and every $\zeta < \ell_{h(m,i)}$ we have

$$Y(\xi, \zeta) = \bigcup_{i' < K} Y^{(\xi)}_{i', j^{(\zeta)}_{i'}} \in D^i(t_1, \ldots, t_{h(m,i)}, K)$$

and such that $j^{(0)}_i, \ldots, j^{(\ell_{h(m,i)}-1)}_i$ forms an arithmetic progression of length $\ell_{h(m,i)}$. Since $h(m,i) \geq m$, $Y(\xi, \zeta)$ is defined for all $\xi < \ell$ and $\zeta < \ell_m^{2^{-(i+1)}}$. Thus we have

$$Y(\xi, \zeta) \in D^i(t_1, \ldots, t_{h(m,i)}, K) \text{ for } \xi < \ell, \; \zeta < \ell_m^{2^{-(i+1)}}.$$

Note that there are at most

$$\ell \cdot \ell_m^{2^{-(i+1)}} \leq \ell_m^{2^{-(i+1)}} \cdot \ell_m^{2^{-(i+1)}} = \ell_m^{2^{-i}}$$

such $Y(\xi, \zeta)$. Thus we may apply the induction hypothesis to these $Y(\xi, \zeta)$ and obtain a sequence

$$Z(\xi, \zeta) \in G^i(t_1, \ldots, t_m, K)$$

with $Z(\xi, \zeta) \sqsubset Y(\xi, \zeta)$ for $\xi < \ell$, $\zeta < \ell_m^{2^{-(i+1)}}$. By the way in which the $Y(\xi, \zeta)$ are defined, for fixed ζ, the position of $Y(\xi, \zeta)$ in $X^{(\xi)}$ is the same as the position of $Y(\eta, \xi)$ in $X^{(\eta)}$ for $\xi, \eta < \ell$. Since the position of each $Z(\xi, \zeta)$ in $Y(\xi, \zeta)$ is the same (by the induction hypothesis), then for fixed ζ, the position of $Z(\xi, \zeta)$ in $Y(\xi, \zeta)$ and $X^{(\xi)}$ is the same as the position of $Z(\eta, \xi)$ in $Y(\eta, \xi)$ and $X^{(\eta)}$ for $\xi, \eta < \ell$ and $\zeta < \ell_m^{2^{-(i+1)}}$. Keep in mind that for $j \leq i$, all the $X^{(\xi)}_j$ are R-equivalent. Thus for $j \leq i$, all the $Z(\xi, \zeta)_j$ with $\xi < \ell, \zeta < \ell_m^{2^{-(i+1)}}$, are R-equivalent. Since $\ell \leq \ell_m^{2^{-(i+1)}}$, we see that $Z(\xi, \zeta)$ is defined for $\xi, \zeta < \ell$. Let

$$Z = \bigcup_{\zeta < \ell} Z(\xi, \zeta)_i, \quad \xi < \ell.$$

As part of the conclusion of Proposition 7.6, we know that $j^{(0)}_i, \ldots, j^{(\ell_{h(m,i)}-1)}_i$ forms an arithmetic progression of length $\ell_{h(m,i)}$. Hence $j^{(0)}_i, \ldots, j^{(\ell-1)}_i$ forms an arithmetic progression of length ℓ. Also by the induction hypothesis, the position of each $Z(\xi, \zeta)$ in $Y(\xi, \zeta)$ is the same. Therefore $Z \in B(t_1, \ldots, t_m, \ell)$. But each $Z(\xi, \zeta) \in G^i(t_1, \ldots, t_m, K)$, so each $Z(\xi, \zeta)_i \in S(t_1, \ldots, t_m)$. Hence $Z \in C(t_1, \ldots, t_m, K)$.

Finally, since each $Y(\xi, \zeta) \in D^i(t_1, \ldots, t_{h(m,i)}, K)$, we have that for $j < i$, $Y(\xi, \zeta)_j \in P(t_1, \ldots, t_{h(m,i)})$. Thus for all $j < i$ and all $\zeta < \ell_m^{2^{-(i+1)}}$, the $Y(0, \zeta)_j$ are R-equivalent. But for $\zeta < \ell_m^{2^{-(i+1)}}$, the position of each $Z(0, \zeta)$ in $Y(0, \zeta)$ is the same. Therefore, for all $j < \ell$ and all $\zeta < \ell_m^{2^{-(i+1)}}$, the $Z(0, \zeta)_j$ are R-equivalent. Hence, since $\ell_m^{2^{-K}} \leq \ell_m^{2^{-(i+1)}}$, we may apply Proposition 7.5

to the sequence $Z(0, \zeta)$, $\zeta < \ell_m^{2^{-K}}$. The conclusion of Proposition 7.5 asserts that for some $\zeta^* < \ell_m^{2^{-K}}$,

$$Z(0, \zeta^*) \in G^{i+1}(t_1, \ldots, t_m, K).$$

But we have already seen that for all $\xi < \ell$, the position of $Z(\xi, \zeta^*)$ in $X^{(\xi)}$ is the same as the position of $Z(0, \zeta^*)$ in $X^{(0)}$. Since all the $X_j^{(\xi)}$ with $j \leq i$ and $\xi < \ell$ are R-equivalent (they are all in $P(t_1, \ldots, t_{m'})$) we see that, for each $j \leq i$, all the $Z(\xi, \zeta^*)_j$ for $\xi < \ell$ are R-equivalent to $Z(0, \zeta^*)_j$. Therefore, for $\xi < \ell$,

$$Z(\xi, \zeta^*) \in G^{i+1}(t_1, \ldots, t_m, K).$$

Lastly, since

$$Z(\xi, \zeta^*) \sqsubset Y(\xi, \zeta^*) \sqsubset X^{(\xi)}, \quad \xi < \ell,$$

the inductive step is complete. This proves Lemma 7.18. ∎

THEOREM 7.5 Szemerédi's Theorem [109]

For every nonnegative integer m and every integer i in $0 \leq i < K$, we have

$$D^i(t_1, \ldots, t_m, K) \neq \emptyset.$$

PROOF   The proof is by induction on i.

Case $i = 0$. Recall that $D^0(t_1, \ldots, t_m, K) = C(t_1, \ldots, t_m, K)$. For $m = 0$, $C(K)$ is certainly nonempty since by definition

$$C(K) = \{X \in B(K) : s^1(X) = K\} = B(K).$$

For $m \geq 1$, t_m has been chosen (see (C) of Section 7.8) so that

$$1 - \beta_{m+1} < \frac{1}{K}.$$

Hence by Lemma 7.14, $C(t_1, \ldots, t_m, K) \neq \emptyset$.

Case $i > 0$. Assume now that the assertion holds for a fixed $i < K - 1$ and all $m \geq 0$. We prove it also holds for $i+1$ and all $m \geq 0$. Let $m' = h(m+1, i)$ and $m'' = h(m'+1, i)$. Let $X \in D^i(t_1, \ldots, t_{m''}, K)$, which is nonempty by the induction hypothesis. By Lemma 7.18, there exists $Y \in G^i(t_1, \ldots, t_{m'+1}, K)$ with $Y \sqsubset X$. We now apply Proposition 7.6 with $\ell = 1$, that is, to the "sequence" consisting of the single term Y, where as usual we write

$$Y_{i'} = \bigcup_{j' < t_{m'+1}} Y_{i',j'} \text{ for } i' < K.$$

The conclusion of Proposition 7.6 asserts that there exists a sequence of arithmetic progressions

$$(j_0^{(\zeta)}, \ldots, j_{K-1}^{(\zeta)}) \in E(t_{m'+1}, K) \text{ for } \zeta < \ell_{m'},$$

such that

$$Y^{(\zeta)} = \bigcup_{i' < K} Y_{i',j_{i'}^{(\zeta)}} \in D^i(t_1, \ldots, t_{m'}, K) \text{ for } \zeta < \ell_{m'},$$

and such that $j_i^{(0)}, \ldots, j_i^{(\ell_{m'}-1)}$ forms an arithmetic progression of length $\ell_{m'}$. Since $m + 1 \leq m'$, we have $\ell_{m+1}^{2^{-i}} \leq \ell_{m'}$, and so we may apply the definition of $h(m+1, i) = m'$ to the sequence $Y^{(\zeta)}$, $\zeta < \ell_{m+1}^{2^{-i}}$. Hence there exists a sequence $Z^{(\zeta)} \in G^i(t_1, \ldots, t_{m+1}, K)$, $\zeta < \ell_{m+1}^{2^{-i}}$, such that $Z^{(\zeta)} \sqsubset Y^{(\zeta)}$ and the position of each $Z^{(\zeta)}$ in $Y^{(\zeta)}$ is the same. Since each $Z^{(\zeta)} \in G^i(t_1, \ldots, t_{m+1}, K)$, we have $Z_i^{(\zeta)} \in S(t_1, \ldots, t_{m+1})$, for $i < K$. Thus, letting $\ell = \ell_{m+1}^{2^{-(i+1)}}$, we have

$$\bigcup_{\zeta < \ell} Z_i^{(\zeta)} \in C(t_1, \ldots, t_{m+1}, \ell)$$

(because $j_i^{(0)}, \ldots, j_i^{(\ell)}$ forms an arithmetic progression). Since we have $Y^{(\zeta)} \in D^i(t_1, \ldots, t_{m'}, K)$, for $\xi < \ell_{m'}$, we see that each $Y_j^{(\zeta)}$, for $j < i$, belongs to $P(t_1, \ldots, t_{m'})$ and consequently all the $Y_j^{(\zeta)}$, for $j < i$ and $\zeta < \ell_{m'}$, are R-equivalent. Since the position of each $Z^{(\zeta)}$ in $Y^{(\zeta)}$ is the same, for each $j < i$, $Z_j^{(\zeta)}$ is R-equivalent to $Z_j^{(\zeta')}$ for $\zeta, \zeta' < \ell_{m'}$. But $i + 1 < K$, so that $\ell \geq \ell_{m+1}^{2^{-K}}$. Thus the hypotheses of Proposition 7.5 are satisfied for the sequence $Z^{(\zeta)}$, $\zeta < \ell$. Proposition 7.5 asserts that there exists a $\xi^* < \ell$, such that $Z^{(\xi^*)} \in G^{i+1}(t_1, \ldots, t_{m+1}, K)$. Finally, we apply Proposition 7.6 to the single configuration $Z^{(\xi^*)}$ (in this case the R-equivalence condition is trivial). The conclusion of Proposition 7.6 implies that

$$D^{i+1}(t_1, \ldots, t_m, K) \neq \emptyset.$$

This completes the inductive step and Theorem 7.5 is proved. ∎

It remains to verify that Theorem 7.5 does in fact show that R contains arbitrarily long arithmetic progressions. According to the theorem there exists an $X \in D^{K-1}(K)$. By definition of $D^{K-1}(K)$, this just means that X is an arithmetic progression in which the first $K - 1$ terms belong to $P(\emptyset) = R$. Since K is arbitrary, R does indeed contain arbitrarily long arithmetic progressions.

The following corollary shows that all $\tau_k = 0$.

COROLLARY 7.3

For all $\varepsilon > 0$ and positive integers k, there exists a number $n_k(\varepsilon)$, such that if $n > n_k(\varepsilon)$ and R is a subset of $[0, n)$ with $|R| \geq \varepsilon n$, then R contains a k-progression.

PROOF   Suppose the corollary is false. Then there exist integers $n_1 < n_2 < \cdots$ and $R_i \subseteq [0, n_i)$ with $|R_i| > \varepsilon n_i$, for $i \geq 1$, such that R_i contains no k-progression. Let $n'_1 < n'_2 < \cdots$ be a subsequence of n_i satisfying $n'_{i+1} \geq 3n'_i$ for all i. Let $d_i = \sum_{j<i} n'_j$ and $R' = \bigcup_{i>0} (R_{n'_i} + \{d_i\})$. Thus

$$\limsup_{n \longrightarrow \infty} |R' \cap [0, n)|/n \geq \varepsilon.$$

By Theorem 7.5, R' must contain a $3k$-progression, say $A = \{a + di : i \in [0, 3k)\}$. Let ℓ satisfy $d_\ell \leq a + (3k-1)d < d_{\ell+1}$. Since $R_{n'_\ell} + \{d_\ell\}$ can contain at most $k - 1$ terms of A, we have $a + 2kd < d_\ell$. But if follows from the definition of n_i and d_i that $d_\ell \geq 3d_{\ell-1}$. Thus

$$a + kd \geq \frac{1}{3}(a + (3k-1)d) > \frac{1}{3}d_\ell \geq d_{\ell-1}.$$

However, this implies

$$\left| A \cap \left(R_{n'_{\ell-1}} + \{d_{\ell-1}\} \right) \right| > k,$$

which is impossible. Therefore the corollary holds.   ∎

7.11   Arithmetic Progressions of Squares

The set of all squares of integers has density 0. Therefore, Szemerédi's Theorem 7.5 says nothing about whether there are arbitrarily long arithmetic progression of squares. The example m^2, $(5m)^2$, $(7m)^2$ shows that there infinitely many 3-progressions of squares. In this section we will determine all 3-progressions of squares. In an exercise, the reader is asked to apply this result to show that there are no 4-progressions of squares. From this exercise and Szemerédi's Theorem 7.5, it follows that the squares have density 0, which we already knew.

Before we determine the 3-progressions of squares we need to determine when a square can be written as the sum of two squares. Of course $z^2 = z^2 + 0^2$, but we are interested in other representations of z^2 as the sum of two squares.

The solutions to $x^2 + y^2 = z^2$ in positive integers are called "Pythagorean triples." Solutions such as 3, 4, 5 and 5, 12, 13 have been known for thousands of years. The ancient Greeks knew that one could construct infinitely many solutions to $x^2 + y^2 = z^2$ by taking

$$x = r^2 - s^2,$$
$$y = 2rs,$$
$$z = r^2 + s^2,$$

where $r > s$ are positive integers. It is surprising that all Pythagorean triples arise this way.

We need one lemma to prove the theorem about Pythagorean triples.

LEMMA 7.19

If m and n are relatively prime positive integers whose product is an exact square, then each of m, n is an exact square.

PROOF   Let p be a prime divisor of m. Let p^α be the highest power of p that divides m. Then p^α divides mn. But $\gcd(m,n) = 1$, so $p \nmid n$ and p^α is the highest power of p that divides mn. Since mn is an exact square, α must be even. The same is true for every prime dividing m, and therefore m is an exact square. Similarly, n is an exact square.   ∎

We now begin to solve $x^2 + y^2 = z^2$. Note first that if x, y, z is one solution, then kx, ky, kz is a solution for any integer k. We will simplify our result with this definition.

DEFINITION 7.10   *A solution x, y, z to $x^2 + y^2 = z^2$ is **primitive** if $\gcd(x, y, z) = 1$.*

It is clear that if d is a positive integer with $d|x$ and $d|y$, then $d^2|x^2 + y^2 = z^2$, and so $d|z$. Similarly, if d divides any two of x, y, z, then d divides the third variable. Thus, if x, y, z is a primitive solution to $x^2 + y^2 = z^2$, then

$$\gcd(x,y) = \gcd(x,z) = \gcd(y,z) = \gcd(x,y,z) = 1.$$

Thus, x and y cannot both be even in a primitive solution. They cannot both be odd either because if they were, then $z^2 \equiv 2 \pmod 4$, which is impossible. As the equation $x^2 + y^2 = z^2$ is symmetric in x and y, we may assume that y is even, and x and z are odd.

We can now state the main theorem about Pythagorean triples.

THEOREM 7.6   Pythagorean Triples

The primitive solutions to $x^2 + y^2 = z^2$ in positive integers with y even are $x = r^2 - s^2$, $y = 2rs$, $z = r^2 + s^2$, where r and s are any integers with $r > s > 0$, $\gcd(r, s) = 1$ and $r \not\equiv s \pmod 2$.

PROOF   As we said above, if x, y, z is a primitive solution to $x^2 + y^2 = z^2$ in positive integers, then x and y must have opposite parity; we are assuming here that y is even and x is odd.

Now write $x^2 + y^2 = z^2$ in the form

$$\left(\frac{z+x}{2}\right)\left(\frac{z-x}{2}\right) = \left(\frac{y}{2}\right)^2.$$

Note that each fraction in parentheses is an integer because x and z are odd, while y is even. If a positive integer d divides both factors on the left side, then it divides their sum, z, and their difference, x. But $\gcd(z, x) = 1$, so d must be 1. Therefore, the two factors on the left side are relatively prime. Since their product is a square, Lemma 7.19 tells us that each is a square. It follows that there are positive integers r and s so that

$$\frac{z+x}{2} = r^2, \quad \frac{z-x}{2} = s^2, \quad \frac{y}{2} = rs.$$

We have $\gcd(r, s) = 1$ because $(z + x)/2$ and $(z - x)/2$ are relatively prime. Clearly $r > s$. Also, r and s have opposite parity because $r + s = z$, which is odd. When we solve for x, y and z in terms of r and s we get the equations in the statement of the theorem. ∎

To construct all integer solutions to $x^2 + y^2 = z^2$, first form all primitive solutions. Allow x and y to be swapped. Then multiply x, y, z by any integer k.

Now we will find the 3-progressions of squares. The squares x^2, y^2, z^2 form an arithmetic progressions of three terms if and only if $y^2 - x^2 = z^2 - y^2$, which is equivalent to $x^2 + z^2 = 2y^2$. As with the equation $x^2 + y^2 = z^2$, we can restrict attention to *primitive solutions*, those with $\gcd(x, y, z) = 1$. As before, if x, y, z is a primitive solution to $x^2 + z^2 = 2y^2$, then

$$\gcd(x, y) = \gcd(x, z) = \gcd(y, z) = \gcd(x, y, z) = 1.$$

Furthermore, if either of x, z were even, then the other would have to be even also because of the "2" in $x^2 + z^2 = 2y^2$. Therefore, both x and z are odd. It follows from considerations modulo 4 that y must be odd, too.

THEOREM 7.7 Three Squares in Arithmetic Progression
If x, y, z is a primitive solution to $x^2 + z^2 = 2y^2$ in positive integers, then there are relatively prime integers r and s, one even and one odd, with $r > s > 0$, so that

$$x = |r^2 - s^2 - 2rs|,$$
$$y = r^2 + s^2,$$
$$z = r^2 - s^2 + 2rs.$$

PROOF   Assume that x, y, z is a primitive solution to $x^2 + z^2 = 2y^2$. With no loss of generality, we may assume that $0 < x < y < z$. (If $x = y$, then $y = z$ and the arithmetic progression is constant. If $0 < z < y < x$, then the progression is descending.) Let $a = (z + x)/2$ and $b = (z - x)/2$. We see that a and b are positive integers, since x and z are odd. Since $a + b = z$, which is odd, one of a, b is even and the other is odd. Furthermore, $a^2 + b^2 = y^2$. If $\gcd(a, b, y) = 1$, we will have a primitive solution a, b, y to $a^2 + b^2 = y^2$. If d is a positive integer dividing both a and b, then d divides their sum $a+b = z$ and difference $a-b = x$. Hence, $d = 1$ since $\gcd(x, z) = 1$. Therefore $\gcd(a, b) = 1$, so we may apply Theorem 7.6. It says there are integers r, s, relative prime, of opposite parity and satisfying $r > s > 0$, so that if b is the even one of a, b, then

$$a = r^2 - s^2,$$
$$b = 2rs,$$
$$y = r^2 + s^2.$$

If instead a is even, then swap a and b in the equations just displayed.

Now use the formulas $z = a+b$ and $x = a-b$ to express x, y and z in terms of r and s. If $b = (z - x)/2$ is even, then we have

$$x = r^2 - s^2 - 2rs,$$
$$y = r^2 + s^2.$$
$$z = r^2 - s^2 + 2rs,$$

while if b is odd (and a is even), then the solution is

$$x = 2rs - r^2 - s^2,$$
$$y = r^2 + s^2.$$
$$z = 2rs + r^2 - s^2,$$

The absolute value bars in the theorem statement provide a uniform treatment of the two cases. ∎

COROLLARY 7.4

All arithmetic progressions $x^2 < y^2 < z^2$ of three squares of integers are given by

$$x = k|r^2 - s^2 - 2rs|,$$
$$y = k(r^2 + s^2),$$
$$z = k(r^2 - s^2 + 2rs),$$

where $r > s > 0$ are any relatively prime integers of opposite parity, and k is any positive integer.

7.12 Exercises

1. Show that for any integer $r > 1$, one can partition the set of positive integers into r sets, none of which contains an infinite arithmetic progression.

2. For which positive integers ℓ, m, r can you compute the $N(\ell, m, r)$ of Statement $S(\ell, m)$? Here "compute" means write the number on a sheet of paper, or at least write an explicit formula for it.

3. Prove Theorem 7.2.

4. Prove that $r_3(n) < cn / \log \log n$ for some constant $c > 0$.

5. Let α and β be real numbers, such that $0 \leq \alpha \leq \beta \leq 1$. Exhibit a set A of positive integers with lower asymptotic density α and upper density β that contains no infinite arithmetic progression.

6. Show that $r_\ell(\ell^2 - \ell - 1) = (\ell - 1)^2 - 1$ when ℓ is prime.

7. Let R be any set of positive integers with positive upper asymptotic density. Prove that for any positive integers t_1, ..., t_m, there is a subset X of R with $X \in B(t_1, \ldots, t_m)$.

8. Show that there are infinitely many 3-progressions of squares of integers with no common divisor > 1.

9. Prove that there are no four squares of integers in arithmetic progression.

10. Prove Sárközy's theorem: For all real $\varepsilon > 0$ there exists a number $N(\varepsilon)$, such that for all integers $n \geq N(\varepsilon)$, a and $d > 0$ we have

$$|\{1 \leq m \leq n : a + md \text{ is a square}\}| < \varepsilon n.$$

Explain why this statement does not follow immediately from the fact that the set of all squares has density 0.

11. Let S be the set of all positive integers that are the sum of two squares. Prove that $d(S) = 0$. Does S contain arbitrarily long arithmetic progressions?

Chapter 8

Applications

The notation $r_s(n)$ now resumes its meaning as the number of ways of writing n as the sum of s squares of integers. Likewise, $\sigma(n)$ again denotes the sum of the positive divisors of n.

This chapter gives several applications of the theory developed in this book. We explore the connection between factoring an integer and finding all its representations as the sum of two squares. We show that it is not necessary to know the factorization of an integer to express it as the sum of s squares when $s \geq 3$. We discover for what values of s and n one can determine $r_s(n)$ easily. Finally, we give three applications of sums of squares of integers to problems far removed from number theory, including microwave radiation, diamond cutting and cryptanalysis.

8.1 Factoring Integers

We will show that if you can express a positive integer as the sum of two squares in two different ways, then you can easily find a proper factor of the number.

Usually, this method is not a useful way to factor a large number. If n is a large number whose factorization is desired, then (a) n might not be the sum of two squares, and (b) even if n is the sum of two squares, it may be very difficult to find even one representation of it as the sum of two squares without already knowing the prime factorization of n. However, sometimes these objections do not apply. For example, if $n = a^2 + 1$, then n is clearly the sum of two squares, and we know one such representation. If we can find a different representation of n as the sum of two squares, then we can factor n easily. See Example 8.3 below, where we factor $2^{128} + 1 = \left(2^{64}\right)^2 + 1$, given a second representation of it as the sum of two squares.

THEOREM 8.1 Factoring Integers

There is a polynomial time algorithm to find a proper factor of a positive integer n, given two distinct representations

$$n = a^2 + b^2, \qquad n = c^2 + d^2,$$

of n as the sum of two squares of integers. By "distinct" we mean that $\{\pm a, \pm b\} \neq \{\pm c, \pm d\}$.

PROOF   If n were prime, then, by Corollary 2.1, n would have only one representation as the sum of two squares. Therefore, n is composite and has a proper factor.

With no loss of generality, we may assume that a, b, c and d are all nonnegative. If any of a, b, c and d were 0, then n would be a square and so $\sqrt{n}$ is a proper factor. Thus, we may assume all of a, b, c and d are positive integers.

We may also assume that $\gcd(a, b) = \gcd(c, d) = 1$ because if either gcd exceeded 1, then it (and also its square) would be a proper factor of n.

After these preliminaries, we claim that both of the numbers $\gcd(ad+bc, n)$, $\gcd(ad - bc, n)$ are proper factors of n. It is well known that we can compute these gcd's in polynomial time in the size of n.

Note that

$$(ad + bc)(ad - bc) = a^2 d^2 - b^2 c^2 = d^2(a^2 + b^2) - b^2(c^2 + d^2) = (d^2 - b^2)n,$$

which shows that n divides $(ad + bc)(ad - bc)$. Either n divides one of these two factors or else both $\gcd(ad + bc, n)$ and $\gcd(ad - bc, n)$ are proper factors of n. Clearly, $ad + bc > ad - bc$ because b and c are positive integers. The equation

$$n^2 = (a^2 + b^2)(c^2 + d^2) = (ac - bd)^2 + (ad + bc)^2$$

shows that $ad+bc \leq n$, and that $ad+bc < n$ unless $ac-bd = 0$. If $ad+bc = n$, then $ac = bd$. In that case, we have $a|bd$. But $\gcd(a, b) = 1$, so $a|d$. Similarly, $d|a$ and so $a = d$. Then $b = c$, and the representation $n = c^2 + d^2$ is the same as $n = a^2 + b^2$, but with a and b swapped. In other words, $n = c^2 + d^2$ is not distinct from $n = a^2 + b^2$. By assumption, this cannot be so. Therefore, $ad + bc < n$ and so $1 < \gcd(ad + bc, n) < n$.

Now $ad - bc < ad + bc < n$. Likewise, $bc - ad < ad + bc < n$, so $ad - bc > -n$. If $ad - bc = 0$, then $a|bc$, so $a|c$; similarly, $c|a$ and $c = a$, so the representations are not distinct, again contrary to hypothesis.

Therefore, both $\gcd(ad + bc) < n$ and $\gcd(ad - bc) < n$, so both are proper factors of n.   ∎

REMARK 8.1   The same trick works for quadratic forms other than $a^2 + b^2$. If we know two distinct representations of n by the same quadratic form $a^2 + Db^2$, then we can factor n easily. See Mathews [69].   ∎

Example 8.1

If we are given $50 = 5^2 + 5^2$ and $50 = 7^2 + 1^2$, then $\gcd(5,5) = 5 > 1$ is a proper factor of 50. If we ignored this, and found $ad + bc = 5 \cdot 7 + 5 \cdot 1 = 40$, then we would have $\gcd(40, 50) = 10$ as a proper factor of 50.

Example 8.2

Suppose we are given $377 = 11^2 + 16^2 = 19^2 + 4^2$. Then $ad + bc = 11 \cdot 4 + 16 \cdot 19 = 348$ and $\gcd(348, 377) = 29$. Also, $ad - bc = 11 \cdot 4 - 16 \cdot 19 = -260$ and $\gcd(-260, 377) = 13$.

Example 8.3

Suppose we wish to factor

$$n = 2^{128} + 1 = 340282366920938463463374607431768211457,$$

and we are told that $n = a^2 + b^2$, where

$$a = 16382350221535464479 \quad \text{and} \quad b = 8479443857936402504.$$

This gives only one representation of n as the sum of two squares. However, we get another such representation for free because of the special form of n:

$$n = 2^{128} + 1 = \left(2^{64}\right)^2 + 1 = 18446744073709551616^2 + 1.$$

We write the latter representation as $n = c^2 + d^2$ with

$$c = 1 \quad \text{and} \quad d = 18446744073709551616.$$

(This order keeps the numbers positive.) Then

$$ad + bc = 302201021862543689419343173647321450568$$

and

$$ad - bc = 302201021862543689402384285931448645560.$$

The gcd's are

$$\gcd(ad + bc, n) = 5704689200685129054721$$
$$\text{and } \gcd(ad - bc, n) = 59649589127497217.$$

One checks easily that these two gcd's are the prime factors of n.

REMARK 8.2   If one is given $\ell > 2$ distinct representations of n as the sum of two squares, then n has at least ℓ odd prime factors, and the method of Theorem 8.1 will lead to $\ell - 1$ different factorizations of n. ∎

8.2 Computing Sums of Two Squares

The previous section explained how to factor a composite integer n, given two distinct representations of n as the sum of two squares. In this section, we reverse the problem and tell how to find one or more representations of n as the sum of two squares, given the prime factorization of n.

Numbers of special form, such as $k^2 + 1$, can be written as the sum of two squares with no work at all. A less obvious example of this phenomenon comes from the Fibonacci numbers, $u_0 = 0$, $u_1 = 1$, $u_{n+1} = u_n + u_{n-1}$ for $n \geq 1$. The well-known identity $u_{2k+1} = u_k^2 + u_{k+1}^2$ provides a representation as the sum of two squares for any Fibonacci number with odd subscript. For example, $34 = u_9 = u_4^2 + u_5^2 = 3^2 + 5^2$.

The algorithms we give for the general case are fast and practical. Their running times are sometimes probabilistic. They work well in practice, although we do not prove before we start that we shall do it, to quote Oliver Atkin.

In this section we will answer these questions, assuming we know the prime factorization of n:

1. How do you tell whether $r_2(n) > 0$?

2. How do you find at least one representation of n as a sum of two squares, if $r_2(n) > 0$?

3. How do you find all representations of n as a sum of two squares, if $r_2(n) > 0$?

The first question is easy to answer using Theorem 2.5. If any prime $q \equiv 3 \pmod 4$ divides n to an odd power, then $r_2(n) = 0$; otherwise, $r_2(n) > 0$.

Now assume that no prime $q \equiv 3 \pmod 4$ divides n to an odd power, so that $r_2(n) > 0$. Let us find a representation of n as the sum of two squares. If n is the sum of two squares, then the prime factors of n are of three types:

1. $2 = 1^2 + 1^2$,

2. $p \equiv 1 \pmod 4$, with $p = a^2 + b^2$,

3. $p \equiv 3 \pmod 4$, with p^e exactly dividing n, where e is even.

In the first two cases, $p = 2$ or $p \equiv 1 \pmod 4$, we can use the identity,

$$(a^2 + b^2)(c^2 + d^2) = (ac \pm bd)^2 + (ad \mp bc)^2, \qquad (8.1)$$

attributed both to Diophantus and Brahmagupta, to combine these representations into a representation of p^e as the sum of two squares. We also have the representation $p^e = \left(p^{e/2}\right)^2 + 0^2$ in the third case because e is even. If we combine the representations of all prime powers p^e in the prime factorization of n using the identity (8.1) again, then we get representations of n. In fact,

by allowing a and b to be swapped, and c and d to be swapped, we obtain all representations of n as the sum of two squares, which answers the third question.

When the identity (8.1) is applied to the powers of 2, one obtains $2^e = \left(2^{(e-1)/2}\right)^2 + \left(2^{(e-1)/2}\right)^2$ when e is odd and $2^e = \left(2^{e/2}\right)^2 + 0^2$ when e is even.

When the identity (8.1) is applied to the even powers of a prime $q \equiv 3 \pmod 4$, it just gives the obvious representation $q^e = \left(q^{e/2}\right)^2 + 0^2$ when e is even.

But when the identity (8.1) is applied to the powers of a prime $p \equiv 1 \pmod 4$, it gives $4(e+1)$ representations of p^e as a sum of two squares. Hence we obtain all of the representations by this simple method.

According to Theorem 2.5, if p is a prime $\equiv 1 \pmod 4$, and e is a positive integer, then $r_2(p^e) = 4(e+1)$. Let us count the "distinct" representations $p^e = x^2 + y^2$ with $0 \le x \le y$. If e is even, then $x = 0$, $y = p^{e/2}$ is one representation and all others have $0 < x < y$ (because p^e cannot be twice a square). If e is odd, then all representations have $0 < x < y$. The representation $p^e = 0^2 + \left(p^{e/2}\right)^2$, when e is even, contributes 4 to $r_2(p^e)$. Each representation $p^e = x^2 + y^2$ with $0 < x < y$ contributes 8 to $r_2(p^e)$. Thus, p^e has $4(e+1)/8 = (e+1)/2$ representations as $p^e = x^2 + y^2$ with $0 < x < y$ (and none with $x = 0$) when e is odd. On the other hand, when e is even, p^e has one representation as $p^e = x^2 + y^2$ with $x = 0 < y$ and $(4(e+1) - 4)/8 = e/2$ representations as $p^e = x^2 + y^2$ with $0 < x < y$. In other words, p^e has $\lfloor (e+2)/2 \rfloor$ "distinct" representations $p^e = x^2 + y^2$ with $0 \le x \le y$.

Example 8.4

Let us find some representations of powers of 5 as the sum of two squares, beginning from $5 = 2^2 + 1^2$. We have

$$5 = 2^2 + 1^2 = 1^2 + 2^2,$$

so

$$5^2 = (2 \cdot 1 + 1 \cdot 2)^2 + (2 \cdot 2 - 1 \cdot 1)^2 = 4^2 + 3^2$$
$$5^2 = (2 \cdot 2 + 1 \cdot 1)^2 + (2 \cdot 1 - 1 \cdot 2)^2 = 5^2 + 0^2.$$

We can write $5^3 = 5 \cdot 5^2$ or $5^2 \cdot 5$. Using one of $2^2 + 1^2$, $1^2 + 2^2$ for 5 and one of $4^2 + 3^2$, $3^2 + 4^2$, $5^2 + 0^2$, $0^2 + 5^2$ for 5^2, the identity (8.1) gives us

$$5^3 = (2 \cdot 3 + 1 \cdot 4)^2 + (2 \cdot 4 - 1 \cdot 3)^2 = 10^2 + 5^2$$
$$5^3 = (2 \cdot 4 + 1 \cdot 3)^2 + (2 \cdot 3 - 1 \cdot 4)^2 = 11^2 + 2^2.$$

We can write $5^4 = 5 \cdot 5^3$, $5^2 \cdot 5^2$ or $5^3 \cdot 5$. Using the expressions already obtained for 5^1, 5^2 and 5^3 as the sum of two squares, the identity (8.1) gives us

$$5^4 = (2 \cdot 11 + 1 \cdot 2)^2 + (2 \cdot 2 - 1 \cdot 11)^2 = 24^2 + 7^2$$
$$5^4 = (2 \cdot 2 + 1 \cdot 11)^2 + (2 \cdot 11 - 1 \cdot 2)^2 = 15^2 + 20^2$$
$$5^4 = (2 \cdot 10 + 1 \cdot 5)^2 + (2 \cdot 5 - 1 \cdot 10)^2 = 25^2 + 0^2.$$

We are left with the problem of representing a prime $p \equiv 1 \pmod 4$, to the first power, as the sum of two squares. There are several ways to do this.

If p is small enough, one can just try all $1 \le a < \sqrt{p}$ to see whether $p - a^2$ is a square. If it equals b^2, then we have $p = a^2 + b^2$. This algorithm has time complexity $O(\sqrt{p})$.

Even slower algorithms use the formulas in these two beautiful theorems. In each theorem, $p = 4k + 1$ is prime, and $p = a^2 + b^2$, where $a \equiv 1 \pmod 4$. Both of these algorithms have time complexity $O(p)$.

THEOREM 8.2 Gauss, pages 90–91 and 168 of [34]
The number a is determined by

$$a \equiv \frac{1}{2}\binom{2k}{k} \pmod p \text{ and } |a| < \frac{p}{2}.$$

Example 8.5

Let $p = 13 = 4 \cdot 3 + 1$. Then

$$a \equiv \frac{1}{2}\binom{2 \cdot 3}{3} = \frac{1}{2}\binom{6}{3} = 10 \equiv -3 \pmod{13}.$$

Thus, $13 = (-3)^2 + b^2$ for some integer b, and we find $13 = (-3)^2 + 2^2 = 3^2 + 2^2$.

THEOREM 8.3 Jacobsthal [58]
The number a is given by the quadratic character sum

$$a = -\sum_{j=1}^{(p-1)/2}\left(\frac{j}{p}\right)\left(\frac{j^2 + 1}{p}\right),$$

Example 8.6

Let $p = 13$. Then $(p - 1)/2 = 6$ and

$$a = -\sum_{j=1}^{6}\left(\frac{j}{13}\right)\left(\frac{j^2 + 1}{13}\right)$$
$$= -(1 \cdot (-1) + (-1) \cdot (-1) + 1 \cdot 1 + 1 \cdot 1 + (-1) \cdot 0 + (-1) \cdot (-1))$$
$$= -3.$$

In fact, there is a probabilistic polynomial time algorithm due to Hermite [54] for expressing a prime $p \equiv 1 \pmod 4$ as the sum of two squares. (His paper immediately followed a paper of Serret [99] on the same subject; hence its title.)

Here is Hermite's algorithm as described by Lehmer [66].

1. Find x with $x^2 \equiv -1 \pmod{p}$ and $0 < x < p/2$.

2. Use the Euclidean algorithm to expand x/p into a simple continued fraction until the denominators of the convergents A'_n/B'_n satisfy $B'_{k+1} < \sqrt{p} < B'_{k+2}$. Then

$$p = (xB'_{k+1} - pA'_{k+1})^2 + (B'_{k+1})^2.$$

The Euclidean algorithm used in the second step is the same one used to compute $\gcd(x, p)$ and is well known to run in polynomial time. We will say more later about the complexity of finding x in the first step.

Here is why the algorithm works. See pages 32–34 of Perron [83] for the needed facts about continued fractions. The theory of continued fractions tells us that

$$\frac{x}{p} - \frac{A'_{k+1}}{B'_{k+1}} = \frac{\vartheta}{B'_{k+1}B'_{k+2}}, \quad \text{with } |\vartheta| < 1.$$

Multiply by pB'_{k+1} and square to get

$$\left(xB'_{k+1} - pA'_{k+1}\right)^2 = \frac{\vartheta^2 p^2}{\left(B'_{k+2}\right)^2} < p$$

because $\sqrt{p} < B'_{k+2}$. Add $\left(B'_{k+1}\right)^2$ to get

$$0 < \left(xB'_{k+1} - pA'_{k+1}\right)^2 + \left(B'_{k+1}\right)^2 < p + \left(B'_{k+1}\right)^2 < 2p,$$

since $B'_{k+1} < \sqrt{p}$. The left side of this inequality is a multiple of p because $x^2 \equiv -1 \pmod{p}$. Therefore it must equal p, and we have expressed p as the sum of two squares.

Brillhart [12] observes that one need not compute the convergents in the second step of the algorithm because the answer appears also in the remainders in the continued fraction. In the second step of his version of the algorithm one performs the Euclidean algorithm on p/x and obtains the remainders r_1, r_2, ..., until the first remainder r_k less than $\sqrt{p}$. Perform one more step of the Euclidean algorithm to get r_{k+1}. Then

$$p = \begin{cases} r_k^2 + r_{k+1}^2 & \text{if } r_1 > 1, \\ x^2 + 1 & \text{if } r_1 = 1. \end{cases}$$

Here is why Brillhart's version of the algorithm works. See pages 32–34 of Perron [83] for the properties of continued fractions.

Assume first that the first remainder $r_1 > 1$. Because $p|(x^2 + 1)$ and $0 < x < p/2$, we have

1. The continued fraction expansion of p/x is palindromic of even length:

$$\frac{p}{x} = [q_0, q_1, \ldots, q_k, q_k, \ldots, q_1, q_0] = \frac{A_{2k+1}}{B_{2k+1}},$$

where the q_i are the quotients. The convergents A'_{n+1}/B'_{n+1} for x/p are the reciprocals of the convergents A_n/B_n for p/x.

2. $A_{2k+1} = p$ and $A_{2k} = x$.

3. $p = A_k^2 + A_{k-1}^2$.

4. The recursion formula for the numerators A_n gives the equations

$$p = q_0 x + A_{2k-1}, \qquad x = q_1 A_{2k-1} + A_{2k-2}, \qquad \dots.$$

The last equations are the same as those for the remainders in the Euclidean algorithm for p/x. Hence, $A_{2k-1} = r_1$, $A_{2k-2} = r_2$, ..., $A_{k+1} = r_{k-1}$, $A_k = r_k$, $A_{k-1} = r_{k+1}$, ... From the third property, we have $p = r_k^2 + r_{k+1}^2$. Since $k > 1$, the first property gives $r_{k-1} = A_{k+1} = B'_{k+2}$, which is $> \sqrt{p}$ by Hermite's version of the algorithm. Therefore, r_k is the first remainder $< \sqrt{p}$.

If $r_1 = 1$, then $p = q_0 x + 1$ and the continued fraction is $[q_0, q_0]$. Therefore, $q_0 = x$ and $p = x^2 + 1$.

Since both versions of the algorithm consist of just the Euclidean algorithm, both have polynomial time complexity *after* we find x.

We now turn to the question of finding a square root x of -1 modulo p. It is here that the algorithm may become probabilistic. First, there is an x with $x^2 \equiv -1 \pmod{p}$ because $p \equiv 1 \pmod 4$. If we knew a quadratic nonresidue q modulo p, then we could let $x \equiv \pm q^{(p-1)/4} \pmod{p}$ with $0 < x < p/2$. In some cases we can exhibit a small quadratic nonresidue modulo p directly. For example, if $p \equiv 5 \pmod 8$, then $q = 2$ is a quadratic nonresidue. Likewise, if $p \equiv 17 \pmod{24}$, then $q = 3$ will work. In the remaining case, $p \equiv 1 \pmod{24}$, we can use this algorithm:

Try successive primes $q = 5, 7, 11, \dots$, until one is found with $q^{(p-1)/4} \not\equiv 1$ or $-1 \pmod{p}$. Then $x \equiv \pm q^{(p-1)/4} \pmod{p}$ with $0 < x < p/2$.

This algorithm works because of Euler's criterion. If q were a quadratic residue modulo p, then $q^{(p-1)/2} \equiv +1 \pmod{p}$ and so $q^{(p-1)/4} \equiv +1$ or $-1 \pmod{p}$. Hence, if $q^{(p-1)/4} \not\equiv 1$ or $-1 \pmod{p}$, then $q^{(p-1)/2} \equiv -1 \pmod{p}$, q is a quadratic nonresidue modulo p, and $x \equiv \pm q^{(p-1)/4} \pmod{p}$ with $0 < x < p/2$ is the square root of 1 that we need.

How many q's might we have to try to get a quadratic nonresidue? Burgess [14] has shown that for every $\varepsilon > 0$ there is a $p_0(\varepsilon)$ so that the least positive quadratic nonresidue modulo p is $< p^{c+\varepsilon}$ for all primes $p > p_0(\varepsilon)$, where $c = 1/(4\sqrt{e}) \approx 0.1516$. A similar bound holds for the number of consecutive integers we must try, beginning at any number, until we find the first quadratic nonresidue. Even these results seem far from the truth. Vinogradov conjectured that for every $\varepsilon > 0$ there is a $p_0(\varepsilon)$ so that the least positive quadratic nonresidue modulo p is $< p^\varepsilon$ for all primes $p > p_0(\varepsilon)$. Although one cannot prove a good upper bound for the least positive quadratic nonresidue, the *average* number of positive integers which must be tried to find a quadratic nonresidue modulo p is 2.

Example 8.7

If $p = 17$, then $x = 4$. The Euclidean algorithm has just one step,

$$17 = 4 \cdot 4 + 1,$$

so $q_0 = x$ and $17 = 4^2 + 1$.

Example 8.8

Let $p = 1217$. Since $p \equiv 17 \pmod{24}$, we can use $q = 3$ as a quadratic nonresidue. This gives $x = q^{(p-1)/4} = 3^{304} \equiv 78 \pmod{p}$. Since $0 < x = 78 < 1217/2$, we do not have to replace x by $p-x$. Note that $x^2 = 78^2 \equiv -1 \pmod{p}$. The first steps of the Euclidean algorithm for $1217/78$ are

$$1217 = 78 \cdot 15 + 47$$
$$78 = 47 \cdot 1 + 31$$
$$47 = 31 \cdot 1 + 16.$$

We have $31 < \sqrt{p} < 47$, so 31 is the first remainder $< \sqrt{p}$. Hence, $p = 1217 = 31^2 + 16^2$.

We remark that 50 is the smallest integer which can be written as the sum of two positive squares in two different ways, $50 = 1^2 + 7^2 = 5^2 + 5^2$, and 65 is the smallest integer which can be written as the sum of two different positive squares in two different ways, $65 = 1^2 + 8^2 = 4^2 + 7^2$.

Finally, it follows easily from Theorem 2.5 that a positive integer n is the sum of two *positive* squares if and only if $n = 2^e n_1 n_2^2$, where e is a nonnegative integer, n_1 is an integer > 1 all of whose prime factors are $\equiv 1 \pmod 4$ and n_2 is the product of zero or more primes $\equiv 3 \pmod 4$.

8.3 Computing Sums of Three Squares

Using Theorem 2.7, it is easy to tell when an integer n is the sum of three squares: Remove all factors of 4 from n, say $n = 4^e m$, where m is not divisible by 4. Then n is the sum of three squares if and only if $m \not\equiv 7 \pmod 8$, that is, $m \equiv 1, 2, 3, 5$ or $6 \pmod 8$. Thus we can easily tell whether $r_3(n) > 0$.

How can we compute a representation of n as a sum of three squares, if there is one? Unlike the case of two squares, we need not know the factors of n. We can solve the problem even for large n whose factorization is unknown.

The proof of Lemma 2.3 gives an explicit correspondence between representations of $4n$ as the sum of three squares and those of n. Thus we may ignore any factors of 4 and just treat the cases $n \equiv 1, 2, 3, 5, 6 \pmod 8$.

Our plan is to subtract a square z^2 from n so that the difference $n - z^2$ is a prime $\equiv 1 \pmod 4$. Then we apply the previous section to this prime to express it as the sum of two squares and get our representation of n.

If $n \equiv 1$ or $5 \pmod 8$, subtract $(2k)^2$ for $k = 0, 1, 2, \ldots$, from n until you hit a prime $n - (2k)^2 = p \equiv 1 \pmod 4$. Use the previous section to write $p = x^2 + y^2$. Then $n = (2k)^2 + x^2 + y^2$.

If $n \equiv 2$ or $6 \pmod 8$, subtract $(2k + 1)^2$ for $k = 0, 1, 2, \ldots$, from n until you hit a prime $n - (2k + 1)^2 = p \equiv 1 \pmod 4$. Use the previous section to write $p = x^2 + y^2$. Then $n = (2k + 1)^2 + x^2 + y^2$.

If $n \equiv 3 \pmod 8$, subtract $(2k + 1)^2$ for $k = 0, 1, 2, \ldots$, from n until you hit *twice* a prime $n - (2k + 1)^2 = 2p$ with $p \equiv 1 \pmod 4$. Use the previous section to write $p = x^2 + y^2$. Then $n = (2k + 1)^2 + (x + y)^2 + (x - y)^2$.

These algorithms are probabilistic in that we cannot say in advance how many squares we will have to try before the difference is prime (or twice a prime in the third case), or even prove that they succeed. The numbers we test for primality will be less than n. The probability that any such number is prime is about $1/\ln n$. The expected number of squares we will have to try is therefore $\ln n$. (A sieve may speed this search by a constant factor.) We need not *prove* that any "prime" we hit really is prime. It suffices to have a probable prime. If the algorithm of the previous section does not produce a representation of a probable prime $\equiv 1 \pmod 4$ as the sum of two squares, then it was not prime, and we return to subtracting more squares.

If we reach a negative number when subtracting squares from n, then n must be small. In this case we start over and use direct search to try to express each $n - x^2$ as $y^2 + z^2$. This takes $O(n)$ steps but doesn't affect the practical running time because we do this only for tiny n.

We saw in the previous section that all the other steps run in probabilistic polynomial time. Thus our algorithms have probabilistic polynomial time complexity.

The construction in the proof of Corollary 2.3 shows that writing a positive integer n as the sum of three triangular numbers is equivalent to writing $8n + 3$ as the sum of three squares, and we just solved the latter problem.

We will give one comprehensive example to illustrate most of the ideas in this section and the previous section.

Example 8.9

Express $n = 139335$ as the sum of three triangular numbers.

Our first task is to write $8n + 3 = 1114683$ as the sum of three squares. Since $8n + 3 \equiv 3 \pmod 8$, we subtract odd squares from it until we reach twice a (probable) prime. We soon find $8n + 3 - 11^2 = 2p$, where $p = 557281$ is probably prime.

Now we have to write $p = 557281$ as the sum of two squares. Since $p \equiv 1 \pmod{24}$ there is no obvious quadratic nonresidue. The first two primes, 2, 3, don't work, and the next two primes, 5, 7, are quadratic residues, too. But the Legendre symbol $(11/p) = -1$ so $q = 11$ is a quadratic nonresidue modulo p.

We compute $x \equiv 11^{(p-1)/4} = 11^{139320} \equiv 252875 \pmod p$, with $0 < 252875 < p/2$, so $x = 252875$ is the solution to $x^2 \equiv -1 \pmod p$ that we can use.

The Euclidean algorithm for p/x (or for $\gcd(p, x)$) begins this way.

$$p = 557281 = 252875 \cdot 2 + 51531$$
$$252875 = 51531 \cdot 4 + 46751$$
$$51531 = 46751 \cdot 1 + 4780$$
$$46751 = 4780 \cdot 9 + 3731$$
$$4780 = 3731 \cdot 1 + 1049$$
$$3731 = 1049 \cdot 3 + 584$$
$$1049 = 584 \cdot 1 + 465.$$

We have $584 < \sqrt{p} < 1049$, so $p = 584^2 + 465^2$.

To write $2p = a^2 + b^2$, we use $a = 584 - 465 = 119$ and $b = 584 + 465 = 1049$. Hence,

$$8n + 3 = 1114683 = 2p + 11^2 = 119^2 + 1049^2 + 11^2 = a^2 + b^2 + c^2,$$

say. Finally, we use the formula from the proof of Corollary 2.3,

$$n = \frac{x^2 + x}{2} + \frac{y^2 + y}{2} + \frac{z^2 + z}{2} = \frac{x(x+1)}{2} + \frac{y(y+1)}{2} + \frac{z(z+1)}{2}$$

(with a new meaning for x), where

$$x = \frac{a - 1}{2} = 59, \qquad y = \frac{b - 1}{2} = 524, \qquad z = \frac{c - 1}{2} = 5,$$

to express $n = 139335$ as the sum of three triangular numbers.

Representation of n as a sum of k k-gonal numbers with $k > 4$ is easy, too, and left as an exercise.

Grosswald, Calloway and Calloway [43] show that there is a finite set S of m integers, such that every integer n is the sum of three positive squares if and only if n is not of the form $n = 4^e n_1$ with integer $e \geq 0$ and $n_1 \equiv 7 \pmod 8$ (in which case n is not the sum of three squares at all) and n is not of the form $n = 4^e n_1$ with integer $e \geq 0$ and $n_1 \in S$ (in which case n is the sum of three squares, but at least one of the squares must be 0^2 in any such representation). They conjecture that $m = 10$ and

$$S = \{1, 2, 5, 10, 13, 25, 37, 58, 85, 130\}.$$

8.4 Computing Sums of Four Squares

By Theorem 2.6, every positive integer is the sum of four squares. One standard proof of this theorem, given in many number theory texts, shows first that every prime is the sum of four squares, and then appeals to Euler's identity,

$$(a^2 + b^2 + c^2 + d^2)(e^2 + f^2 + g^2 + h^2) = w^2 + x^2 + y^2 + z^2,$$

where

$$w = ae + bf + cg + dh, \qquad x = af - be \pm ch \mp dg,$$
$$y = ag \mp bh - ce \pm df, \qquad z = ah \pm bg \mp cf - de,$$

which shows that the product of two numbers, each of which is the sum of four squares, is also the sum of four squares.

One might surmise from this proof that it is necessary to know the prime factorization of n in order to find a representation of n as the sum of four squares. However, we will show that there is a probabilistic polynomial time algorithm to find such a representation, without knowing the factorization of n.

First remove any factors of 4 from n. We assume this has been done and so $n \equiv 1, 2$ or $3 \pmod 4$.

If $n \equiv 1$ or $2 \pmod 4$, then one of the four squares can be taken to be 0. Use results of the previous section to write n as the sum of three squares.

In case $n \equiv 2 \pmod 4$, an alternative approach is to write $n = p + q$, where p and q are primes with $p \equiv q \equiv 1 \pmod 4$. Write p and q each as the sum of two squares by Brillhart's variation of Hermite's algorithm.

If $n \equiv 3 \pmod 4$, then $n - 1^2 \equiv 2 \pmod 4$ is the sum of three squares, so use results of the previous section to get $n - 1 = x^2 + y^2 + z^2$.

The situation for sums of four or more positive squares was settled by Pall.

THEOREM 8.4 Pall [82] Sums of Four or More Positive Squares
Let $B = \{1, 2, 4, 5, 7, 10, 13\}$. For $s \geq 6$, every integer n is the sum of s positive squares, except for $1, 2, \ldots, s - 1$ and $s + b$ with $b \in B$. For $s = 5$ the same statement holds, but with $b \in B \cup \{28\}$. For $s = 4$ the same statement holds, but with $b \in B \cup \{25, 37\}$, except for $n = 4^e n_1$ with $n_1 \in \{2, 6, 14\}$.

If you want to find all representations of a large integer n as the sum of four squares, there probably are too many of them to list. However, see the next section for how to count them, that is, compute $r_4(n)$.

If you want the squares to satisfy other conditions, you are on your own. However, sums of squares of primes has been done. See for example Liu and Zhan [68] and its references.

Sprague [106] has shown that every integer > 128 is the sum of (some number of) distinct squares. In fact, his proof shows that there are exactly 32 integers, $2, 3, 6, 7, 8, 11, \ldots, 128$, that are not the sum of distinct squares.

8.5 Computing $r_s(n)$

For what values of s and n can we compute $r_s(n)$ in (probabilistic) polynomial time?

One purpose of formulas proved in this book is to express $r_s(n)$ in terms of easily computed functions. Some of these functions, like divisor functions, are easy to compute provided you know the factorization of n.

The case $s = 1$ is trivial.

The case $s = 2$ is easy for any n that you can factor. Theorem 2.5 expresses $r_2(n)$ as $4T(n)$, where $T(n)$ is a multiplicative function with the value 1 for $n = 2^e$ or $n = q^{2e}$, 0 for $n = q^{2e+1}$, and $e+1$ for $n = p^e$, where $q \equiv 3 \pmod 4$ and $p \equiv 1 \pmod 4$. The value of $r_2(n)$ depends only on the number of primes $\equiv 1 \pmod 4$ dividing n, their multiplicities, and whether any prime $\equiv 3 \pmod 4$ divides n to an odd power. It does not depend on the actual values of these primes.

Example 8.10

Find $r_2(n)$ for $n = 2^3 p_1^2 p_2^3 p_3^4 q^6$, where $q \equiv 3 \pmod 4$ and the $p_i \equiv 1 \pmod 4$ are distinct primes.

The power of 2 is irrelevant. Since q^6 exactly divides n, it contributes 1, rather than 0, as a factor of $r_2(n)$. The p_i's contribute factors of $2+1$, $3+1$ and $4+1$ to $r_2(n)$. Therefore, $r_2(n) = 4T(n) = 4 \cdot 3 \cdot 4 \cdot 5 = 240$.

Computing $r_3(n)$ is more troublesome. Gauss' theorem expresses $r_3(n)$ in terms of a class number. For small n, the class number may be computed efficiently by listing all the reduced forms. One can also count the representations of n as the sum of three squares by generating all of them, if n is small. These algorithms take exponential time in $\log n$. Computing $r_3(n)$ is essentially equivalent to computing the class number of a quadratic field. Cohen [18] describes subexponential time algorithms for doing this in both the imaginary (Hafner and McCurley [44]) and real (Buchmann) cases. These algorithms, which are too complicated to present here, allow one to compute $r_3(n)$ when n is as large as about 50 decimal digits. But we don't know how to compute $r_3(n)$ when n has 200 decimal digits, even if the prime factorization of n is known. (However, we can tell whether $r_3(n) = 0$, even for very large n, by Theorem 2.7.)

Formulas (6.12) and (6.13) allow one to compute the approximate values of $r_5(n)$ and $r_7(n)$, respectively, when n is large.

For $s = 4, 6, 8, \ldots, 16$ and 24, $r_s(n)$ can be expressed in terms of "divisor" functions, which are easy to compute provided the factorization of n is known. For example,

$$\text{if } n = \prod_{p \text{ prime}} p^\alpha, \text{ then } \sigma_k(n) \overset{\text{def}}{=} \sum_{d|n} d^k = \prod_{p \text{ prime}} \frac{p^{k(\alpha+1)} - 1}{p^k - 1}.$$

Example 8.11

Find $r_4(n)$ when

$$n = 2381^3 \cdot 87881 \cdot 354047 = 419985305104760206787.$$

Since n is odd, Jacobi's theorem, Theorem 2.6, gives

$$r_4(n) = 8\sigma(n) = 8\sigma_1(n) = 8(1 + 2381 + 2381^2 + 2381^3)(1 + 87881)(1 + 354047)$$
$$= 3361341898424587272192.$$

Example 8.12

Find $r_8(n)$ when $n = 2381 \cdot 87881 = 209244661$.
Since n and all of its divisors d are odd, Formula (6.4) tells us that

$$r_8(n) = 16\sigma_3(n) = 16(1 + 2381^3)(1 + 87881^3) = 14658284253348758555 3087424.$$

One theme of Chapter 6 was computing $r_s(n)$ approximately. Hardy's Theorem 6.1 tells us that the singular series $\rho_s(n)$ is a good approximation for $r_s(n)$ for $s > 4$, and $\rho_s(n)$ may be computed approximately by Theorems 6.3 and 6.4. These theorems (and others for $s = 3$ and 4) say, roughly, that when $s > 2$, $\rho_s(n)$ is approximately a constant times $n^{(s/2)-1}$ times a factor close to 1 that depends on n. This is easy to see geometrically, for $r_s(n)$ counts the points with integer coordinates on the surface of the sphere with radius $\sqrt{n}$ in s-dimensional space. The volume of this sphere is approximately equal to the number of lattice points within it, and this is the total number of representations of integers $\leq n$ as the sum of s squares, that is,

$$\left| \left\{ (x_1, \ldots, x_s) \in \mathbb{Z}^s : \sum_{i=1}^{s} x_i^2 \leq n \right\} \right| = \sum_{k=0}^{n} r_s(k).$$

Thus the average value of $r_s(k)$ for $0 \leq k \leq n$ is just the volume of the sphere with radius $\sqrt{n}$ in s-dimensional space divided by n. According to page 9 of [20] this volume is $\pi^{s/2}(\sqrt{n})^s/((s/2)\Gamma(s/2))$, so the average value of $r_s(k)$ for $0 \leq k \leq n$ is

$$\frac{2(\pi n)^{s/2}}{ns\Gamma(s/2)}.$$

Compare this volume with the coefficient of the singular series in Equation (6.17).

8.6 Resonant Cavities

Baltes and Kneubühl [8] studied microwave radiation in finite cavities. They define a cavity as a bounded open connected set G with boundary F in Euclidean space $\mathbb{R}^3$ with volume V. It is known that if G is empty (that is, without matter), has perfectly reflecting walls and F is smooth, then the free electromagnetic oscillations $\mathbf{E}$, $\mathbf{H}$ in G obey the Helmholtz equations

$$\left(\nabla^2 + K^2\right) \mathbf{E} = \mathbf{0} \text{ and } \left(\nabla^2 + K^2\right) \mathbf{H} = \mathbf{0},$$

and the divergence conditions

$$\nabla \cdot \mathbf{E} = 0, \qquad \nabla \cdot \mathbf{H} = 0$$

throughout G, and the boundary conditions

$$\mathbf{n} \times \mathbf{E} = \mathbf{0}, \qquad \mathbf{n} \cdot \mathbf{H} = 0$$

on F, where $\mathbf{n}$ is the normal vector to F.

Under these conditions, one can show that the oscillations have an infinite number of eigenfrequencies, ν_n. They are all real, positive, and they have no finite accumulation point. The corresponding eigenfunctions $\mathbf{E}_n$ and $\mathbf{H}_n$ form complete orthonormal sets, are continuous in $G \cup F$ and have continuous derivatives in G.

Baltes and Kneubühl [8] studied several shapes for the cavity G. From our point of view, the most interesting shape is the cube, whose analysis relies on facts about sums of squares. They let $N(\nu)$ denote the number of modes with frequencies $\nu_n \leq \nu$ and wrote

$$N(\nu) = \sum_{\nu_n < \nu} G_n + \frac{1}{2} \sum_{\nu_n = \nu} G_n, \tag{8.2}$$

where G_n is the finite weight (multiplicity) of the resonance ν_n. A hundred years ago, Jeans found the approximation

$$N(\nu) \approx \frac{8\pi}{3} V \left(\frac{\nu}{c}\right)^3,$$

where c is the speed of light.

Consider a cube-shaped cavity with edge L and volume $V = L^3$. Let W be the parameter

$$W = \frac{2L}{\lambda} = \frac{2L\nu}{c},$$

where λ is the wavelength, and let $Q = W^2$. We shall also write $W_n = 2L\nu_n/c$. Then the resonances are determined by

$$Q = W^2 = n_1^2 + n_2^2 + n_3^2,$$

where the n_i are integers. We may assume that all $n_i \geq 0$.

Baltes and Kneubühl [8] separated these resonances into E- and H-type modes. They note that one obtains nonvanishing E- and H-type modes when all three n_i are positive. However, if $n_1 = 0$ or $n_2 = 0$, then the E-type solution vanishes. In case $n_3 = 0$, the H-type solution is zero. Finally, if more than one of the three n_i are zero, then both solutions vanish. Thus the total multiplicities are

$$G_n = G(n_1, n_2, n_3) = \begin{cases} 2 & \text{if no } n_i \text{ is } 0, \\ 1 & \text{if one } n_i \text{ is } 0, \\ 0 & \text{if two or three } n_i \text{ are } 0. \end{cases}$$

Combining this formula with (8.2), we get

$$N(\nu) = \sum_{(a)} G(n_1, n_2, n_3) + \frac{1}{2} \sum_{(b)} G(n_1, n_2, n_3),$$

where

$$\begin{aligned} (a): \quad & n_1^2 + n_2^2 + n_3^2 < Q, \\ (b): \quad & n_1^2 + n_2^2 + n_3^2 = Q. \end{aligned}$$

Define $F_s(Q)$ to be the number of lattice points inside the s-dimensional sphere with center at the origin and radius Q, where each lattice point on the surface of the sphere counts as one-half point. Then

$$F_s(Q) = \sum_{(c)} 1 + \frac{1}{2} \sum_{(d)} 1,$$

where

$$\begin{aligned} (c): \quad & n_1^2 + \cdots + n_s^2 < Q, \\ (d): \quad & n_1^2 + \cdots + n_s^2 = Q. \end{aligned}$$

In terms of $F_3(Q)$, with $Q = W^2$ and $W = 2L\nu/c$, the number of modes is

$$N(\nu) = \frac{1}{4} F_3(Q) - \frac{3}{2} \sqrt{Q} + \frac{1}{2}.$$

Baltes and Kneubühl [8] also considered the two-dimensional analogue of this problem with

$$Q = n_1^2 + n_2^2, \qquad \text{with all } n_i > 0.$$

For this problem the number of modes is

$$N^{(2)}(Q) = \frac{1}{4} F_2(Q) - \sqrt{Q} + \frac{1}{4}.$$

Another interesting problem considered in their article is the three-dimensional scalar Dirichlet problem in which

$$Q = n_1^2 + n_2^2 + n_3^2, \qquad \text{with all } n_i > 0.$$

For this problem the number of modes is

$$N^{(3)}(Q) = \frac{1}{8}F_3(Q) - \frac{3}{8}F_2(Q) + \frac{3}{4}\sqrt{Q} - \frac{1}{8}.$$

It follows from the definition of $F_s(Q)$ that

$$F_s(Q) = \sum_{n=0}^{Q-1} r_s(n) + \frac{1}{2}r_s(Q).$$

The smallest value of ϑ for which the estimate

$$F_2(Q) = \pi Q + O(Q^{\vartheta+\varepsilon})$$

holds for all $\varepsilon > 0$ is known to satisfy

$$\frac{1}{4} \le \vartheta < \frac{12}{37}.$$

Thus we have

$$N^{(2)}(Q) = \frac{\pi}{4}Q - \sqrt{Q} + O(Q^{\vartheta+\varepsilon}).$$

Counting the lattice points inside the sphere of radius Q centered at the origin leads to the estimate

$$F_3(Q) = \frac{4\pi}{3}Q^{3/2} + O(Q^{\mu})$$

for $\mu = 1$. Several number theorists have reduced the exponent μ. Vinogradov showed that we may take $\mu = 19/28 + \varepsilon$. This shows that the total number of modes is

$$N(\nu) = \frac{\pi}{3}Q^{3/2} + O(Q^{19/28+\varepsilon})$$

for the electromagnetic problem, and

$$N^{(3)}(Q) = \frac{\pi}{6}Q^{3/2} - \frac{3}{8}Q + O(Q^{19/28+\varepsilon}) \tag{8.3}$$

for the three-dimensional scalar Dirichlet problem.

Exercise 19 of Chapter 2 shows that five-sixths of the positive integers can be written as the sum of three squares. Hence the average distance between consecutive integers with this property is $\Delta Q = \Delta(W^2) = 6/5$. Therefore, $\Delta W = (6/5)/2W = 3/5W$. Using Equation (8.3) one finds that the mean degeneracy of an eigenfrequency W is $\overline{G}(W) = 3\pi W/5$.

In (8.3), only the lattice points (n_1, n_2, n_3) with all three $n_i > 0$ are counted. Therefore, we must discard those Q which can be represented as the sum of three squares if and only if at least one n_i is zero. This problem was considered by Grosswald, Calloway and Calloway [43] and discussed in Section

8.3. Exercise 11 of this chapter shows that when these numbers are discarded, the asymptotic density is unchanged at 5/6, so that (8.3) remains correct.

Essentially the same formulas arise in a problem in nonrelativistic quantum statistical mechanics of an ideal gas at low temperature. The quantum statistician is interested in the asymptotic form of $N^{(3)}(Q)$ with more precision than the first term $\pi Q^{3/2}/6$, and in particular with the difference $\Delta W = (6/5)/2W = 3/5W$. Thus, (8.3) is important in statistical mechanics as well as in electromagnetism. See [7] for more about how the theorem of Grosswald, Calloway and Calloway [43] is used in statistical mechanics.

8.7 Diamond Cutting

Why does a round brilliant-cut diamond have 56 facets? Note that according the table at the end of Section 2.9, we have

$$r_3(1) + r_3(2) + r_3(3) + r_3(4) + r_3(5) = 6 + 12 + 8 + 6 + 24 = 56$$

and diamonds are three-dimensional solids. Is this a coincidence?

A diamond is composed of pure carbon. Carbon has valence 4, which means that each carbon atom is bonded directly to exactly four other carbon atoms to form a lattice. This lattice has the structure of a "face-centered cube." This "cube" is the basic building block of a diamond. It is replicated throughout a diamond crystal, with vertices at points having three integer coordinates.

There are eight carbon atoms contained in each face-centered cube of a diamond. The atoms at the eight vertices, at $(0,0,0)$, $(0,0,1)$, $(0,1,0)$, $(1,0,0)$, $(1,1,0)$, $(1,0,1)$, $(0,1,1)$ and $(1,1,1)$, each count as 1/8 atom, for a total of one atom, because one-eighth of each of these atoms lies inside one of the eight cubes that meet at each vertex. The atoms at the centers of the six faces of the cube, at $(\frac{1}{2},\frac{1}{2},0)$, $(\frac{1}{2},0,\frac{1}{2})$, $(0,\frac{1}{2},\frac{1}{2})$, $(\frac{1}{2},\frac{1}{2},1)$, $(\frac{1}{2},1,\frac{1}{2})$ and $(1,\frac{1}{2},\frac{1}{2})$, each count as one-half atom, for a total of three atoms, because one-half of each of these atoms lines inside each of the two cubes having a face in common. There are also four carbon atoms in the interior of each cube, centered at coordinates $(\frac{1}{4},\frac{1}{4},\frac{1}{4})$, $(\frac{1}{4},\frac{3}{4},\frac{3}{4})$, $(\frac{3}{4},\frac{1}{4},\frac{3}{4})$ and $(\frac{3}{4},\frac{3}{4},\frac{1}{4})$. This gives a total of

$$\frac{8}{8} + \frac{6}{2} + 4 = 8$$

carbon atoms per basic cube. This cube is repeated by displacing it by any integer vector until the boundary of the diamond crystal is reached. In a diamond, the edge length of this basic face-centered cube is 3.56 Å = 3.56 · 10^{-10} meter. This length is the unit length for the coordinate system. Figure 8.1 shows the location of the centers of the carbon atoms in a single face-centered cube. The double lines in the figure represent the chemical bonds between adjacent carbon atoms.

A plane $hx + ky + \ell z = 1$ in $\mathbb{R}^3$ intercepts the axes at $x = 1/h$, $y = 1/k$ and $z = 1/\ell$ (but there is no intercept if the coefficient vanishes). Physicists like to

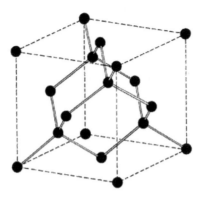

Figure 8.1 The Face-Centered Cube in a Diamond.

describe the plane $hx + ky + \ell z = 1$ and its orientation by its "Miller indices" (h, k, ℓ), just as mathematicians specify the orientation by giving the "normal vector" to the plane. (Physicists denote a negative number $-h$ by $\bar{h}$ in a Miller index, so that they would write $(1, \bar{2}, 3)$ for the plane $x - 2y + 3z = 1$, but we will use the minus signs.)

Example 8.13

The Miller indices $(1, 0, 0)$, $(0, -1, 0)$, $(0, 0, 2)$ and $(1, -2, 1)$ represent the planes $x = 1$, $y = -1$, $2z = 1$ and $x - 2y + z = 1$, respectively.

Let $\{h; k; \ell\}$ denote the set of all planes

$$\pm h'x \pm k'y \pm \ell'z = 1$$

with independent $\pm$ signs and h', k', ℓ' any permutation of h, k, ℓ.

Example 8.14

The notation $\{1; 0; 0\}$ represents the six planes $x = \pm 1$, $y = \pm 1$, $z = \pm 1$, with independent $\pm$ signs.

The notation $\{1; 1; 0\}$ represents the twelve planes $\pm x \pm y = 1$, $\pm x \pm z = 1$, $\pm y \pm z = 1$, with independent $\pm$ signs.

The numbers 1, 2, 3, 4, 5 and 6 each have exactly one representation as $a^2 + b^2 + c^2$, not counting sign changes and permutations of a, b, c. Thus, if $h^2 + k^2 + \ell^2 = n \in \{1, 2, 3, 4, 5, 6\}$, then $\{h; k; \ell\}$ represents all $r_3(n)$ planes $ax + by + cz = 1$ with any integers a, b, c satisfying $a^2 + b^2 + c^2 = n$.

The planes specified by Miller indices are called *lattice planes*. Generally, the lattice planes having small Miller indices (in absolute value) are the planes in which the carbon atoms are most densely packed. When a crystal is cleaved (split), it separates into two pieces on a lattice plane with small Miller indices because relatively few chemical bonds span this boundary and so the cleaving requires the least energy. For a diamond, with its face-centered cube crystal structure, the eight $\{1; 1; 1\}$ planes have the densest packing of carbon atoms of any lattice plane. When naturally occurring (uncut) diamonds are examined under a microscope, the $\{1; 1; 1\}$ planes are the most common surfaces observed. Usually, only one cleavage plane is prominent. It is easily identified by its transparency and smoothness.

Figure 8.2   Side View of a Round Brilliant-Cut Diamond.

Here is a table of some representations of integers $1 \le n \le 5$ as the sum of three squares, and the planes they represent.

$n = h^2 + k^2 + \ell^2$	$\{h; k; \ell\}$	$r_3(n) = \|\{h; k; \ell\}\|$
$1 = 1^2 + 0^2 + 0^2$	$\{1; 0; 0\}$	6
$2 = 1^2 + 1^2 + 0^2$	$\{1; 1; 0\}$	12
$3 = 1^2 + 1^2 + 1^2$	$\{1; 1; 1\}$	8
$4 = 2^2 + 0^2 + 0^2$	$\{2; 0; 0\}$	6
$5 = 2^2 + 1^2 + 0^2$	$\{2; 1; 0\}$	24
	Total	56

Could it be that the 56 facets of a round brilliant-cut diamond are planes with the smallest possible Miller indices, that is, the lattice planes parallel to $\{1; 0; 0\}$, $\{1; 1; 0\}$, $\{1; 1; 1\}$, $\{2; 0; 0\}$ and $\{2; 1; 0\}$?

Unfortunately, this beautiful guess is wrong.

First of all, a round brilliant-cut diamond actually has one or two other facets not counted in the 56. The flat octagonal surface at the top of a round brilliant-cut diamond is called the "table" and is not one of the 56 facets. Many round brilliant-cut diamonds have the point at the bottom cut off to form another planar surface, called the "culet," parallel to the table. If the facets were lattice planes, what would the Miller indices of the table and culet be?

So a round brilliant-cut diamond actually has 57 or 58 facets. It also has a thin circular curved surface called the "girdle" around the middle. The girdle separates the upper and lower facets. If a diamond is set, it is usually held in place by the girdle.

The next reason the guess is wrong comes from the way a diamond reflects and refracts light. A typical light beam enters the diamond through the table or an upper facet, where it is bent and its colors are separated. Then it is reflected totally inside the diamond by one or two lower facets. Finally, it exits the diamond through the table or an upper facet, where it is refracted once more, and we see the diamond "sparkle." The angles between the facets and the plane through the table must lie in certain limited ranges in order for this refraction and total reflection to work properly. As shown in Figure 8.3, if a diamond is too tall (2.) or too short (3.), light will exit through a lower facet and the gem will not sparkle. Part 1. of Figure 8.3 shows the correct light path. It turns out that a diamond cut on lattice planes with small Miller

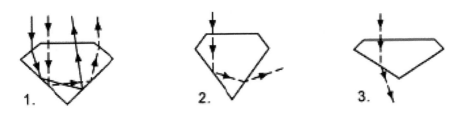

Figure 8.3   Light Ray Paths in a Diamond.

indices would not have facets with the correct angles to make it sparkle. In 1919, Marcel Tolkowsky published a book [111] in which he analyzed which angles between facets would make a round brilliant-cut diamond sparkle the most. The angles he recommended were based on the reflection and refraction properties of diamond and were chosen to maximize the brilliancy, fire and color of the crystal. These angles simply are not the angles between lattice planes with small Miller indices.

Finally, when crystals, including diamonds, are cleaved on lattice planes with small Miller indices, these surfaces are fragile and material tends to flake off from them. When a diamond is cut into a gem stone, its facets are polished (with diamond dust) rather than being cleaved, and the facets are chosen to be not lattice planes. Even though diamond is the hardest mineral, its lattice planes would still flake when it is handled, giving it a poor appearance.

8.8   *Cryptanalysis of a Signature Scheme*

Recall the RSA [91] public key cryptosystem. A user, Alice, who wishes to receive messages with the system, chooses two large primes p and q and lets $n = pq$. Thus, $\phi(n) = (p - 1)(q - 1)$. She chooses a random integer $1 < e < n$, relatively prime to $\phi(n)$, and computes an integer $1 < d < n$ with

$ed \equiv 1 \pmod{\phi(n)}$. She keeps p, q and d secret, but makes n and e public, for example, on her web page or in a directory of RSA users.

When Bob wants to send a secret message to Alice he breaks it into pieces $0 \le m < n$ and enciphers each piece by raising it to the power e modulo n. He knows e and n because they are public.

When Alice receives the enciphered message $c = m^e \bmod n$, she deciphers it by raising it to the d power modulo n. This works because of Euler's theorem, $m^{\phi(n)} \equiv 1 \pmod{n}$, so that

$$c^d \equiv (m^e)^d \equiv m^{de} \equiv m^1 = m \pmod{n},$$

since $ed \equiv 1 \pmod{\phi(n)}$.

The RSA system separates secrecy and authentication. In the use just described, Alice has no assurance that m came from Bob; anyone could have enciphered m and sent it to her. Let us reverse the roles and suppose that Alice wants to send an authentic message to Bob, so that he will be certain that it came from her. Suppose for simplicity that the message is not secret. Alice is happy that anyone can read the message and know that it came from her.

Alice can sign a message $1 < m < n$ by raising it to the power d, say $s = m^d \bmod n$. When Bob, or anyone else, receives s, he can read the message and check that Alice signed it by finding Alice's public keys n and e, and raising s to the power e modulo n. As for encryption, Euler's theorem shows that $m \equiv s^e \pmod{n}$. (Alice can sign a secret message to Bob by first signing it with her secret key d and then enciphering s with Bob's public key.)

The RSA signature scheme just described has a flaw, called the *multiplicative property*, which allows some signatures to be forged. If a malicious observer Eve learns two or more signatures of messages signed by Alice, then she can multiply them modulo n to construct a valid signature of a message Alice did not sign. For example, if Eve collects Alice's signatures s_1 and s_2 of messages m_1 and m_2, respectively, then she can forge the signature $s_3 = s_1 s_2 \bmod n$ of the message $m_3 = m_1 m_2 \bmod n$. It doesn't matter that m_3 is unlikely to be meaningful. The mere fact that Eve can construct Alice's valid signature of a message Alice did not actually sign is considered a flaw.

In 1991, ISO Standard 9796-1 addressed this flaw by limiting messages to the interval $0 < m < \sqrt{n}$ and introducing a redundancy function $\mu(m)$ with these properties:

1. $\mu : \{0, 1, \ldots, \lfloor \sqrt{n} \rfloor\} \longrightarrow \{0, 1, \ldots, n-1\}$,

2. μ is publicly known, one-to-one and easy to compute,

3. given $\mu(m)$, it is easy to find m, and

4. a random x in $0 \le x < n$ has probability about $1/\sqrt{n}$ of being $\mu(m)$ for any $0 \le m < \sqrt{n}$.

Alice would sign a message $0 < m < \sqrt{n}$ by signing $\mu(m)$, that is, compute $s = (\mu(m))^d \bmod n$. The recipient would check Alice's signature on $\mu(m)$ and then recover m from $\mu(m)$.

The four properties show that the use of μ repairs the flaw in the signature scheme because, given two messages $0 < m_1, m_2 < \sqrt{n}$, it is unlikely that there is a message $0 < m_3 < \sqrt{n}$ for which

$$\mu(m_1)\mu(m_2) \equiv \mu(m_3) \pmod{n}.$$

Given signatures for messages m_1 and m_2, Eve could still construct a signature for some number $x = \mu(m_1)\mu(m_2) \bmod n$ in $0 < x < n$, but it is unlikely (the probability is about $1/\sqrt{n}$) that $x = \mu(m_3)$ for any message $0 < m_3 < \sqrt{n}$.

The ISO Standard 9796-1 used Hamming codes in the definition of μ. The Standard worked well for eight years, but in 1999 it was partially broken and, shortly thereafter, completely broken. Coppersmith, Halevi and Jutla [21] found a way to construct triples of messages $0 < m_1, m_2, m_3 < \sqrt{n}$ with $\mu(m_1)\mu(m_2) \equiv \mu(m_3) \pmod{n}$. In the same paper they suggested five alternate redundancy functions to replace the μ they had broken.

The next year Girault and Misarsky [37] found attacks for all five of these countermeasures. The attack on μ_5 from [21] interests us because it uses the representation of integers as the sum of two squares.

The redundancy function μ_5 was defined [21] for $0 \leq m < \sqrt{n}$ by $\mu_5(m) = (m^2 + \delta) \bmod n$, where δ is a (public) constant modulo n. Assume that δ is about as big as $n/2$.

Girault and Misarsky [37] attacked this countermeasure by letting $A = 2(n - \delta)$, so that A is about as large as n, and assuming that A has at least two distinct representations as the sum of two squares, say,

$$A = w^2 + x^2 = y^2 + z^2 \quad \text{with} \quad 0 < w, x, y, z < \sqrt{n}.$$

They proved the following theorem and corollary.

THEOREM 8.5 Multiplicative Property of Redundancy Function
If $A = 2(n - \delta) = w^2 + x^2 = y^2 + z^2$ with $0 < w, x, y, z < \sqrt{n}$, then

$$\mu_5(w)\mu_5(y) \equiv \mu_5(x)\mu_5(z) \pmod{n}.$$

PROOF We have $\mu_5(m) = (m^2 + \delta) \bmod n$ when $0 \leq m < \sqrt{n}$. Thus

$$\begin{aligned}
\mu_5(w)\mu_5(y) &\equiv (w^2 + \delta)(y^2 + \delta) \pmod{n} \\
&\equiv (A - x^2 + \delta)(A - z^2 + \delta) \pmod{n} \\
&\equiv (-x^2 - \delta)(-z^2 - \delta) \pmod{n} \\
&\equiv (x^2 + \delta)(z^2 + \delta) \pmod{n} \\
&\equiv \mu_5(x)\mu_5(z) \pmod{n}.
\end{aligned}$$

∎

COROLLARY 8.1

If signatures using μ_5 are known for any three of the four messages w, x, y, z, where $A = 2(n - \delta) = w^2 + x^2 = y^2 + z^2$ with $0 < w, x, y, z < \sqrt{n}$, then one can compute the signature for the fourth one.

PROOF   If three of the four numbers $\mu_5(w)$, $\mu_5(x)$, $\mu_5(y)$, $\mu_5(z)$ are known, then the fourth one can be computed from the congruence in Theorem 8.5.
∎

The attack described in Corollary 8.1 is an unacceptable flaw in the signature scheme using μ_5.

Example 8.15

Suppose Alice uses $n = 733549 = 827 \cdot 887$ as her RSA modulus. Suppose $\delta = 325256$. Then

$$A = 2 \cdot (n - \delta) = 2 \cdot (733549 - 325256) = 816586 = 2 \cdot 281 \cdot 1453.$$

Since the odd prime factors of A are $\equiv 1 \pmod 4$, we can apply Hermite's algorithm of Section 8.2 and find

$$281 = 5^2 + 16^2, \qquad 1453 = 3^2 + 38^2.$$

The identity (8.1) gives us

$$\frac{A}{2} = (16 \cdot 3 + 5 \cdot 38)^2 + (16 \cdot 38 - 3 \cdot 5)^2 = 238^2 + 593^2, \qquad \text{and}$$

$$\frac{A}{2} = (16 \cdot 38 + 5 \cdot 3)^2 + (16 \cdot 3 - 5 \cdot 38)^2 = 623^2 + 142^2.$$

We use $2 = 1^2 + 1^2$ and another application of (8.1) to obtain

$$A = (593 + 238)^2 + (593 - 238)^2 = 831^2 + 355^2 \text{ and}$$
$$A = (623 + 142)^2 + (623 - 142)^2 = 765^2 + 481^2.$$

Then Theorem 8.5 gives

$$\mu_5(831)\mu_5(765) \equiv \mu_5(355)\mu_5(481) \pmod n.$$

Girault and Misarsky [37] gave a similar example with a 512-bit modulus n. Hermite's algorithm works fine even for numbers of that size.

8.9   Exercises

1. The Fibonacci number $u_{101} = 573147844013817084101$ has exactly two prime factors. Use the information that $u_{101} = a^2 + b^2$, where $a = 9844590449$ and $b = 21822737750$, to find them.

2. Write 5^5 and 5^6 as the sum of two squares in all distinct ways.

3. Prove that a positive integer n is the sum of two positive squares if and only if $n = 2^e n_1 n_2^2$, where e is a nonnegative integer, n_1 is an integer > 1 all of whose prime factors are $\equiv 1 \pmod 4$ and n_2 is the product of zero or more primes $\equiv 3 \pmod 4$.

4. Show that if $p = 4k + 1$ is prime, then one can compute the binomial coefficient $\binom{2k}{k}$ modulo p in probabilistic polynomial time. (Note that you are not required to compute the binomial coefficient, which may be huge; only its remainder modulo p is required.)

5. Show that Jacobsthal's quadratic character sum in Theorem 8.3 can be evaluated in probabilistic polynomial time.

6. The number 593914915675537 is prime. Express it as the sum of two squares.

7. Write 8888888888888888888888888888883 as the sum of three squares.

8. Write 8888888888888888888888888888887 as the sum of four squares.

9. Write 8888888888888888888888888888887 as the sum of three triangular numbers.

10. Give a probabilistic polynomial time algorithm which, given integers $k > 4$ and $n > 0$, will find a representation of n as a sum of k k-gonal numbers.

11. Use the theorem quoted from Grosswald, Calloway and Calloway [43] to show that the asymptotic density of the set of positive integers that are the sum of three *positive* squares is $5/6$.

12. Find the 32 positive integers that are not the sum of distinct squares.

13. Show that infinitely many positive integers cannot be written as the sum of four positive squares.

14. Prove that every integer > 169 can be written as the sum of five positive squares.

15. Show that the $\{1; 1; 1\}$ planes have the densest packing of carbon atoms of any lattice plane in a diamond.

16. Alice uses RSA with the modulus $n = 373322303 = 17203 \cdot 21701$ and uses the redundancy function $\mu_5(m) = (m^2 + 156486298) \bmod n$ when signing messages. Exhibit four messages with the property that if Eve can persuade Alice to sign any three of them, then Eve can compute Alice's signature on the fourth message.

References

[1] M. Abramowitz and I. Stegun, editors. *Handbook of Mathematical Functions*. Dover, New York, 1965.

[2] L. V. Ahlfors. *Complex Analysis*. McGraw-Hill, New York, 1966.

[3] A. N. Andrianov. On the representation of numbers by certain quadratic form and the theory of elliptic curves. *Am. Math. Soc. Translations, Ser. 2*, 66:191–203, 1968.

[4] A. O. L. Atkin and J. Lehner. Hecke operators on $\Gamma_0(m)$. *Math. Ann.*, 185:134–160, 1970.

[5] P. Bachmann. *Die Arithmetik der quadratishen Formen*. Teubner, 1898.

[6] P. Bachmann. *Niedere Zahlentheorie*. Teubner, 1910.

[7] H. P. Baltes, P. K. J. Draxl, and E. R. Hilf. Quadratsummen und geweise Randwert probleme der mathematischen Physik. *J. Reine Angew. Math.*, 268/269:410–417, 1974.

[8] H. P. Baltes and F. K. Kneubühl. Thermal radiation in finite cavities. *Helv. Phys. Acta*, 45:481–529, 1972.

[9] F. A. Behrend. On sequences of integers containing no arithmetic progression. *Časopis Mat. Fis. Praha.*, 67:235–239, 1938.

[10] R. E. Borcherds. Monstrous moonshine and monstrous Lie superalgebras. *Invent. Math.*, 109:405–444, 1992.

[11] Z. I. Borevich and I. R. Shafarevich. *Number Theory*. Academic Press, New York, 1966.

[12] J. Brillhart. Note on representing a prime as a sum of two squares. *Math. Comp.*, 26:1011–1013, 1972.

[13] J. Buhler, R. Crandall, R. Ernvall, T. Metsänkylä, and M. Shokrollaki. Irregular primes and cyclotomic invariants to 12 million. *J. Symb. Comput.*, 11:1–8, 2000.

[14] D. A. Burgess. A note on the distribution of residues and non-residues. *J. London Math. Soc.*, 38:253–256, 1963.

[15] L. Carlitz. Notes on irregular primes. *Proc. Am. Math. Soc.*, 5:329–331, 1954.

[16] D. X. Charles. Computing the Ramanujan tau function. Preprint, March 2003. Available at http://www.cs.wisc.edu/~cdx/CompTau.pdf.

[17] S. Chowla. There exists an infinity of 3-combinations of primes in A. P. *Proc. Lahore Phil. Soc.*, 6:15–16, 1944.

[18] H. Cohen. *A Course in Computational Algebraic Number Theory.* Springer-Verlag, New York, 1996.

[19] H. Cohn. *A Second Course in Number Theory.* John Wiley & Sons, New York, New York, 1962.

[20] J. H. Conway and N. J. A. Sloan. *Sphere Packings, Lattices, and Groups.* Springer-Verlag, New York, Berlin, 1999.

[21] D. Coppersmith, S. Halevi, and C. Jutla. ISO 9796-1 and the new forgery strategy (Working Draft). Research Contribution to P1363, 1999. Available at http:grouper.ieee.org/groups/1363/contrib.html.

[22] P. Deligne. *Formes modulaires et représentations ℓ-adiques,* volume 179 of *Séminaire Bourbaki Exposé 355, Lecture Notes in Mathematics,* pages 139–172. Springer-Verlag, Berlin, 1971.

[23] P. Deligne. La conjecture de Weil, I. *Publ. Math. IHES,* 43:273–307, 1974.

[24] L. E. Dickson. *History of the Theory of Numbers.* Reprinted by Chelsea, New York, 1952.

[25] G. L. Dirichlet. Sur l'équation $t^2 + u^2 + v^2 + w^2 = 2m$, [Extrait d'une lettre de M. LeJeune Dirichlet à M. Liouville]. *J. de Mathématiques,* 1(2):210–214, 1856.

[26] H. M. Edwards. *Riemann's Zeta Function.* Academic Press, New York, 1974.

[27] M. Elchler. Quaternary quadratische Formen und die Riemannsche Vermutung für die Kongruenzzeta funktion. *Arch. Math.,* 5:355–366, 1954.

[28] M. Eichler. *The basis problem for modular forms and the traces of the Hecke operators,* in *Modular functions of one variable, I,* volume 320 of *Springer Lecture Notes,* pages 75–151. Springer-Verlag, Berlin, 1973.

[29] P. Erdős and S. S. Wagstaff, Jr. The fractional parts of the Bernoulli numbers. *Ill. J. Math.,* 24:104–112, 1980.

[30] P. Erdős and P. Turán. On some sequences of integers. *J. London Math. Soc.,* (2) 11:261–264, 1936.

[31] N. J. Fine. *Basic Hypergeometric Series and Applications.* Number 27 in Mathematical Surveys and Monographs. Am. Math. Soc., Providence, R.I., 1988.

[32] R. Fricke and F. Klein. *Vorlesungen über die Theorie der elliptischen Modulfunktionen.* Vol. 2, Bibliotheca Mathematica Teubneriana, no. 11, Johnson Reprint Corporation, New York, 1966.

[33] H. Furstenberg. Ergodic behaviour of diagonal measures and a theorem of Szemerédi on arithmetic progressions. *J. d'Analyse Math.,* 31:204–256, 1977.

[34] C. F. Gauss. *Werke,* volume II. Königlichen Gesellschaft der Wissenschaften, Göttingen, 1863.

[35] C. F. Gauss. *Disquisitiones Arithmeticae.* Yale University Press, New Haven, English edition, 1966.

[36] S. Gelbart. *Weil's Representation and the Spectrum of the Metaplectic Group*, volume 530 of *Lecture Notes in Mathematics*. Springer-Verlag, Berlin, New York, 1976.

[37] M. Girault and J.-F. Misarsky. Cryptanalysis of countermeasures proposed for repairing ISO 9796-1. In B. Preneel, editor, *Advances in Cryptology—EUROCRYPT 2000*, volume 1807 of *Lecture Notes in Computer Science*, pages 81–90, Springer-Verlag, Berlin, 2000.

[38] W. L. Glaisher. On the number of representations of a number as a sum of $2r$ squares, where $2r$ does not exceed eighteen. *Proc. London Math. Soc. (2)*, 5:479–490, 1907.

[39] W. T. Gowers. A new proof of Szemerédi's theorem. *Geom. and Funct. Anal.*, 11:465–588, 2001.

[40] R. L. Graham and B. L. Rothschild. A short proof of van der Waerden's theorem on arithmetic progressions. *Proc. Am. Math. Soc.*, 42:385–386, 1974.

[41] B. Green and T. Tao. The primes contain arbitrarily long arithmetic progressions. Available at `arXiv:math.NT/0404188` v3 2 Jun 2005.

[42] E. Grosswald. *Representations of Integers as Sums of Squares*. Springer-Verlag, New York, New York, 1985.

[43] E. Grosswald, A. Calloway, and J. Calloway. The representation of integers by three positive squares. *Proc. Am. Math. Soc.*, 10:451–455, 1959.

[44] J. Hafner and K. McCurley. A rigorous subexponential algorithm for computation of class groups. *J. Am. Math. Soc.*, 2:837–850, 1989.

[45] H. Halberstam and K. F. Roth. *Sequences, vol. I*. Oxford University Press, Oxford, England, 1966.

[46] H. Hamburger. Über einige Beziehungen, die mit der Funktionalgleichung der Riemannschen ζ-Funktion äquivalent sind. *Math. Ann.*, 85:129–140, 1922.

[47] G. H. Hardy. On the representation of a number as the sum of any number of squares, and in particular of five or seven. *Proc. Natl. Acad. Sci. U.S.A.*, 4:189–193, 1918.

[48] G. H. Hardy. On the representation of a number as the sum of any number of squares, and in particular of five. *Trans. Am. Math. Soc.*, 21:255–284, 1920.

[49] G. H. Hardy. *Ramanujan; Twelve Lectures on Subjects Suggested by his Life and Work*. Cambridge Univ. Press, Cambridge, 1940.

[50] G. H. Hardy. *Ramanujan*. Reprinted by Chelsea, New York, 1959.

[51] G. H. Hardy and J. E. Littlewood. Some problems of 'Partitio Numerorum': IV. The singular series in Waring's problem and the value of the number $G(k)$. *Math. Zeit.*, 12:161–181, 1922.

[52] G. H. Hardy and E. M. Wright. *An Introduction to the Theory of Numbers*. Clarendon Press, Oxford, England, Fifth edition, 1979.

[53] E. Hecke. Herleitung des Euler-Produktes der Zetafunktion und einiger L-Reihen aus ihrer Funktionalgleichung. *Math. Ann.*, 119:266–287,

1944.

[54] C. Hermite. Note au sujet de l'article précédent. *J. Math. Pures Appl.*, page 15, 1848. The title refers to Serret [99].

[55] Y. Ihara. Hecke polynomials as congruence zeta functions in the elliptic modular case. *Ann. Math., Ser. 2*, 85:267–295, 1967.

[56] K. Iwasawa. *Lectures on p-adic L-functions*. Annals of Math. Studies No. 74. Princeton Univ. Press, Princeton, New Jersey, 1971.

[57] C. G. J. Jacobi. Démonstration élémentaire d'une formule analytique remarquable. *J. de Mathématiques*, 7:85–109, 1842. Originally published in German in *J. Math.*, 21:13–32, 1840.

[58] E. Jacobsthal. Anwendungen einer Formel aus der Theorie der quadratischen Reste. *J. Math.*, 132:238–245, 1907.

[59] G. J. Janusz. *Algebraic Number Fields*. Academic Press, New York, 1973.

[60] K. L. Jensen. Om talteoretiske Egenskaber ved de Bernoulliske Tal. *Nyt Tidsskrift Für Mathematik, Afdeling B*, 26:73–83, 1915.

[61] W. Johnson. *p*-adic proof of congruences for the Bernoulli numbers. *J. Number Theory*, 7:251–265, 1975.

[62] A. Khinchin. *Three Pearls of Number Theory*. Graylock, 1952.

[63] M. Knopp. *Modular Functions in Analytic Number Theory*. Markham, Chicago, 1970.

[64] E. Landau. *Elementary Number Theory*. Reprinted by Chelsea, New York, 1955.

[65] A. M. Legendre. *Théorie des Nombres, Tome I*. Gauthiers-Villars, Paris, France, 1830.

[66] D. H. Lehmer. *Computer technology applied to the theory of numbers*, volume 6 of *MAA Studies in Mathematics*, pages 117–151. Math. Assn. Am., 1969.

[67] W. Li. New forms and functional equations. *Math. Ann.*, 212:285–315, 1975.

[68] J. Liu and T. Zhan. Distribution of integers that are sums of three squares of primes. *Acta Arith.*, 98:207–228, 2001.

[69] G. B. Mathews. *Theory of Numbers, Part I*. Deighton, Bell & Co., Cambridge, 1892.

[70] F. I. Mautner. Spherical functions over *p*-adic fields, I. *Am. J. Math.*, 80:441–457, 1958.

[71] T. Metsänkylä. Distribution of irregular primes. *J. Reine Angew. Math.*, 282:126–130, 1976.

[72] H. L. Montgomery. Distribution of irregular primes. *Ill. J. Math.*, 9:553–558, 1965.

[73] L. J. Mordell. On the representation of a number as a sum of an odd number of squares. *Trans. Cambridge Phil. Soc.*, 23:361–372, 1919.

[74] L. J. Mordell. On the representation of numbers as a sum of 2r squares. *Quart. J. of Math.*, 48:93–104, 1920.

[75] M. B. Nathanson. *Additive Number Theory: The Classical Bases*.

Springer-Verlag, New York, New York, 1996.

[76] M. Newman. Subgroups of the modular group and sums of squares. *Am. J. Math.*, 82:761–778, 1970.

[77] D. Niebur. A class of nonanalytic automorphic functions. *Nagoya Math. J.*, 52:133–145, 1973.

[78] N. Nielsen. *Traité Élémentaire des Nombres de Bernoulli.* Gauthiers-Villars, Paris, France, 1923.

[79] I. Niven, H. S. Zuckerman, and H. L. Montgomery. *An Introduction to the Theory of Numbers.* John Wiley, New York, Fifth edition, 1991.

[80] K. Ono. *The web of modularity: Arithmetic of the coefficients of modular forms and q-series*, volume 102 of *CBMS Regional Conference Series in Mathematics.* Am. Math. Soc., Providence, R.I., 2004.

[81] H. H. Ostmann. *Additive Zahlentheorie.* Springer-Verlag, Berlin, New York, 1956.

[82] G. Pall. On sums of squares. *Am. Math. Monthly*, 40:10–18, 1933.

[83] O. Perron. *Die Lehre von den Kettenbrüchen.* Reprinted by Chelsea, New York, Second edition, 1950.

[84] H. Petersson. Über die Entwicklungskoeffizienten der Automorphen Formen. *Acta Math.*, 58:169–215, 1932.

[85] H. Petersson. *Modulfunktionen und Quadratisch Formen.* Number 100 in Ergebnisse der Mathematik und ihrer Grenzgebeite. Springer-Verlag, New York, 1982.

[86] H. Rademacher. Zur Theorie der Modulfunktionen. *J. Reine Angew. Math.*, 167:312–326, 1931.

[87] H. Rademacher. *Topics in Analytic Number Theory.* Springer-Verlag, Berlin, New York, 1973.

[88] S. Ramanujan. On certain arithmetical functions. *Trans. Camb. Phil. Soc.*, 22(9):159–184, 1916.

[89] R. A. Rankin. Sets of integers containing not more than a given number of terms in arithmetical progression. *Proc. R. Soc. Edinburgh, Sect A*, 65:332–344, 1960/61.

[90] R. A. Rankin. On the representation of a number as the sum of any number of squares, and in particular of twenty. *Acta Arith.*, 7:399–407, 1961.

[91] R. L. Rivest, A. Shamir, and L. Adleman. A method for obtaining digital signatures and public-key cryptosystems. *Comm. A. C. M.*, 21(2):120–126, 1978.

[92] K. F. Roth. On certain sets of integers. *J. London Math. Soc.*, 28:104–109, 1953.

[93] L. Saalschütz. *Vorlesungen über die Bernoullischen Zahlen.* Springer-Verlag, Berlin, 1893.

[94] R. Salem and D. C. Spencer. On sets of integers which contain no three terms in arithmetical progression. *Proc. Natl. Acad. Sci. U.S.A.*, 28:561–563, 1942.

[95] P. Samuel. *Algebraic Theory of Numbers.* Hermann, Paris, 1970.

[96] A. Selberg. Harmonic analysis and discontinuous groups in weakly sym-
      metric Riemannian spaces with applications to Dirichlet series. *J. In-
      dian Math. Soc.*, 20:47–87, 1956.

[97] J.-P. Serre. *Représentations Linéaires des Groupes Finis*. Hermann,
      Paris, 1971.

[98] J.-P. Serre. *Congruences et formes modulaires (d'après H. P. F.
      Swinnerton-Dyer)*, volume 317 of *Séminaire Bourbaki Exposé 416, Lec-
      ture Notes in Mathematics*, pages 319–338. Springer-Verlag, Berlin,
      1973.

[99] J. A. Serret. Sur un théorème rélatif aux nombres entières. *J. Math.
      Pures Appl.*, pages 12–14, 1848.

[100] G. Shimura. A reciprocity law in non-solvable extensions. *J. Reine
      Angew. Math.*, 221:209–220, 1966.

[101] G. Shimura. *Introduction to the Arithmetic Theory of Automorphic
      Functions*. Princeton Univ. Press, Princeton, New Jersey, 1971.

[102] G. Shimura. On modular forms of half integral weight. *Ann. Math.*,
      97:440–481, 1973.

[103] C. L. Siegel. Die Funktionalgleichungen einer Dirichletschen Reihen.
      Math. Zeit., 63:363–373, 1956.

[104] H. J. S. Smith. *Report on the Theory of Numbers*. Reprinted by Chelsea,
      New York, 1965.

[105] J. Laurie Snell. A conversation with Joe Doob. *Statistical Sci-
      ence*, 12:301–311, 1997. Also available at http://www.dartmouth.edu/
      ~chance/Doob/conversation.html.

[106] R. Sprague. Über Zerlegungen in ungleiche Quadratzahlen. *Math. Zeit.*,
      51:289–290, 1948.

[107] H. P. F. Swinnerton-Dyer. *On ℓ-adic representations and congruences
      for coefficients of modular forms*, in *Modular Functions of One Variable,
      III*, volume 350 of *Springer Lecture Notes*, pages 1–55. Springer-Verlag,
      Berlin, 1973.

[108] E. Szemerédi. On sets of integers containing no four elements in arith-
      metic progression. *Acta Math. Acad. Sci. Hung.*, 20:89–104, 1969.

[109] E. Szemerédi. On sets of integers containing no k elements in arithmetic
      progression. *Acta Arith.*, 27:199–245, 1975.

[110] J. Tits. Symmetry. *Am. Math. Monthly*, 107(5):454–461, 2000.

[111] M. Tolkowsky. *Diamond Design*. E. & F. N. Spon, Ltd., 57 Haymarket,
      S.W. 1, London, 1919.

[112] J. V. Uspensky. On Jacobi's theorems concerning the simultaneous
      representation of numbers by two different quadratic forms. *Trans.
      Am. Math. Soc.*, 30:385–404, 1928.

[113] J. V. Uspensky and M. A. Heaslet. *Elementary Number Theory*.
      McGraw-Hill, New York, 1939.

[114] A. van der Poorten. A proof that Euler missed ... Apéry's proof of the
      irrationality of $\zeta(3)$. *Math. Intelligencer*, 1:195–203, 1978/79.

[115] H. S. Vandiver. On Bernoulli's numbers and Fermat's last theorem.

Duke Math. J., 3:569–584, 1937.

[116] G. N. Watson. *A Treatise on the Theory of Bessel Functions*. Cambridge Univ. Press, Cambridge, Second edition, 1966.

[117] A. Weil. On some exponential sums. *Proc. Natl. Acad. Sci. U.S.A.*, 34:204–207, 1948.

[118] A. Weil. Über die Bestimmung Dirichletscher Reihen durch Funktion-algleichungen. *Math. Ann.*, 168:149–156, 1967.

[119] A. Weil. Sur les sommes des trois et quatre carrés. *L'Enseign. Math. (2)*, 20:215–222, 1974.

[120] J. R. Wilton. Congruence properties of Ramanujan's function $\tau(n)$. *Proc. London Math. Soc.*, (2) 31:1–10, 1929.

[121] J. R. Wilton. A note on Ramanujan's arithmetical function $\tau(n)$. *Proc. Camb. Phil. Soc.*, 25:121–129, 1929.

Index

354   *Index*

Milton Keynes UK
Ingram Content Group UK Ltd.
UKHW021635071024
449327UK00020BA/1307